AF393584

# PROTOPLASMA
*Supplementum 2*

M. Tazawa (ed.)

# *Cell Dynamics*

## *Volume Two*
*Molecular Aspects of Cell Motility*
*Cytoskeleton in Cellular Structure and Activity*

*Springer-Verlag Wien New York*

Prof. MASASHI TAZAWA, Ph.D.
Department of Botany, Faculty of Science,
University of Tokyo, Tokyo, Japan



With 126 Figures

ISSN 0934-8727

ISBN-13:978-3-7091-9013-5     e-ISBN-13:978-3-7091-9011-1
DOI: 10.1007/978-3-7091-9011-1

# Contents


*Molecular Aspects of Cell Motility*

**Yano, M., Mioh, H., Shimizu, H.**: Motion-Controlled ATP Hydrolysis in Reconstituted Streaming from Rabbit Skeletal Muscle . . . . . . . . . . . . . . . . . . . . . . . . . . . . . . . . . . 3

**Yamashiro-Matsumura, S., Ishikawa, R., Matsumura, F.**: Purification and Characterization of 83 kDa Nonmuscle Caldesmon from Cultured Rat Cells: Changes in Its Expression upon L 6 Myogenesis . . . . 9

**Schleicher, M., Wallraff, E., Gerisch, G., Isenberg, G.**: Construction and Analysis of *Dictyostelium* Mutants with Defects in Actin-Binding Proteins . . . . . . . . . . . . . . . . . . . . . . . . . . 22

**Higashi-Fujime, S.**: Actin-Induced Elongation of Fibers Composed of Cytoplasmic Membrane from *Nitella* 27

**Kohama, Kazuhiro, Shoda, Miwa, Murayama, Kimie, Okamoto, Yoh**: Domain Structure of *Physarum* Myosin Heavy Chain . . . . . . . . . . . . . . . . . . . . . . . . . . . . . . . . . . . . . . . . . . . 37

**Ogihara, S.**: Immunoelectron Microscopic Localization of the *Physarum* 36,000-Dalton Actin Binding Protein on the Surface of Vesicular Structures in the Plasmodium . . . . . . . . . . . . . . . . 48

**Inoué, S.**: Manipulating Single Microtubules . . . . . . . . . . . . . . . . . . . . . . . . . . 57

*Cytoskeleton in Cellular Structure and Activity*

**Lancelle, S. A., Hepler, P. K.**: Cytochalasin-Induced Ultrastructural Alterations in *Nicotiana* Pollen Tubes 65

**Klein, H. P., Köster, Bärbel, Stockem, W.**: Pinocytosis and Locomotion of *Amoebae*. XVII. Different Morphodynamic Forms of Endocytosis and Microfilament Organization in *Amoeba proteus* . . . . 76

**Hiramoto, Yukio, Kaneda, Isamu**: Diffusion of Substances in the Cytoplasm and Across the Nuclear Envelope in Egg Cells . . . . . . . . . . . . . . . . . . . . . . . . . . . . . . . . . . . . . . . 88

**Kakimoto, T., Shibaoka, H.**: Cytoskeletal Ultrastructure of Phragmoplast-Nuclei Complexes Isolated from Cultured Tobacco Cells . . . . . . . . . . . . . . . . . . . . . . . . . . . . . . . . . . 95

**Yonemura, S., Tsukita, S., Tsukita, S., Mabuchi, I.**: Structural Analysis of the Sea Urchin Egg Cortex Isolated on a Substratum . . . . . . . . . . . . . . . . . . . . . . . . . . . . . . . . . . 104

**Wayne, Randy, Tazawa, M.**: The Actin Cytoskeleton and Polar Water Permeability in Characean Cells . 115

**McBeath, Elena, Fujiwara, Keigi**: A New Class of Photoactivated Fixatives for Immunocytochemistry . . 131

**Loewy, A. G., Dollahon, N., Klainer, P., Wolfe, K.**: An Insoluble Matrix of the Nerve Cytoskeleton . . 137

**Shimada, O., Ishikawa, H., Wakabayashi, K.**: Role of Microtubules in Hormone Secretory Function of the Rat Anterior Pituitary . . . . . . . . . . . . . . . . . . . . . . . . . . . . . . . . . . . 145

**Index of Keywords** . . . . . . . . . . . . . . . . . . . . . . . . . . . . . . . . . . . . . 159

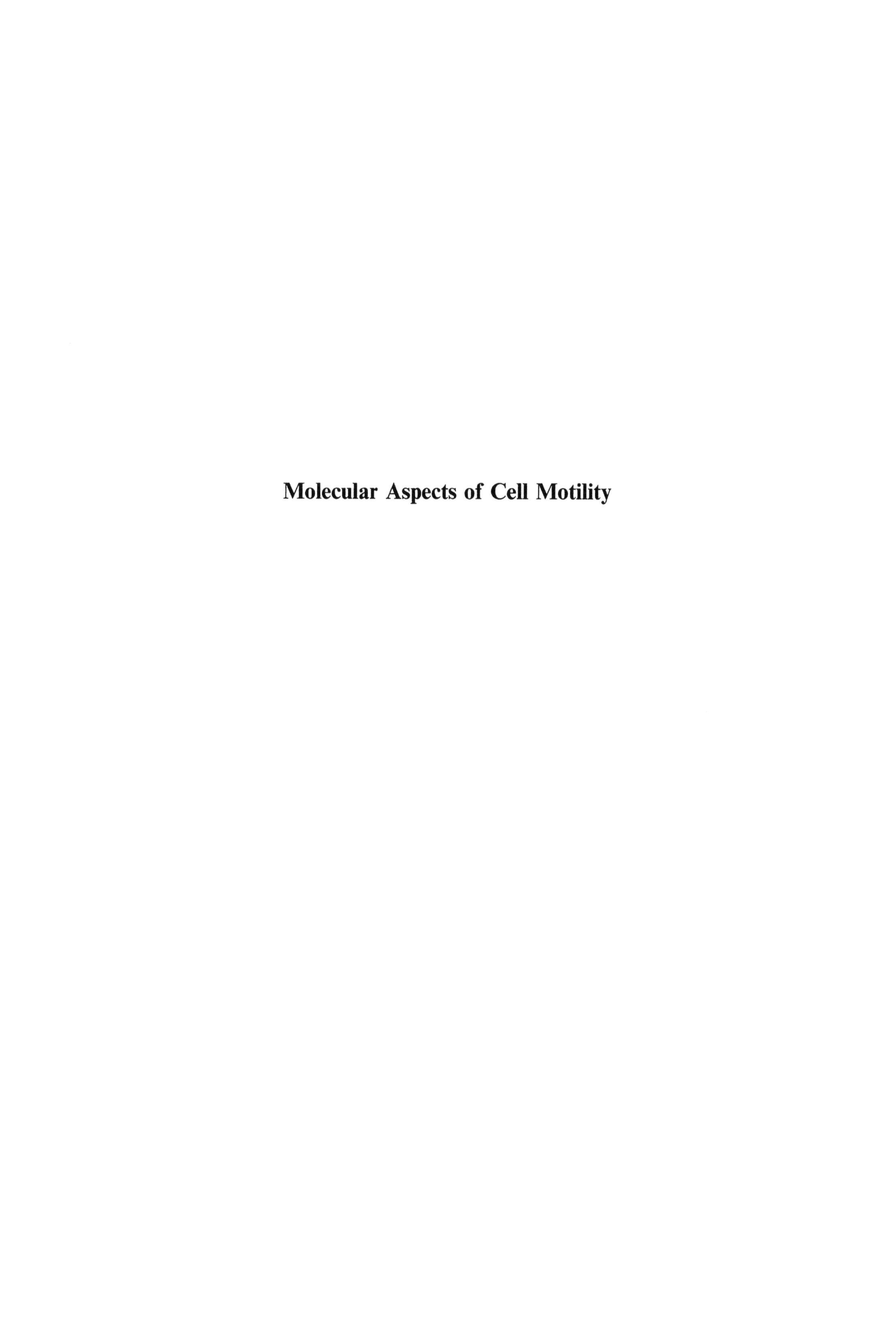

# Molecular Aspects of Cell Motility

Protoplasma (1988) [Suppl 2]: 3–8


# Motion-Controlled ATP Hydrolysis in Reconstituted Streaming from Rabbit Skeletal Muscle

M. Yano*, H. Mioh, and H. Shimizu

Faculty of Pharmaceutical Sciences, University of Tokyo

Received January 6, 1988
Accepted March 16, 1988

Dedicated to Professor Dr. Noburo Kamiya on the occasion of his 75th birthday

## Summary

Recently, we described an apparatus termed "stream cell" in which circular streaming could be induced by the interaction of heavy-meromyosin with F-actin purified from rabbit skeletal muscle. Since the streaming was observed only under conditions similar to those during muscle contraction, we concluded that it was active and driven by ATPase. The most typical feature of this system was that the acto-heavymeromyosin ATPase activity was uniquely determined by the macroscopic streaming velocity. In order to explain the chemo-mechanical conversion observed in the active streaming, dynamic cooperativity among force generators was required. Dynamic cooperativity during chemo-mechanical conversion implies that myosin crossbridges do not always act as independent force generators. For further study of the energy conversion in actomyosin system, we report here a new streaming system "grid stream system", in which not only the velocity but also the direction of streaming can be controlled externally. Upon increasing the positive streaming velocity, the ATPase activity first increases and then decreases. On the contrary, at negative streaming velocities, corresponding to lengthening, the ATPase activity is not affected. The molecular mechanism of energy conversion will be discussed in relation to those observations.

*Keywords:* Actomyosin; Reconstituted system; Myosin crossbridge; Energy conversion; Dynamic cooperativity.

## 1. Introduction

For the study of chemo-mechanical coupling, it is necessary to simultaneously measure both chemical and mechanical processes in steady states. Several years ago, we reported reconstitution of steady streaming in a circular slit, in which circulating flow could be in-

duced by the interaction of heavy meromyosin (HMM) with attached polarized actin from rabbit skeletal muscle (Yano 1978). The reconstituted system indicates that heavy meromyosin can be assembled onto F-actin from solution, not necessarily in filamentous form, and produce a vectorial force utilizing the energy of ATP hydrolysis (Yano *et al.* 1982, Yano *et al.* 1983). Since it is very difficult to analyze the chemical change accompanying mechanical work in a living muscle, a reconstituted motile system is useful for studying the molecular mechanism of the chemo-mechanical coupling of energy conversion. Of course, the physiological significance of these reconstituted systems depends on the extent to which they are compatible with chemical and physical properties of contracting muscle. Indeed, reconstituted motility can only be observed under physiological conditions (Yano *et al.* 1978). The most typical feature of this system was that the acto-HMM ATPase activity was uniquely determined by the streaming velocity, the magnitude of which was the order of diffusive displacement of small molecules in aqueous solutions, *i.e.*, up to about 20 μm/sec at 20 °C. The ATPase activity at higher streaming velocities is higher than that at slower velocities, suggesting a Fenn's effect (Fenn 1924) in that system. An analysis of the relation between chemical and mechanical properties during active streaming led to the notion that the extent of selforganization of macroscopic streaming (the streaming velocity) enslaves the molecular dynamics of the ATPase activities of the individual myosin molecules (Yano and Shimizu 1978, Shimizu and Yano 1978, Shimizu 1979).

* Correspondence and Reprints: Faculty of Pharmaceutical Sciences, University of Tokyo, Hongo 7-3-1, Bunkyo-ku, Tokyo 113, Japan.

This means that streaming could alter the ATPase mechanism introducing spontaneous acceleration as a rate-determining step. That is, the individual myosin molecules work together through the macroscopic streaming. We named this "dynamic cooperativity" since the information on the molecular dynamics of the HMM molecules is mediated by the streaming fluid, resulting in higher efficiency in the mechano-chemical conversion. We concluded that the slaving relationship holds only when dynamic cooperativity occurs in the molecular dynamics of energy conversion. In our system, as a function of the conditions, *e.g.*, ATPase concentration, ionic strength and temperature, the streaming velocity can be uniquely determined.

For further study of the slaving relation in the actomyosin system, we report here a new streaming system, the *grid stream system*, in which acto-HMM from rabbit skeletal muscle is used, and in which not only the velocity but also the direction of streaming can be controlled externally. Using the grid stream system, we can obtain the information on the molecular mechanism at not only beyond the maximum shortening velocity of muscle but at various lengthening speed. In addition, it was possible to obtain information on the muscle shortening. In light of our results, we propose a new control mechanism for the molecular mechanism of chemo-mechanical energy conversion in living motile system.

## 2. Material and Methods

### 2.1. Materials

Rabbit skeletal muscle myosin was prepared by the method of Perry (1955) and heavy meromyosin prepared by limited digestion of purified myosin with chymotrypsin (Weeds and Pope 1977). G-actin was purified by the method of Spudich and Watt (1971). Native tropomyosin was obtained according to the method of Ebashi and Ebashi (1965). The protein concentration was determined by the Biuret reaction and the method of Lowry *et al.* (1951).

### 2.2. F-actin Fixation

The two sides of grids for electron microscopy (3 mm in diameter, Veco Grid, Ni, #100 or #75) were coated by using photo-resist (photo-resist 747, Kodak), which makes the surface hydrophobic. After being rinsed several times with distilled water, the photo-resist-coated grids were further coated with poly-L-lysine from a solution of 5 mg/ml in 0.4 N potassium hydroxide for 3 hours at 0 °C. Grids were floated on the surface of the various solutions in order to obtain F-actin attached to only one side. The grids were washed with 2 mM imidazole buffer, pH 7.4 (0.1 M KCl, 1 mM MgCl$_2$). To obtain grids with actin of different polarities, the grids were incubated on actin solutions in the absence or presence of cytochalasin B (Fig. 1). The grids were incubated on actin solutions containing 0.2 mg/ml G-actin, 0.2 mM ATP, 0.2 mM ATP, 0.1 M KCl, 1 mM MgCl$_2$, and

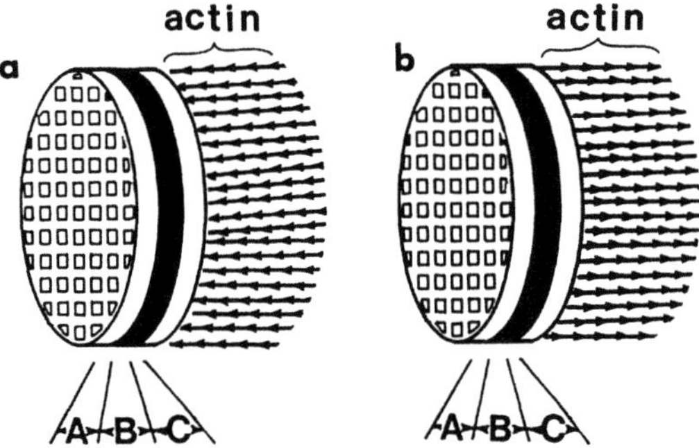


Fig. 1. A schematic representation of the stream grid. *a* Normal type. *b* Reverse type with cytochalasin B. *A* Grid, *B* photo-resist *C* poly-L-lysine

2 mM imidazole buffer, pH 7.4 for 1 hour at 0 °C (Fig. 1 *a*). The amount of fixed F-actin was 3–4 µg/cm$^2$. Actin (40 µg/ml) was also polymerized in the presence of 40 µM cytochalasin B (added from 2 mM stock solution in DMSO) with 0.15 M KCl, 1 mM MgCl$_2$, 0.2 mM ATP and 2 mM imidazole buffer (pH 7.4). After incubation for 3 hours at 20 °C, 4–5 µg/cm$^2$ of F-actin was fixed on the grids (Fig. 1 *b*) (Tsukita *et al.* 1984). F-actin prepared in the presence of cytochalasin B did not show any significant difference in the potentiation of ATPase activity of HMM from that of F-actin polymerized in the absence of cytochalasin B.

### 2.3. Rectangular Stream Cell

A new stream cell system was designed to permit control of the streaming velocity from the outside (see Fig. 2). A precision peristatic pump permitted precise control of flow rate. The stream cell used was composed of a ceiling and a sole plate 2 × 5 × 1 cm. Between the ceiling and sole plate there is a chamber 3 mm in diameter and 3 cm in length. In order to avoid interference of the attached actin filaments with the opposite sides of neighboring grids placed in the chamber, a metal spacer was inserted between the grids. For measurements, about 400 grids and spacers were placed in a row. The system was kept at 20 °C throughout the experiments. Immediately

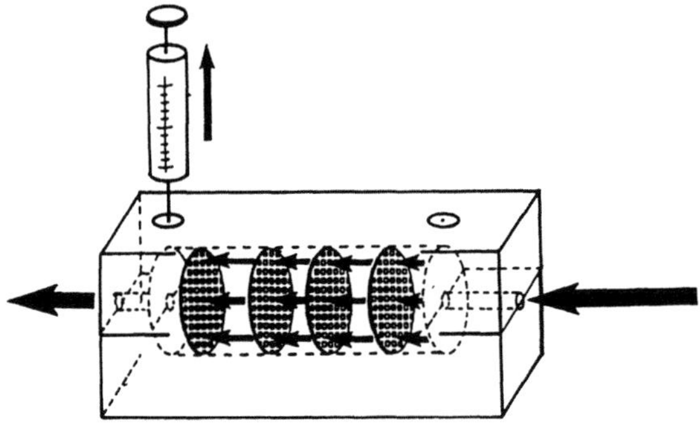

Fig. 2. A schematic illustration of the grid stream cell. The grid cell is composed of a ceiling and a sole plate of 2 × 5 × 1 cm. Between the ceiling and the sole plate there is a chamber with 3 mm in diameter and 3 cm in length. In order to avoid squeezing the attached actin filaments by the adjacent grid, a metal spacer was inserted between the grids. For our measurements, about four hundred grids and spacers were placed in a row

before injection of the solution, HMM solution (2 mg/ml HMM) and ATP solution (both containing 70 mM KCl, 5 mM $MgCl_2$, 0.1 mM $CaCl_2$, 20 mM MOPS, pH 7.0) were mixed in 1 : 1 proportions. The streaming was observed with an optical microscope equipped with night vision (Ikegami CTC9000) in order to avoid turbulence caused by the heat from the lamp. For measuring the ATPase activity, two holes sealed with silicon rubber, at intervals of 2 cm, at both ends of the chamber were drilled in the ceiling, and the samples were aspirated from there.

### 2.4. ATPase Activity

The ATPase activity was determined by the method of Anner and Moosmayer (1975) with slight modifications. Immediately after withdrawal of 10 µl of sample from the stream cell, it was transferred to 1 ml of ice-cold 4% trichloroacetic acid, and the mixture was rapidly mixed and kept below 4 °C until color development. After mixing all the color reagents, the solution was incubated for half an hour at 20 °C and the optical density was measured at 623 nm.

## 3. Results

### 3.1. Polarity of the Attached Actin Filaments

We have reported that under suitable conditions, G-actin of rabbit skeletal muscle polymerizes with a specific polarity on the surface of poly(L-lysine)-coated teflon. This technique was applied to fix F-actin and its polarity was verified using the electron microscopy. The arrowhead structure of HMM and F-actin and its polarity was verified using the electron microscope. Almost all actin filaments grew with arrowheads pointing toward the grid surface, indicating predominantly barbed end elongation (Fig. 3 a). This will be called the "normal" F-actin grid. Under our conditions, 3–4 µcm² F-actin was fixed to the surface of the grids. In the presence of 40 µM cytochalasin B, the elongation of actin filaments at the barbed ends was inhibited, so that most of the actin filaments pointed away from the surface of the grids, while a small number of the actin filaments were bidirectional (Fig. 3 b). The amount of fixed F-actin was 4–5 µg/cm². Hereafter we call this grid the "CB F-actin" grid.

### 3.2. Observation of Active Streaming

After fixing actin filaments to grids in the absence and presence of cytochalasin B, one portion of them was placed in chamber of the stream cell, without external pressure difference or external force. Using either grid, active streaming was observed in a specific direction with a velocity of about 1 µm/sec in the presence of a sufficient amount of HMM and ATP. The spontaneous flow was induced in the direction of the tail of the actin arrowheads, which is the same direction of motion as that of the crossbridge movement on muscle fiber thin

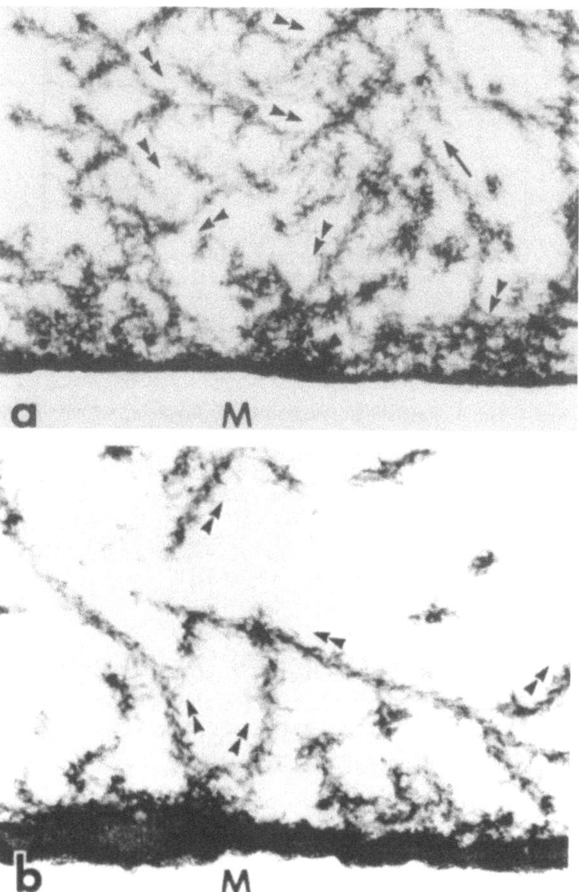



Fig. 3. *a* An electron micrograph of the surface of a normal stream grid and *b* a CB stream grid

filaments, and continued for 40–50 min. That is, with F-actin fixed to the grid with arrowheads pointing toward the surface of grid, the fluid moved from the side of the grids opposite from that carrying attached F-actin. On the contrary, in the case of the CB F-actin grid, the fluid flowed in the opposite direction. This spontaneous flow could not be observed in the absence of ATP and was regulated by calcium if troponin and tropomyosin were attached to actin filaments on the grids.

### 3.3. The Effects of the Streaming Velocity on ATPase Activity

To avoid aggregation of actin filaments on the surface of the grid, a flow was always driven from the side of the grid opposite to that carrying attached F-actin. In the case of the normal F-actin grids, the direction of the flow was the same as that of movement of crossbridges in shortening muscle, so it was defined as pos-

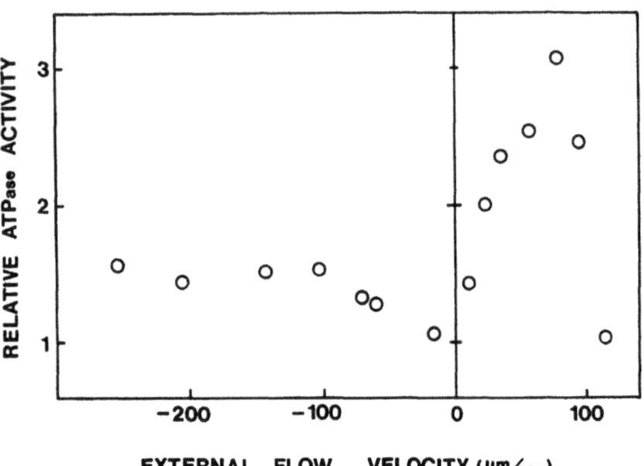

Fig. 4. The ATPase activity dependent on streaming velocity in the grid stream cell. The streaming velocity was evaluated by dividing the flowing volume per time unit by the total cross section of the openings of the grids. In order to compare the both cases quantitatively, the measured ATPase activities were normalized using the equation; Relative ATPase Activity = (measured ATPase activity)/ (ATPase activity of acto-HMM in solution)

itive flow. When the CB F-actin grids were placed in the chamber, the direction of the flow was opposite to that of crossbridges in shortening muscle, and we defined it as negative flow. The external flow did not cause turbulent flow even at maximum flow velocity (300 μm/sec), as far as we examined by studying the flow of bromphenol blue in the solution. Two kinds of grids with different sizes of meshes were used: one was the #100 (Veco Grid) with 100 μm meshes, and the other was #75, one side of a mesh of which was two times larger than the former. As shown in Fig. 4, using #100 grids, upon increasing the velocity of positive flow, the ATPase activity first increased and then decreased. The optimum flow velocity for enhancement of ATPase activity was 80–90 μm/sec. This striking phenomenon was not affected by the sequence of measurements. We also verified that the size of the meshes the grid (#100 to #75) and the amount of fixed actin filaments did not qualitatively influence on the biphasic characteristics of the ATPase activity (data not shown). On the contrary, in the case of negative flow velocity, the ATPase activity did not strongly depend on the streaming velocity, but remained almost constant. However, the ATPase activity was slightly enhanced by the streaming, possibly due to the existence of a small fraction of bidirectional actin filaments. In order to compare both cases quantitatively, the measured ATPase activities should be normalized, because the amount of fixed actin filament was different in each experiment. For this purpose, it is reasonable to measure the degree of the enhancement caused by the flow, normalized by

the ATPase activity measured at stationary streaming. However, it is very difficult to stop the streaming completely due to its spontaneity; therefore the ATPase activity in the homogeneous solution containing the same amount of actin and HMM as the grid cell was used as a calibration factor instead of the grid cell itself. The measured ATPase activities were normalized using the following equation:

Relative ATPase activity = (Measured ATPase)/(ATPase activity of Acto-HMM in solution)

Evidently the normalized ATPase activities showed at extreme asymmetry on both sides of the velocity of endogenous streaming.

## 4. Discussion

In the present study, we have successfully reconstituted a new streaming system, the grid cell, in which active streaming could be induced under physiological conditions by the interaction of heavy meromyosin with F-actin with two different polarities. Controlling the flow velocity from the outside, the grid cell made it possible to study how the two kinds of vectorial flows were coupled with ATP hydrolysis. The direction of the spontaneous streaming and the direction opposite to it corresponded to shortening and lengthening of muscle, respectively. The fact that, within the velocity range of shortening muscle, the motion-dependent ATPase activity of the reconstituted system corresponded to that of contracting muscle, shows that this system is physiologically relevant. In addition, the grid cell has enabled us to obtain information on chemo-mechanical coupling at higher velocities than the maximum shortening speed of muscle, and at negative velocities corresponding to the lengthening of muscle. Therefore, our system is potentially useful for studying the molecular mechanism of chemo-mechanical coupling in actomyosin.

The system must be carefully designed to prevent turbulent flow, which might complicate the flow-dependent ATPase activity of actomyosin. Judging from the Reynold's number of applied flow, it is reasonable that the flow was laminar, even at maximum streaming velocity (300 μm/sec). By studying a dye flowing in the solution, it was further verified that there was no turbulent flow.

For the quantitative analysis of the results from the stream cell, it was necessary to confirm that the structure of the actin filaments attached to the surface of the grids was intact throughout the series of experiments. The amount of F-actin attached to the surface was unchanged within experimental error before and after the experiments, even if the solution in the cell flowed at the highest streaming velocity, the characteristic flow dependence of the ATPase activity was maintained. The sequence of the flow velocity changes did not affect the characteristic ATPase profile. Accordingly, these observations ensured that the structure of the actin filaments attached to the surface of the grids was not perturbed irreversibly by the external flow.

When actin filaments were well organized with a specific polarity, the ATPase activity in the stream cell varied depending on the flow velocity even under the same chemical conditions. On the contrary, in the case of a homogeneous solution containing F-actin and heavy meromyosin without fixing actin filaments, the ATPase activity remained unchanged even when the solution was squeezed through the grid cell. As shown in Fig. 4, when the flow velocity is in the physiological range, the acto-HMM ATPase activity is an increasing function of the velocity. However it decreases when the velocity exceeds about 80–90 $\mu$m/sec. Furthermore, the negative flow does not almost influence the ATPase activity. This fact indicates that the physiological relative motion of actin and HMM slaves the ATPase activity in the grid cell. This peculiar behaviour of the ATPase activity in the grid cell would be closely related to the molecular mechanism of the chemo-mechanical conversion and would lead to a new control mechanism in enzyme reactions.

Here we suppose the new control mechanism termed "dissipative control". For simplicity, the cycle of ATP hydrolyzed by myosin can be written as follows,

$$\text{M} \cdot \text{ATP} \rightleftharpoons \begin{array}{c} \nearrow \text{M} \cdot \text{ADP} \cdot \text{Pi} \searrow \\ \text{ATP} \end{array} \text{M} + \text{ADP} + \text{Pi}$$

Since the reaction of ATP hydrolysis can not proceed until the free energy of the ATP molecule is consumed up, the reaction rate is controlled by the dissipation rate of the free energy of ATP molecule. As a matter of course, the dissipation control for chemical reaction may work on condition that the reaction rate is sufficiently slow. It is well known that $\text{M} \cdot \text{ADP} \cdot \text{Pi}$, which is a long life intermediate of ATP hydrolysis cycling, is the force generating state. Then the decomposition rate of the intermediate is proportional to the rate of the heat production and the performed work. Roughly speaking, the total force generation is proportional to the total number of the force generating intermediates, which is inversely proportional to the life time of the

intermediate. When the external load is lighter than the total force generation, myosin molecules in the force generating state move relatively on actin filaments with larger power and larger energy dissipation, resulted in shorter life time of the intermediates. Then, the total force generation is reduced in the motile system. In this way, the motile system generates force well balancing to the external load and reaches a steady state.

Finally, we refer to an important problem, the chemical detailed balance in the case of the chemo-mechanical coupling in the motile system. It is widely accepted that the chemical detailed balance is well maintained as long as the system is in local equilibrium. That is, enzymes enhance the forward reaction and the reverse one, equivalently. The relation in such cases is written as follows;

$$\frac{K_f (\text{without enzyme})}{K_r (\text{without enzyme})} = \frac{K_f (\text{with enzyme})}{K_r (\text{with enzyme})}$$
$$= \text{constant (for usual reaction)}$$

where $K_f$ and $K_r$ are the rate constant of the forward reaction and the reverse reaction, respectively. However, in the grid cell, the chemo-mechanical coupling is unidirectional and coupled only with one specific vectorial flow, and then the reaction rate of the ATP hydrolysis is enhanced during that specific vectorial flow but not during the opposite flow. It means that in a motile system the forward reaction is accelerated specifically when the motile system performs mechanical work practically. The extent of the enhancement closely depends on the power rate, so the chemo-mechanical energy conversion is efficiently controlled. We may reasonably conclude that the chemical detailed balance, the ratio of the rate constant of the forward reaction to the reverse one, does not always hold in the chemo-mechanical coupling system.

$$\frac{K_f (\text{chemo-mechanical})}{K_r (\text{chemo-mechanical})} > \frac{K_f (\text{usual})}{K_r (\text{usual})} = \text{constant}$$

Where $K_f$(chemo-mechanical) and $K_r$(chemo-mechanical) denote the apparent rate constant for the forward reaction and the reverse reaction in the motile system, respectively. A quantitative treatment will appear elsewhere.

## References

ANNER B, MOOSMAYER M (1975) Rapid determination of inorganic phosphate in biological systems by a highly sensitive photometric method. Anal Biochem 65: 305–309

EBASHI S, EBASHI F (1964) A new protein component participating in the superprecipitation. J Biochem 55: 604–613

FENN WO (1924) The relation between the work performed and the energy liberated in muscular contraction. J Physiol (Lond) 58: 373–395

LOWRY OH, ROUSEBROUGH NJ, FARR AL, RANDALL RJ (1951) Protein determination with the Folin phenol reagent. J Biol Chem 193: 265–275

PERRY SV (1955) Myosin adenosinetriphosphatase. In: COLOWICH SP, KAPLAN NO (eds) Methods in enzymology, vol 2. Academic Press, New York, pp 582–588

SHIMIZU H, YANO M (1978) Studies of the chemo-mechanical conversion in artificially produced streamings III. Dynamic co-operativity—A new cooperativity in actomyosin systems with a polarized arrangement of F-actin. J Biochem 84: 1093–1102

— — (1979) Dynamic cooperativity of molecular processes in active streaming, muscle contraction, and subcellular dynamics. Adv Biophys 13: 195–278

SPUDICH JA, WATT S (1971) The regulation of rabbit skeletal muscle contraction. I. Biochemical studies of the interaction of the tropomyosin-troponin complex with actin and the proteolytic fragments of myosin. J Biol Chem 246: 4866–4871

TSUKITA S, TSUKITA S, ISHIKAWA H (1984) Bidirectional polymerization of G-actin on the human erythrocyte membrane. J Cell Biol 98: 1102–1110

YANO M (1978) Observation of steady streamings in a solution of Mg-ATP and acto-heavy meromyosin from rabbit skeletal muscle. J Biochem 83: 1203–1204

— YAMADA T, SHIMIZU H (1978) Studies of the chemomechanical conversion in artificially produced streamings I. Reconstitution of a chemo-mechanical system from acto-HMM of rabbit skeletal muscle. J Biochem 84: 277–283

— SHIMIZU H (1978) Studies of the chemo-mechanical conversion in artificially produced streamings II. An order-disorder phase transition in the chemo-mechanical conversion. J Biochem 84: 1087–1092

— YAMAMOTO Y, SHIMIZU H (1982) An actomyosin motor. Nature 299: 557–559

— — — (1983) Actomyosin motor of active fragments of myosin and F-actin from rabbit skeletal muscle. In: Structure and function in non-muscle cells. Academic Press, Australia. pp 201–208

WEEDS AG, POPE B (1977) Studies on the chymotryptic digestion of myosin. Effects of divalent cations on proteolytic susceptibility. J Mol Biol 111: 129–257

Protoplasma (1988) [Suppl. 2]: 9–21


# Purification and Characterization of 83 kDa Nonmuscle Caldesmon from Cultured Rat Cells: Changes in Its Expression upon L 6 Myogenesis

S. Yamashiro-Matsumura, R. Ishikawa, and F. Matsumura*

Department of Biochemistry, Busch campus, Rutgers University, Piscataway, New Jersey

Received January 18, 1988
Accepted April 15, 1988

Dedicated to Professor Dr. Noburo Kamiya on the occasion of his 75th birthday

## Summary

We have purified and characterized nonmuscle caldesmon with Mr of 83,000 (83 kDa protein) from cultured rat cells, and have examined its intracellular localization in rat fibroblasts as well as L 6 myoblasts and myotubes. Like smooth muscle caldesmon, it is heat-resistant, binds to calmodulin-Sepharose in a calcium dependent manner, and binds to actin in a $Ca^{2+}$/calmodulin dependent manner. Further, affinity-purified polyclonal antibodies raised against 83 kDa protein cross-react with smooth muscle caldesmon from chicken gizzard, suggesting that they are immunologically related proteins. Actin binding of 83 kDa protein is saturated at an approximate molar ratio of 6 actin monomers to one 83 kDa protein molecule, and actin bundling is fully observed at an approximate molar ratio of 8 actin monomers to one 83 kDa protein in the absence of reducing agents. The protein is found to be a phosphoprotein, which probably explains the presence of four isoelectric variants around pI 7 on high resolution two dimensional gels. Indirect immunofluorescence has shown that 83 kDa protein is localized in stress fibers of L 6 myoblasts, but disappears after fusion of L 6 myoblasts into myotubes. Biochemical analyses have confirmed this observation. SDS gel analyses have revealed that the microfilament fraction isolated from L 6 myoblasts contains 83 kDa protein as one of major components while the same protein is not detected in the microfilament fraction isolated from L 6 myotubes.

*Keywords:* Actin-binding protein; Calmodulin; Microfilaments; Nonmuscle caldesmon; Tissue cultured cells.

*Abbreviations:* PBS phosphate-buffered saline (137 mM NaCl, 2.5 mM KCl, 1.5 mM $KH_2PO_4$, 8 mM $Na_2PO_4$ (pH 7.3)); SDS PAGE sodiumdodecylsulfate-polyacrylamide gel electrophoresis; PMSF phenylmethylsulfonyl fluoride; DTT dithiothreitol; Buffer A, 2.5 mM EDTA, 0.2 mM PMSF, 0.5 mM DTT, 20 mM Tris/HCl

---

* Correspondence and Reprints: Department of Biochemistry, Busch campus, Rutgers University, P.O. Box 1059, Piscataway, NJ 08854, U.S.A.

(pH 8.0); Buffer B, 50 mM Tris acetate buffer, 3 mM $MgCl_2$, 1.5 mM $CaCl_2$, 1 mM DTT, 0.2 mM PMSF (pH 7.5); Buffer C, 10 mM sodium phosphate buffer, 0.5 mM DTT, 0.2 mM PMSF (pH 7.0).

## 1. Introduction

Caldesmon is an actin-binding, calmodulin-binding protein first identified in chicken gizzard smooth muscle (Sobue *et al.* 1981). The protein from gizzard contains two polypeptides of 138,000 and 140,000. Its binding to actin is regulated through the $Ca^{2+}$-dependent interaction with calmodulin (Sobue *et al.* 1981, Bretscher 1984). In the presence of micromolar concentrations of $Ca^{2+}$, caldesmon binds to calmodulin and such caldesmon/calmodulin complex does not bind to actin. On the other hand, caldesmon does not bind to calmodulin in the absence of $Ca^{2+}$, and calmodulin-free caldesmon binds to actin.

Recently immunologically cross-reactive forms of caldesmon have been identified in many nonmuscle cells including cultured cells although nonmuscle caldesmon-like proteins show much lower molecular weights (ranging from 70,000 to 80,000 vs 140,000 of smooth muscle caldesmon) than those of smooth muscle caldesmon (Koji-Owada *et al.* 1984, Bretscher and Lynch 1985). These studies have also shown that nonmuscle caldesmon is localized in stress fibers as well as membrane ruffles. Subsequent studies (Sobue *et al.* 1985 b, Dingus *et al.* 1986) on purified caldesmon from nonmuscle cells have revealed that actin-binding of these polypeptides is similar to that of smooth muscle caldesmon. While these studies suggest the important roles of nonmuscle caldesmon in the microfilament organization and motility, further studies are needed to elucidate the properties and functions of nonmuscle caldesmon. For example, it is still not clear whether nonmuscle caldesmon has actin-bundling activity.

We have previously reported the characterization of microfilaments isolated from cultured rat cells by use of monoclonal antibodies to tropomyosin (Matsumura *et al.* 1983). During the analysis of molecular components of the isolated microfilaments, we found that one protein with a Mr of 83,000 was a heat-stable, actin-binding protein like tropomyosin or caldesmon (Matsumura *et al.* 1983). Further, we have found that the actin-binding of 83 kDa protein is regulated by $Ca^{2+}$/calmodulin like caldesmon.

In this paper, we describe the purification and characterization of the caldesmon-like 83 kDa protein, as well as its intracellular localization in cultured rat cells including L 6 myoblasts and myotubes. We find that 83 kDa protein makes actin filaments into bundles in the absence of reducing agents. This property of 83 kDa nonmuscle caldesmon is similar to that of smooth muscle caldesmon reported by Lynch *et al.* (1987) but inconsistent with those of other nonmuscle caldesmons previously reported (Sobue *et al.* 1985 b, Dingus *et al.* 1986). We also show that 83 kDa protein is a phosphoprotein with four isoelectric variants on high resolution two-dimensional gels. Further we have examined localization of 83 kDa protein in both L 6 myoblasts and myotubes. In L 6 myoblasts, the 83 kDa nonmuscle caldesmon is found in stress fibers, as well as in membrane ruffles, which is similar to the localization of 83 kDa protein in rat fibroblasts. The protein, however, disappears when L 6 myoblasts fuse into myotubes.

## 2. Materials and Methods

### 2.1. Cell Culture

Cultured cells used in the present study were L 6 myoblasts and myotubes, normal rat kidney cells, REF-52 cells (an established rat embryo cell line) and an SV 40 transformed REF-52 (REF-4A). Cell lines were maintained in Delbecco's modified Eagle's medium containing 10% fetal calf serum in an atmosphere of 5% $CO_2$ and 95% air at 37 °C unless otherwise specified. Fetal calf serum was replaced with newborn calf serum for culture of L 6 myoblasts and myotubes. For culture of REF-4A cells in a large scale, cells were grown in large square plates (245 × 245 × 20 mm, Nunc) in the Delbecco's modified Eagle's medium containing 10% calf serum.

### 2.2. Isolation of Microfilaments from Cultured Cells

The details of isolation method of microfilaments were described in the previous paper (Matsumura *et al.* 1983). Briefly, monolayer cells (10–20 plates of 100 mm culture dish) were washed 2–3 times with PBS and then extracted for 2 min with Triton-glycerol solution (0.1 M PIPES, 5 mM $MgCl_2$, 0.2 mM EGTA, 0.05% Triton X-100, 4 M glycerol) to stabilize the cytoskeleton. After washing with PBS, cell residues were collected and homogenized in the presence of 5 mM PMSF and 5 mM ATP. After centrifugation, 1/50 volume of ascites fluid of a monoclonal antibody, IV 15, was added to the supernatant to induce aggregation of microfilaments into bundles. The resultant microfilament bundles were collected by low speed centrifugation away from other elements of cell extracts and washed with PBS. The final pellet was suspended in 50–100 µl of PBS and used as a microfilament fraction.

For *in vivo* labelling, cells in one of 100 mm cultured dishes were labeled for 16 hours either with 250 µCi of $S^{35}$-methionine (1110 Ci/mmol) in methionine-free Dulbecco's modified Eagle's medium containing 2.5% fetal calf serum, or with 1 mCi of $^{32}P$ orthophosphoric acid (10 mCi/ml) in phosphate-free medium containing 5% dialyzed fetal calf serum.

### 2.3. Purification of 83 kDa Protein from Cultured Rat Cells

REF-4A cells grown in large square plates were washed 3 times with PBS and stored at —70 °C. Cells in wet weight of 120 g were homogenized with a Waring Blender (three times, 15 s each) in 500 ml of buffer A containing 0.1 M NaCl. The homogenates were stirred at 4 °C for 1 hour and the first extract was obtained by centrifugation at 16,000 × **g** for 15 min. The pellet was re-extracted for 10 min in 50 ml of buffer A containing 0.1 M NaCl and the second extract was obtained by centrifugation at 16,000 × **g** for 15 min. The first and second extracts were combined and heated in a boiling water bath for 15 min, cooled on ice for 30 min, and then centrifuged at 16,000 × **g** for 15 min. The heat-stable supernatant was fractionated by adding ammonium sulfate powder between 0–28 g per 100 ml of the supernatant. The precipitates were recovered by centrifugation at 16,000 × **g** for 15 min and dialyzed for 40 hours with three changes against 4 liters of 5 mM $NaHCO_3$ containing 0.5 mM DTT and 0.2 mM PMSF. The solution was clarified by centrifugation at 17,300 × **g** for 10 min and applied to a DE-52 (Whatman) ion-exchange column (2.5 × 10 cm) equilibrated in buffer A. The column was washed with 2 column volumes of buffer A and developed with a linear gradient of NaCl (0–500 mM) in buffer A (300 ml/chamber). Column fractions were monitored by SDS-PAGE as shown in Fig. 1 *A* and the 83 kDa protein eluted at about 100 mM NaCl were pooled. After addition of 1.3 mM (final conc.) $CaCl_2$, the pooled fractions of the DE-52 column were directly applied to a Calmodulin-Sepharose (Pharmacia) column (0.8 × 18 cm) equilibrated with buffer B. The column was washed with 5 column volumes of buffer B containing 0.15 M NaCl and eluted with the buffer B except that 1.5 mM $CaCl_2$ was replaced with 4 mM EGTA. Column fractions were monitored by SDS-PAGE (Fig. 1 *B*). The 83 kDa protein eluted with EGTA-containing solution were pooled and directly applied to a Hydroxylapatite (Bio-Rad) column (1.2 × 15 cm) equilibrated with buffer C. The column was developed with a linear gradient of phosphate (10–250 mM) containing 0.5 mM DTT and 0.2 mM PMSF (100 ml/chamber). Column fractions were monitored by SDS-PAGE (Fig. 1 *C*) and the 83 kDa protein fractions eluted at 200 mM phosphate were pooled, dialyzed against 20 mM imidazole buffer of pH 7.0 containing 100 mM KCl and 0.5 mM DTT, and concentrated by Centricon 30 (Amicon).

### 2.4. F-Actin Binding Assay

Purified 83 kDa protein at varying concentrations (final concentration 0–0.29 mg/ml) was mixed with F-actin (0.5 mg/ml) in 70 µl of

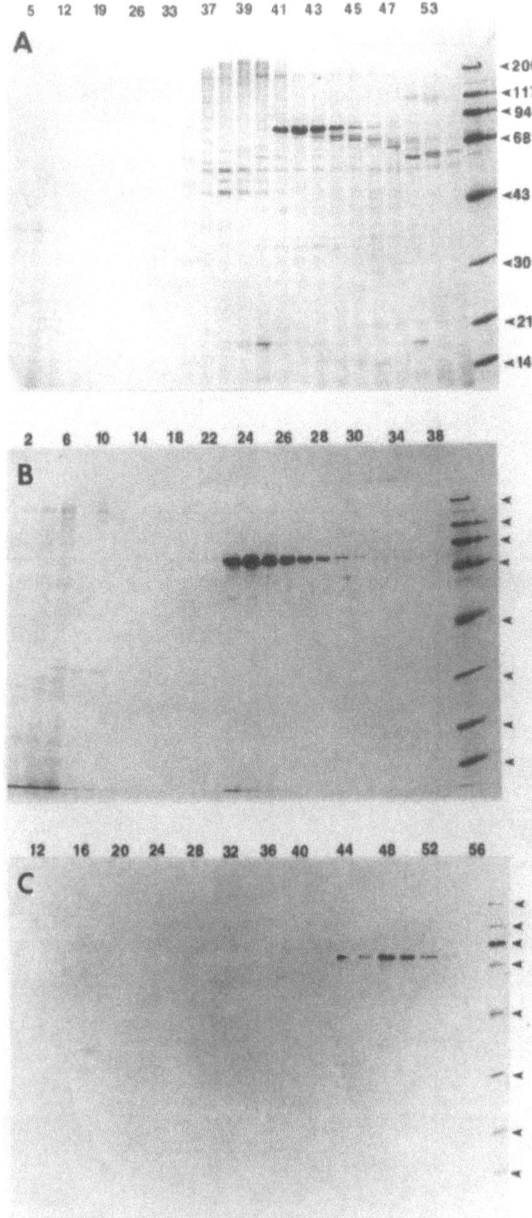

Fig. 1. Purification of 83 kDa protein. *A* DEAE-cellulose column chromatography. Ammonium sulfate (0–28 g per 100 ml of heat-stable extract) precipitated proteins were loaded on a DE-52 ion-exchange column equilibrated with buffer A. The proteins were eluted with 600 ml (total) of a linear NaCl gradient (0–500 mM). NaCl gradient started at fraction No. 23. Fractions of 7.5 ml were collected and analyzed by SDS-PAGE. 83 kDa protein (fraction No. 41–44) eluted at about 100 mM NaCl were pooled. *B* Calmodulin-Sepharose affinity column chromatography. After addition of 1.3 mM CaCl₂, the DE-52 column fractions were directly applied to a Calmodulin-Sepharose column equilibrated with buffer B. The column was washed with buffer B containing 0.15 M NaCl and eluted at fraction No. 21 with the buffer B except that 1.5 mM CaCl₂ was replaced with 4 mM EGTA. Column fractions of 1.8 ml were collected and monitored by SDS-PAGE. Fractions from No. 23 to 28

100 mM KCl, 20 mM imidazole HCl buffer and 0.5 mM DTT (pH 7.0). In some experiments, DTT was omitted. After incubation for 90 min at room temperature, the mixture was centrifuged in a Beckman Airfuge at 140,000 × g (28 p.s.i.) for 20 min. The 83 kDa protein alone was not precipitated under these conditions. Both supernatants and pellets were suspended in an equivalent volume of SDS sample buffer, run with SDS-PAGE, and then, the amounts of actin and 83 kDa protein were quantified as described previously (YAMASHIRO-MATSUMURA and MATSUMURA 1985).

The effects of calmodulin on the binding of 83 kDa protein to actin were examined in the following conditions; 0.5 mg/ml of F-actin, 0.18 mg/ml of 83 kDa protein, 0.18 mg/ml of calmodulin, 20 mM imidazole buffer of pH 7.0, 0.1 M KCl and 0.5 mM DTT. The concentrations of free $Ca^{2+}$ were regulated using 2 mM EGTA-$Ca^{2+}$ buffer calculated according to AMOS et al. (1976).

### 2.5. F-Actin Bundling Assay by Low Speed Centrifugation

F-actin at 0.5 mg/ml was incubated with varying amounts (0–0.258 mg/ml) of 83 kDa protein in 50 µl of 100 mM KCl, 20 mM Imidazole-HCl, pH 7.0, with or without 2 mM MgCl₂, or with or without 0.5 mM DTT. After incubation for 90 min at room temperature, the mixtures were centrifuged for 15 min at 12,000 × g with an Eppendorf centrifuge model 5412. Under these conditions, F-actin alone remained in the supernatant while bundles of F-actin were precipitated. Supernatants were carefully separated from the pellets and both fractions were suspended in an equivalent volume of SDS sample buffer. Samples were run on SDS-PAGE and quantified as described previously (YAMASHIRO-MATSUMURA and MATSUMURA 1985). The pellets were also examined by an electron microscopy using the negative staining technique.

### 2.6. Antibody Production

The hydroxylapatite column purified protein was further purified by preparative SDS-PAGE and used to raise polyclonal antibodies. Several rabbits were immunized by subcutaneous injection of the protein. For the first injection, 150 µg of the 83 kDa protein was emulsified in 1 ml of Freund's complete adjuvant and four boosts were done with 100 µg of protein emulsified in Freund's incomplete adjuvant over a 2-month period. The antibody was affinity-purified essentially following the methods of OLMSTED (1981) and of SMITH and FISHER (1984) with slight modification as described in the previous paper (YAMASHIRO-MATSUMURA and MATSUMURA 1985).

### 2.7. Other Procedures

Protein concentration was determined by the method of LOWRY et al. (1951) using bovine serum albumin as a standard. SDS-PAGE was performed essentially as described by BLATTER et al. (1972) using

were pooled for 83 kDa protein. *C* Hydroxylapatite column chromatography. Calmodulin-Sepharose column fractions were directly applied to a Hydroxylapatite column equilibrated with buffer C. The proteins were eluted with 200 ml of a linear gradient of phosphate (10–250 mM) containing 0.5 mM DTT and 0.2 mM PMSF. Fractions (2 ml each) were monitored by SDS-PAGE and the 83 kDa protein (indicated by an arrow) eluted at 200 mM phosphate was pooled, dialyzed against 20 mM imidazole buffer of pH 7.0 containing 100 mM KCl and 0.5 mM DTT. The positions of molecular mass markers (from top to bottom: 200 kDa, 117 kDa, 94 kDa, 68 kDa, 43 kDa, 30 kDa, 21 kDa, and 14 kDa) are indicated on the right

12.5% polyacrylamide except that the buffer system of Laemmli (1970) was used. Samples were dissolved in equal volumes of 2 × SDS sample buffer at a final concentration of 1% SDS, 50 mM DTT, 40 mM Tris-HCl, pH 6.8, 7.5% glycerol and 0.0005% bromphenol blue. Sucrose density gradient centrifugation for the determination of sedimentation coefficients was done according to the procedure of Martin and Ames (1961). The samples were loaded on the 5–20% (w/v) sucrose gradients in 20 mM imidazole buffer (pH 7.0) containing 100 mM KCl and 0.5 mM DTT, and centrifuged in a SW 50.1 Ti rotor at 38,000 rpm for 20 hours. Protein standards included chymotrypsinogen ($S_{20,w}$ = 2.6) and aldolase ($S_{20,w}$ = 7.4). The Stokes radius was determined by gel filtration with Sephacryl S-300 according to the method of Siegel and Monty (1966). Western blot was performed according to the method of Towbin *et al.* (1979). To prepare total cell lysates, cells (one 100 mm culture dish, $10^6$–$10^7$ cells) were quickly washed 3 times with PBS and extracted by addition of 300 μl of hot SDS sample buffer. The extracts were homogenized with a syringe by several passages through a needle (gauge no. 28). The homogenates were heated for 3 min at 100 °C and used immediately for immunoblotting. Immunoprecipitation was performed as described (Matsumura *et al.* 1983). Immunofluorescence was performed as described (Yamashiro-Matsumura and Matsumura 1986). Two-dimensional gel electrophoresis was performed according to O'Farrell (1975), and samples were prepared after Garrels (1979 a). For amino acid composition, purified 83 kDa protein was hydrolyzed for 16 hours at 115 °C in 6 N HCl, 0.2% phenol and the hydrolysate was run on a Beckman 121 M analyzer. Actin was prepared from rabbit skeletal muscle as described previously (Matsumura *et al.* 1983). Smooth muscle caldesmon was prepared from chicken gizzard or bovine aorta by the method described by Bretscher (1984).

## 3. Results

### 3.1. Purification of 83 kDa Protein

We have purified 83 kDa nonmuscle caldesmon by three steps of column chromatography. The order of chromatography is designed in such a way that each column fraction can be directly applied on the next column without dialysis, thus allowing us to avoid lengthy dialysis. Approximately 1.3 mg of 83 kDa protein was purified from 120 g (wet weight) of cells within 6 days. A typical purification pattern is shown in Fig. 1.

The protein is susceptible to proteases. If heat-treatment was omitted or done at 80 °C instead of 100 °C, the protein tended to be degraded during the column purification.

### 3.2. Amino Acid Composition

Table 1 shows the amino acid composition of 83 kDa protein. For comparison, the amino acid composition of bovine smooth muscle caldesmon is also included. In general, the compositions of these two proteins are similar to each other.

Table 1. *Amino acid composition of 83 kDa protein and smooth muscle caldesmon*

|      | 83 kDa protein | Smooth muscle caldesmon |
|------|----------------|-------------------------|
| Asx  | 65.2  | 80.5  |
| Ser  | 56.7  | 49.0  |
| Gly  | 53.7  | 52.5  |
| Glx  | 151.9 | 338.1 |
| Thr  | 27.7  | 54.8  |
| Ala  | 61.2  | 101.4 |
| Val  | 27.5  | 52.5  |
| Met  | 11.6  | 14.0  |
| Tyr  | 7.3   | 2.3   |
| Ile  | 12.7  | 26.8  |
| Leu  | 44.2  | 58.3  |
| Phe  | 15.0  | 17.5  |
| His  | 5.5   | 8.9   |
| Lys  | 76.3  | 165.6 |
| Arg  | 72.8  | 143.4 |

Number of residues are calculated assuming a molecular weight of 83,000 for 83 kDa protein and 140,000 for bovine aorta smooth muscle caldesmon. Tryptophan, cysteine and proline not determined

### 3.3. Actin Binding Properties

We have measured stoichiometry of the actin binding of 83 kDa protein by high speed centrifugation (150,000 × **g** for 20 min) in the conditions of 20 mM imidazole buffer of pH 7.0, 100 mM KCl and 0.5 mM

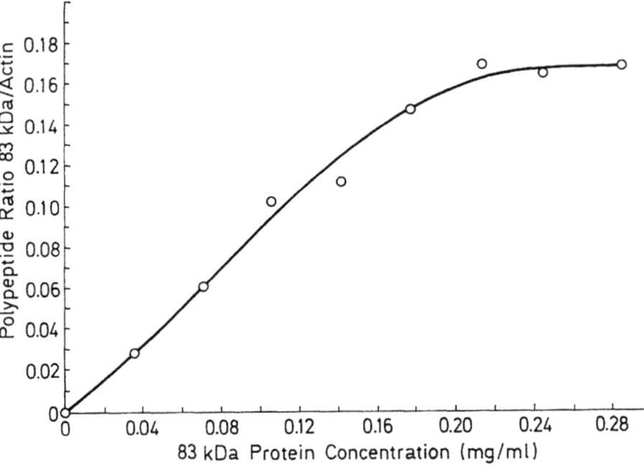

Fig. 2. Saturation of F-actin with 83 kDa protein. 83 kDa protein at the indicated concentration was mixed with F-actin (final concentration 0.5 mg/ml) and adjusted to 100 mM KCl, 20 mM imidazole buffer of pH 7.0 and 0.5 mM DTT. After incubation for 90 min at room temperature, the mixtures were centrifuged at 140,000 × **g** (28 p.s.i.) for 20 min in a Beckman Airfuge. Both supernatants and pellets were adjusted to equivalent volumes of SDS sample buffer and run on 12.5% acrylamide gels. The binding of 83 kDa protein was determined by densitometry as described previously (Yamashiro-Matsumura and Matsumura 1985). Saturation of binding is observed at a molar ratio of one 83 kDa protein to six actin molecules

DTT. As Fig. 2 shows, saturation is achieved at an approximate polypeptide ratio of one 83 kDa protein to six actin monomers.

The Stokes radius and sedimentation coefficient of 83 kDa protein in the same salt conditions were determined as 60.5 A and 3.5, respectively (Yamashiro-Matsumura and Matsumura 1988). These values suggest that the 83 kDa protein is a monomeric, asymmetric protein with a native molecular weight of 87,000. Thus the molar ratio of 83 kDa protein to actin for saturation should be about 1:6. These results are inconsistent with those of the previous report (Sobue *et al.* 1985 b). It showed that nonmuscle caldesmon from bovine adrenal medula is a tetramer with a molecular weight of 300,000 and that it saturates actin at the molar ratio of one nonmuscle caldesmon per 12–14 actin monomers. The reason for this difference is currently unknown but similar discrepancy is reported on smooth muscle caldesmon (Bretscher 1984, Lynch *et al.* 1987, Sobue *et al.* 1981, 1985 c).

Like smooth muscle caldesmon, the binding of 83 kDa protein to actin is regulated by $Ca^{2+}$/calmodulin (data not shown). In the presence of $Ca^{2+}$, calmodulin reduces actin binding of 83 kDa protein to 20–30% of the control. In the absence of $Ca^{2+}$, on the other hand, calmodulin does not inhibit the binding of 83 kDa protein to actin. Similar to the case of smooth muscle caldesmon, high concentrations of calmodulin, 5–10 molar excess of calmodulin over 83 kDa protein are required for the release of 83 kDa protein from actin.

### 3.4. Actin Bundling Activity

The morphology of 83 kDa protein/F-actin complex was examined by an electron microscopy using the negative staining technique. In the presence of 0.5 mM DTT, the structure of F-actin complexed with a saturating amount of 83 kDa protein appeared same as that of F-actin. On the other hand, when 83 kDa protein was incubated with F-actin in the absence of reducing agents, numerous bundles of F-actin were observed. This effect of DTT on the actin bundling activity is similar to that of smooth muscle caldesmon recently reported by Lynch *et al.* (1987).

The bundling effect was quantitatively examined by low speed centrifugation where bundles of F-actin but not free F-actin were pelletable. As Fig. 3 shows, full bundling was observed at an approximate polypeptide ratio of one 83 kDa protein to 8 actin molecules. The bundling effect was enhanced by the presence of magnesium ions.

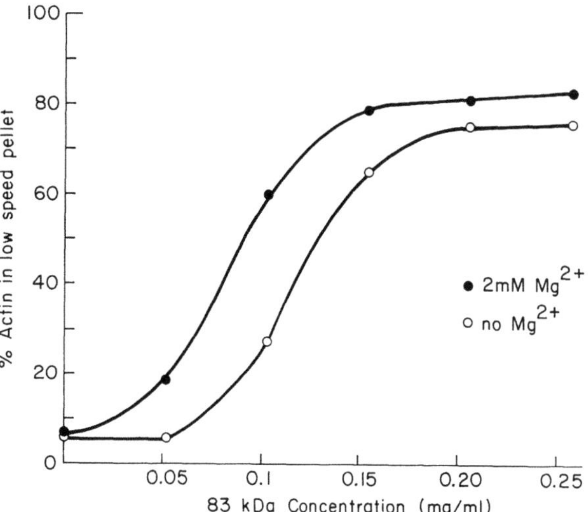


Fig. 3. F-actin bundling activity of 83 kDa protein in the absence of reducing agents. Actin bundling activity of 83 kDa protein in the absence of reducing agent was assayed by low speed centrifugation as described in Materials and Methods. Full bundling was observed at a molar ratio of one 83 kDa protein to 8 actin molecules. The addition of 2 mM MgCl$_2$ enhances the bundling activity of 83 kDa protein

### 3.5. Phosphorylation of 83 kDa Protein

As described previously, 83 kDa protein is enriched in microfilament fraction isolated from rat cultured cells using anti-tropomyosin monoclonal antibodies (Matsumura *et al.* 1983). When the microfilament fraction as well as total cell lysates of L 6 myoblasts were analysed on high resolution 2 D gels (run by Dr. J. I. Garrels, the Cold Spring Harbor Lab), the 83 kDa protein was found to be focused as four spots with different isoelectric points around 7.0 (Fig. 4 *A* and *B*). We have examined if phosphorylation may explain these variations in isoelectric points.

REF-4 A cells were first labeled *in vivo* with $^{32}$P-phosphate and microfilaments were isolated as described (Matsumura *et al.* 1983). Because 83 kDa protein is heat-stable like tropomyosin, microfilament fraction was heat-treated to further purify 83 kDa protein. Figure 5 *B* shows an autoradiograph of two-dimensional gel analyses of heat-stable proteins of isolated microfilaments. Although our two-dimensional gel system only resolves two spots for 83 kDa protein, these two spots are labeled with $^{32}$P-phosphate. It also appears that a spot with a more acidic isoelectric point has more incorporation of phosphate, which probably explains the difference in isoelectric points of these two spots.

Immunoprecipitation was performed using rabbit anti-

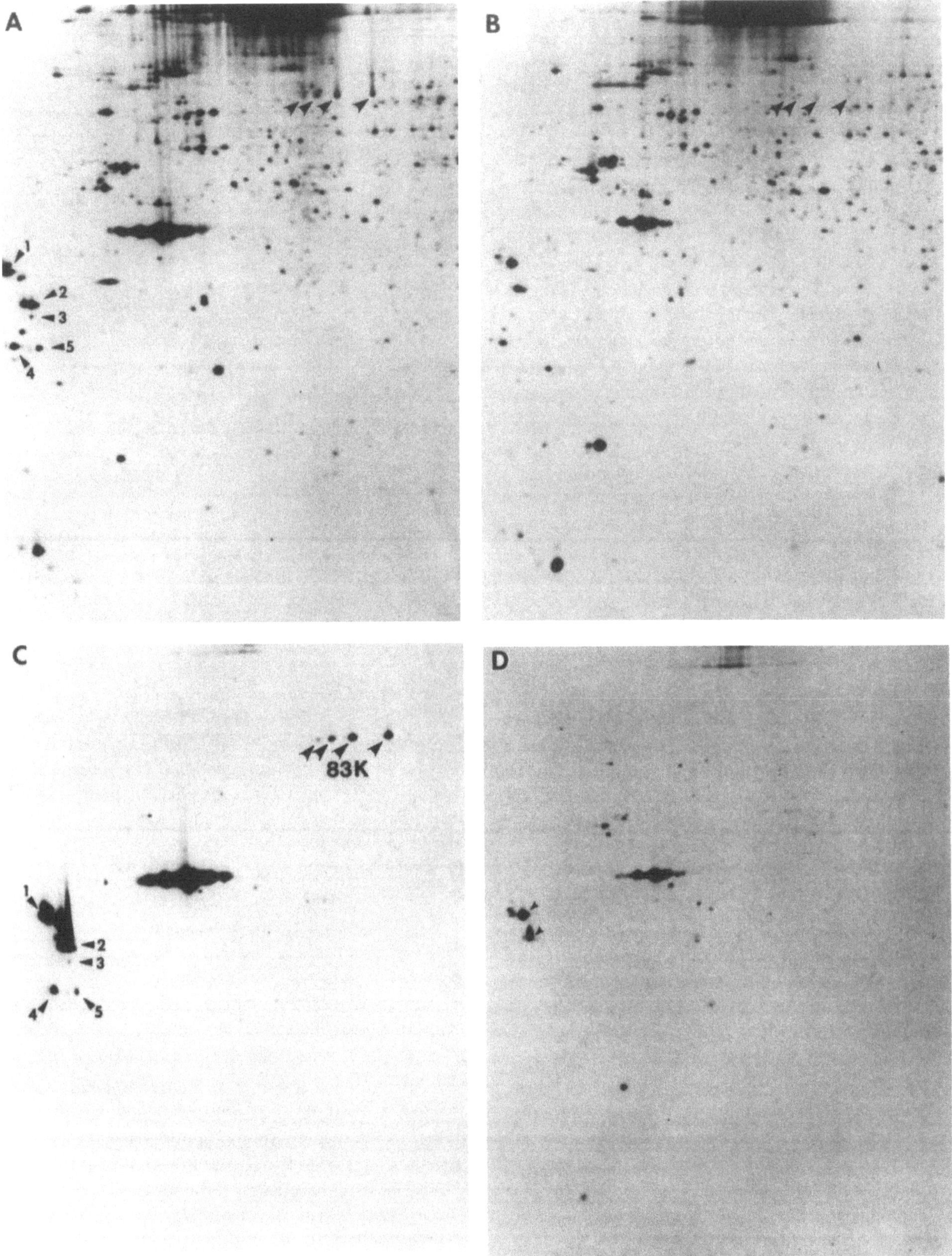

Fig. 4. High resolution two-dimensional gel analysis of 83 kDa protein. *A* Total cell lysates prepared from [35]S-methionine-labeled L 6 myoblasts were analyzed by two dimensional gel electrophoresis with pH 3–10 ampholytes for the first dimension (*left*, acidic; *right*, basic) and 10% acrylamide for the second dimension (these high resolution gels were run by Dr. G. I. GARRELS at the Cold Spring Harbor Laboratory). Four arrows without numbers indicate isoelectric variants of 83 kDa proteins: Arrows with numbers from one to five are multiple isoforms of tropomyosin (MATSUMURA *et al.* 1983). *B* Total cell lysates of L 6 myotubes. *C* Microfilament fraction isolated from L 6 myoblasts. *D* Microfilament fraction isolated from L 6 myotubes

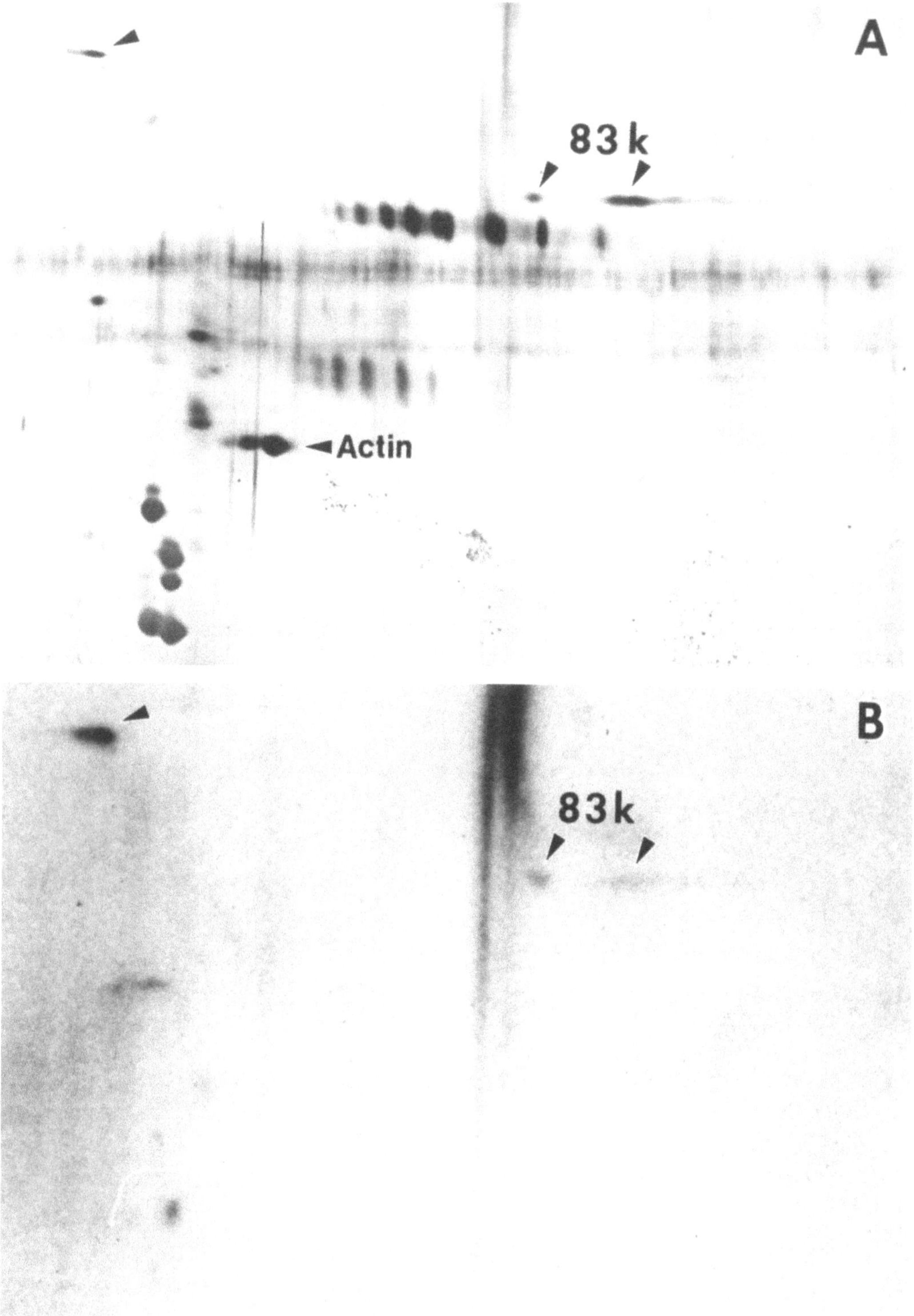



Fig. 5. Phosphorylation of 83 kDa protein. Microfilament fraction was prepared from $^{32}$P orthophosphoric acid-labeled REF-4 A cells and then, heat treated at 100 °C for 10 min to recover heat stable proteins in the supernatant. The supernatant was analyzed by two-dimensional gel electrophoresis with pH 3–10 ampholytes for the first dimension (*left*, acidic; *right*, basic) and 12.5% acrylamide for the second dimension. After electrophoresis, gels were silver-stained (*A*) and autoradiographed (*B*). Both 83 kDa protein and tropomyosin are major heat stable components of isolated microfilaments (*A*) although the efficiency of silver staining was low to compare other contaminated proteins such as IgM heavy chain (focused as multiple spots just under 83 kDa protein). Unlike the high resolution two-dimensional gels shown in Fig. 4, our gel system resolved 83 kDa protein as only 2 spots. Autoradiography (*B*), however, clearly shows that these spots of 83 kDa protein are phosphorylated. The very acidic spot with a higher Mr than 83 kDa protein (indicated by an arrow) is an unidentified phosphoprotein but not smooth muscle caldesmon

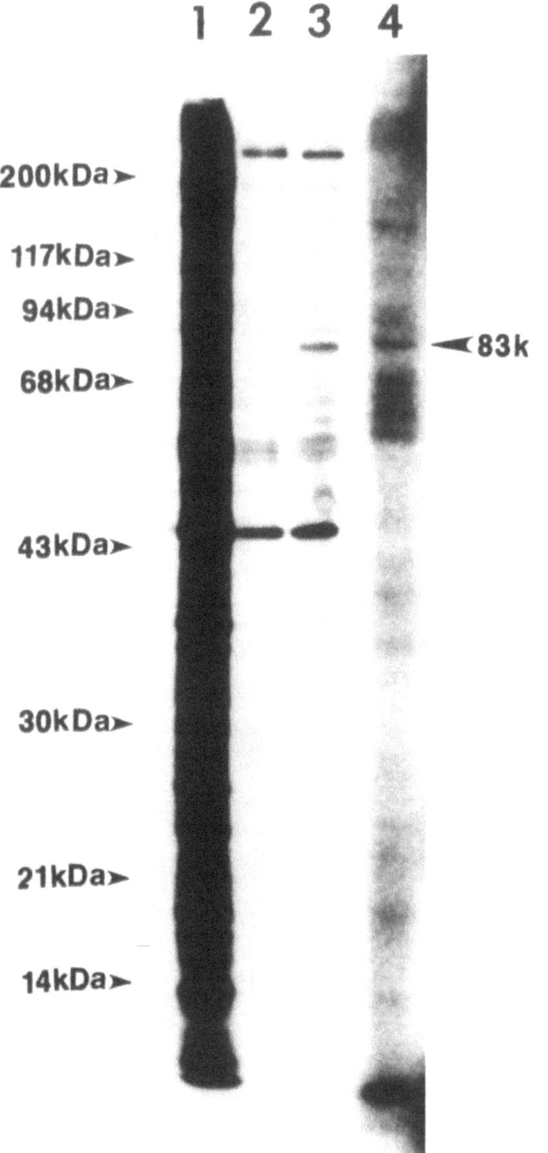



Fig. 6. Immunoprecipitation of 83 kDa protein from [35]S-methionine or [32]P-phosphate labeled cell by rabbit antiserum to 83 kDa protein. Lane *1*, total cell extracts labeled with [35]S-methionine; lane *2*, immunoprecipitates with preimmune serum from [35]S-methionine labeled cell extracts; lane *3*, immunoprecipitates with immune serum from [35]S-methionine labeled cell extracts; lane *4*, immunoprecipitates with immune serum from [32]P-phosphate labeled cell extracts. Arrow indicates 83 kDa protein immunoprecipitated

serum against 83 kDa protein to see if immunoprecipitated protein contains phosphate. REF-4 A cells were labeled *in vivo* with either [35]S-methionine or [32]P-phosphate. As lane 3 of Fig. 6 shows, the antiserum immunoprecipitated 83 kDa protein from [35]S-methionine labeled cell lysates. When [32]P-phosphate labeled lysates

were used instead, the immunoprecipitated 83 kDa protein was found to be labeled with phosphate (lane 4), confirming that 83 kDa protein is a phosphoprotein.

### 3.6. Intracellular Localization of 83 kDa Protein in L6 Myoblasts and Myotubes

Polyclonal rabbit antibody against 83 kDa protein was affinity-purified in order to examine localization of the protein in cultured rat cells including L 6 myoblasts and myotubes. Figure 7 shows the specificity of the

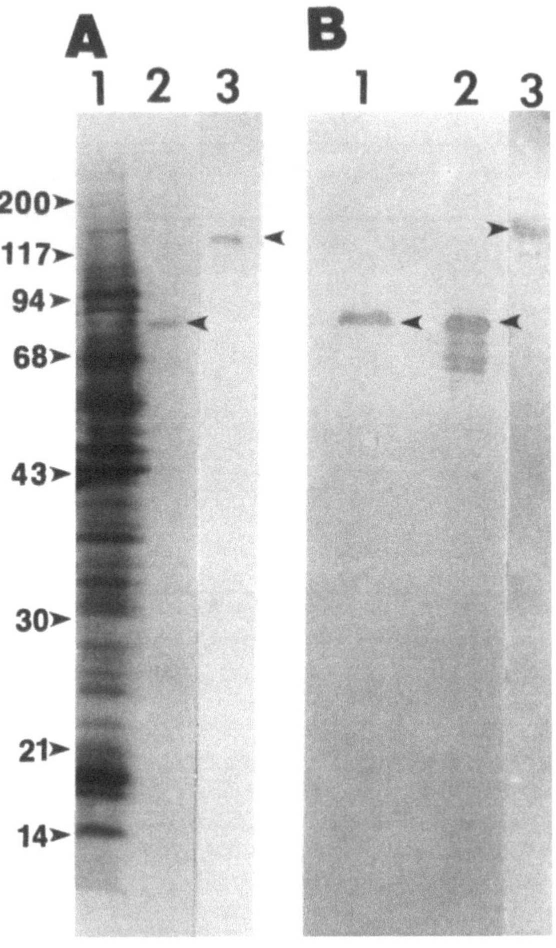



Fig. 7. Immunoblot analysis of 83 kDa protein with affinity-purified polyclonal antibody to 83 kDa protein. Total cell lysates of normal rat kidney cells (lane *1* of *A* and *B*), purified 83 kDa protein (lane *2* of *A* and *B*), and purified chicken gizzard smooth muscle caldesmon (lane *3* of *A* and *B*) were electrophoresed, transferred to nitrocellulose paper, and stained with Amido Black (*A*) or stained for immunoreactivity with the affinity-purified antibodies against 83 kDa protein (*B*). The antibody was reactive with both purified 83 kDa protein (lane 2 of *B*) and smooth muscle caldesmon (lane 3 of *B*), suggesting that they are related protein. However, the antibody reacted only with a polypeptide corresponding to 83 kDa protein when total cell lysates of cultured rat cells were tested (arrow in lane 1 of *B*)

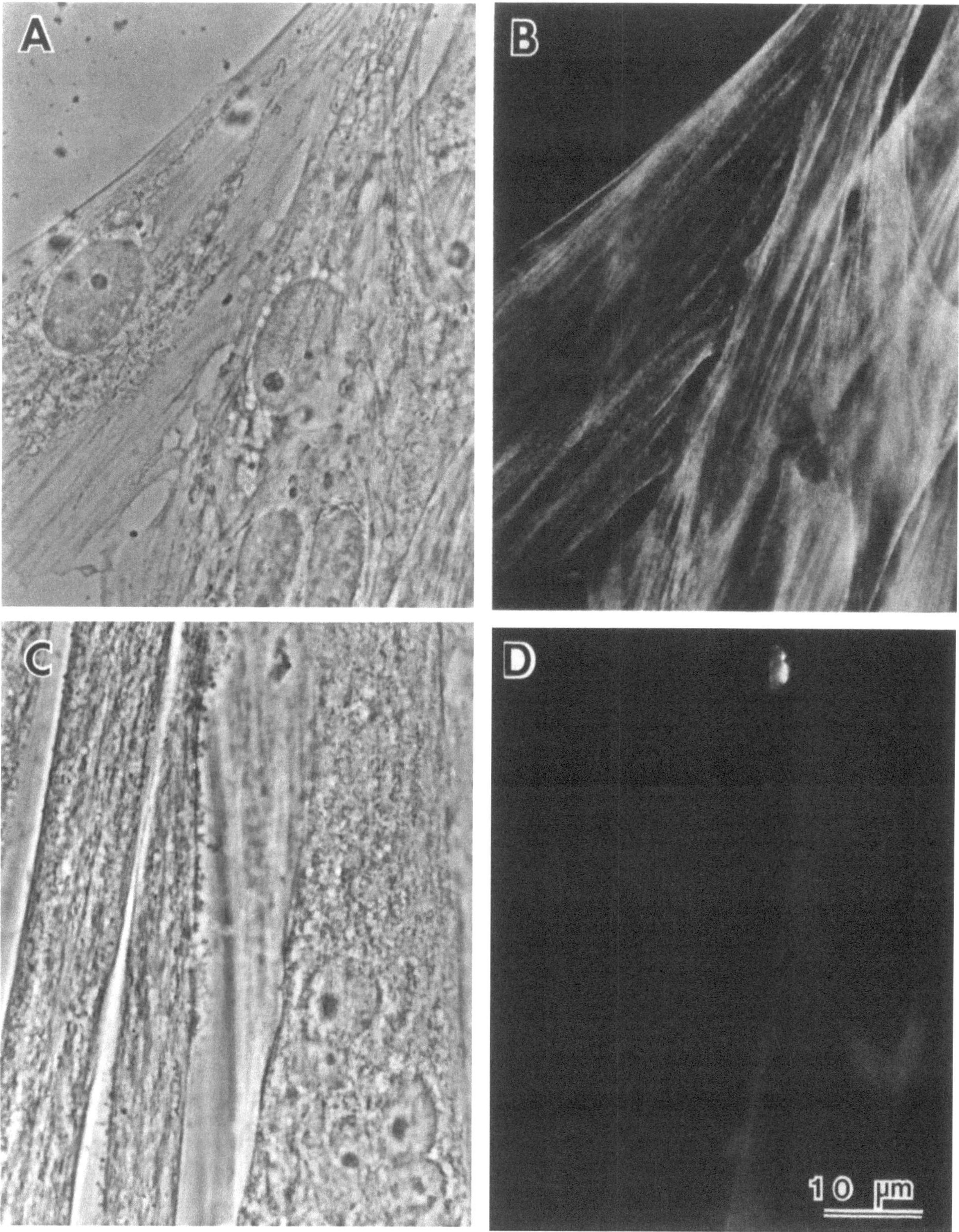

Fig. 8. 83 kDa protein disappears when L 6 myoblasts fuse into myotubes. Both myoblasts (*A* and *B*) and myotubes (*C* and *D*) were stained with the affinity-purified antibody against 83 kDa protein. *A*, *C* Phase-contrast; *B*, *D* Immunofluorescence. Note that myotubes are not stained with 83 kDa protein

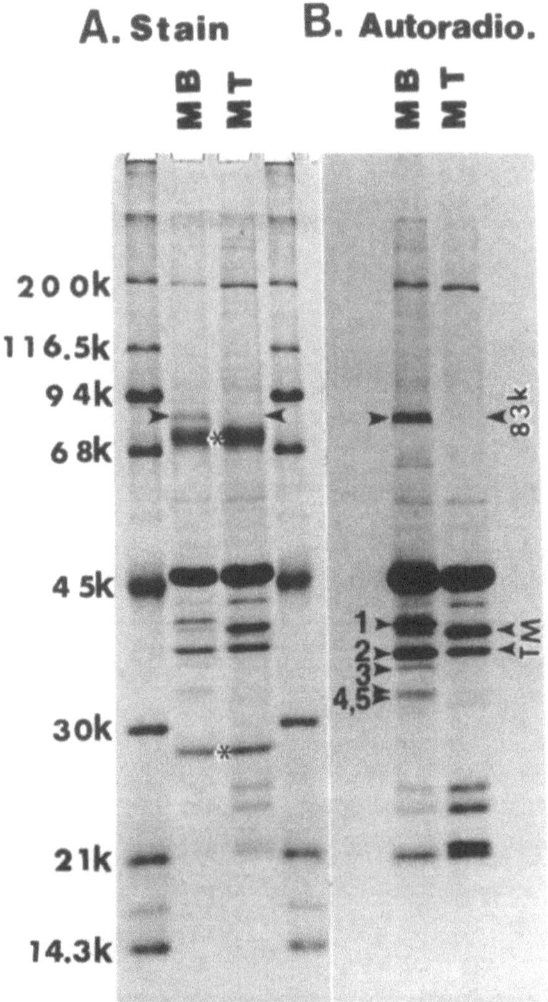



Fig. 9. Comparison of components of microfilaments between L 6 myoblasts and myotubes. Microfilaments were isolated from both L 6 myoblasts and myotubes labeled *in vivo* with ³⁵S-methionine as described previously (Matsumura *et al.* 1983). Each microfilament preparation was analyzed by SDS-PAGE. After electrophoresis, gels were stained with Coomassie blue (*A*) and autoradiographed (*B*). *MB* Myoblast microfilaments; *MT* myotube microfilaments. Note that 83 kDa protein (indicated by arrow with 83 k) is present in microfilaments from myoblasts but absent in microfilaments from myotubes. Asterisks in the stained gel of *A* show heavy and light chains of immunoglublin IgM used for the preparation of microfilaments. Numbers (*1–5*) in lane *MB* of *B* show multiple isoforms of nonmuscle type of tropomyosin present in L 6 myoblasts and arrows in lane *MT* of *B* show muscle type tropomyosin newly expressed in myotubes

antibody to 83 kDa protein by Western blot. The antibody reacts with a band corresponding to 83 kDa protein when total cell lysates of cultured rat cells were examined (lane 1). No bands with higher Mr than 83,000 were detected in the total cell lysates. The antibody, however, cross-reacts with caldesmon purified

from chicken gizzard smooth muscle (lane 3), suggesting that these two proteins are immunologically related. Furthermore, this antibody decorated stress fibers, as well as membrane ruffles and the basal portions of microspikes in cultured rat fibroblasts. The localization of 83 kDa protein is very similar to that of other nonmuscle caldesmons (Koji-Owada 1984, Bretscher and Lynch 1985, Dingus *et al.* 1986).

Figure 8 shows the intracellular localization of 83 kDa protein in both L 6 myoblasts and myotubes. As expected, 83 kDA protein is present in myoblasts (Fig. 8 *A* and *B*) and its localization is similar to that in fibroblasts. When L 6 myoblasts fused into myotubes, the staining with anti-83 kDa protein antibody was reduced to a background level as shown in Fig. 8 *C* and *D*. These observations suggest that the synthesis of 83 kDa nonmuscle caldesmon is shut down in myotubes. The results appear to be inconsistent with the report by Ngai and Walsh (1985) that skeletal and cardiac muscle contain smooth muscle caldesmon. The reason for this discrepancy is currently unknown.

The absence of nonmuscle caldesmon in myotubes was confirmed by biochemical analyses. First we have analyzed total protein pattern of myotubes by high resolution 2 D gel electrophoresis. As described earlier, 83 kDa protein is present as four spots on 2 D gel in total cell lysates of L 6 myoblasts (Fig. 4 *A*). In contrast, the total cell lysates of L 6 myotubes (Fig. 4 *C*) contain little of 83 kDa protein (a very small amount of the protein found in the Fig. 4 *C* is probably derived from a small population of L 6 myoblasts contaminated in L 6 myotube preparation).

Second, we have isolated microfilaments from both L 6 myoblasts and myotubes to see their protein compositions. As Fig. 9 shows, SDS-PAGE has shown that microfilaments from L 6 myoblasts contain 83 kDa protein as one of the major constituents while microfilaments from myotubes lack the same component. Two dimensional gel analyses shown in Fig. 4 *B* and *D* also confirmed the above result. It is also worth to note the changes in the tropomyosin composition from nonmuscle type to muscle type as shown in Figs. 4 and 9.

## 4. Discussion

We have concluded that 83 kDa protein is a nonmuscle caldesmon of cultured rat cell as judged from the following criteria. These include (a) the binding to calmodulin-Sepharose column in a calcium dependent way (Fig. 1 *B*); (b) the binding to actin through $Ca^{2+}$/ calmodulin interaction; (c) stability to heat treatment;

(d) an elongated shape of molecule, and (e) immunological cross-reactivity (Fig. 7).

Because smooth muscle caldesmon is very susceptible to proteases, it may be possible that 83 kDa protein is a proteolytic product of smooth muscle caldesmon *during preparation*. However, our polyclonal anti-83 kDa protein antibody failed to detect any band higher than 83 kDa protein in total cell lysates of cultured rat cells, while the same antibody cross-reacted with smooth muscle caldesmon (Fig. 7). In addition, the immunoblot of total cell lysates (Fig. 7 *B*, lane 1) discretely shows 83 kDa protein band without proteolytic fragments while the immunoblot of purified 83 kDa protein (Fig. 7 *B*, lane 2) shows many proteolytic fragments. This suggests that our preparation of total cell lysates had minimum proteolytic activities.

Other laboratories, however, reported that anti-smooth muscle caldesmon antibody detected an immunoreactive band corresponding to the mobility of smooth muscle caldesmon in nonmuscle cells (BRETSCHER and LYNCH 1985, KAKIUCHI *et al.* 1983, NGAI and WALSH 1985). Although the reason for this discrepancy is currently unclear, similar discrepancy was reported for platelets (KAKIUCHI *et al.* 1983, DINGUS *et al.* 1986). 83 kDa nonmuscle caldesmon and smooth muscle caldesmon appear to share most of properties, although the size of the 83 kDa protein is about half of smooth muscle caldesmon. This is not surprising as judged from the recent results on domain mapping of smooth muscle caldesmon from several laboratories (FUJII *et al.* 1987, SZPACENKO and DABROWSKA 1986, RISEMAN and BRETSCHER 1986). These laboratories have shown that 20–40 kDa fragments include both actin-binding and calmodulin binding sites. In addition, SZPACENKO and DABROWSKA (1986) have presented evidence that these fragments can even inhibit ATPase activity of actomyosin. At present, we do not know why smooth muscle caldesmon has almost twice larger molecule than nonmuscle caldesmon. Future comparative studies on smooth muscle and nonmuscle caldesmon should have insight into the functional differences between these muscle and nonmuscle caldesmons.

The functions of 83 kDa protein in cultured cells are yet to be determined. Because smooth muscle caldesmon has been reported to inhibit both ATPase activity and superprecipitation of actomyosin, it is suggested to play a regulatory role in smooth muscle contraction (SOBUE *et al.* 1982, SOBUE *et al.* 1985 a, NGAI and WALSH 1984, MARSTON and SMITH 1985, DABROWSKA *et al.* 1985, HORIUCHI *et al.* 1986, BRETSCHER 1986, SMITH *et al.* 1987). Our preliminary data have shown that 83 kDa nonmuscle caldesmon, like smooth muscle caldesmon, can also inhibit tropomyosin-enhanced ATPase activity of skeletal muscle actomyosin, although the inhibition is only observed at 50 mM KCl concentration but not at 100 mM KCl concentration unlike smooth muscle caldesmon. These results suggest that 83 kDa nonmuscle caldesmon may also play a regulatory role in the motility and contraction of nonmuscle cells as smooth muscle caldesmon does in smooth muscle contraction.

Another important function of 83 kDa nonmuscle caldesmon is its stimulating effect on binding of cultured cell tropomyosin to actin (YAMASHIRO-MATSUMURA and MATSUMURA 1988). This property is especially important because tropomyosin isoforms with lower Mr values (low Mr tropomyosins) from cultured cells showed a very low affinity to actin by themselves (MATSUMURA and YAMASHIRO-MATSUMURA 1985). This regulatory function for the binding of tropomyosin to actin by 83 kDa protein may be related to the co-localization of 83 kDa protein with tropomyosin in stress fibers. Because tropomyosin binding to actin appears to stabilize structure of actin filaments, the 83 kDa protein may regulate the organization of microfilament in cultured cells through stabilization of microfilament structure.

It is interesting to note that 83 kDa nonmuscle caldesmon disappears upon L 6 myogenesis. L 6 myoblasts differentiate into skeletal muscle through fusion. Expression of many contractile proteins are concomitantly changes from nonmuscle type to muscle type (DEVLIN and EMERSON 1978, GARRELS 1979 b, MONTARRAS *et al.* 1982). It appears that regulatory system for contraction is also changed during myogenesis. In myoblasts, phoshporylation of myosin light chain seems to regulate actomyosin activity while troponin-tropomyosin system replaces it after differentiation into skeletal muscle. Thus, the disappearance of nonmuscle caldesmon in myotubes may reflect the changes in the regulatory system during myogenesis. Further studies are needed to elucidate the functions of nonmuscle caldesmon in the regulation of motility and organization of microfilaments in myogenesis.

## Acknowledgements

We thank Dr. G. I. GARRELS (Cold Spring Harbor Laboratory) for running high resolution two-dimensional gels of 83 kDa protein. This work is supported by grants from the National Cancer Institute (CA 42742), Biomedical Sciences Support Programs and Charles and Johanna Busch Fund at Rutgers. S. Y-M is supported by a postdoctoral fellowship from Charles and Johanna Busch Fund.

## References

Amos WB, Rontledge LM, Weis-Fogh T, Yew FF (1976) The spasmoneme and calcium-dependent contraction in connection with specific calcium binding proteins. In: Duncan CJ (ed) Calcium in biological systems. Cambridge Univ Press, Cambridge, pp 273–301

Blatter DP, Garner F, Van Slyke K, Bradley A (1972) Quantitative electrophoresis in polyacrylamide gels of 2–40%. J Chromatogr 64: 147–155

Bretsher A (1984) Smooth muscle caldesmon. J Biol Chem 259: 12873–12880

— Lynch W (1985) Identification and localization of immunoreactive forms of caldesmon in smooth and nonmuscle cells: A comparison with the distributions of tropomyosin and alpha-actinin. J Cell Biol 100: 1656–1663

— (1986) Thin filament regulatory proteins of smooth- and nonmuscle cells. Nature 321: 726–727

Dabrowska R, Goch A, Galazkiewicz B, Osinska H (1985) The influence of caldesmon and ATPase activity of the skeletal muscle actomyosin and bundling of actin filaments. Biochim Biophys Acta 842: 70–75

Devlin RB, Emerson CP Jr (1978) Coordinate regulation of contractile protein synthesis during myoblast differentiation. Cell 13: 599–611

Dingus J, Hwo S, Bryan J (1986) Identification by monoclonal antibodies and characterization of human platelet caldesmon. J Cell Biol 102: 1748–1757

Fujii T, Imai M, Rosenfeld GC, Bryan J (1987) Domain mapping of chicken gizzard caldesmon. J Biol Chem 262: 2757–2763

Garrels JI (1979a) Two-dimensional gel electrophoresis and computer analysis of proteins synthesized by clonal cell lines. J Biol Chem 254: 7961–7977

— (1979b) Changes in protein synthesis during myogenesis in a clonal cell line. Dev Biol 73: 134–152

Horiuchi KY, Miyata H, Chacko S (1986) Modulation of smooth muscle actomyosin ATPase by thin filament associated proteins. Biochem Biophys Res Commun 136: 962–968

Kakiuchi R, Inui M, Morimoto K, Kanda K, Sobue K, Kakiuchi S (1983) Caldesmon, a calmodulin-binding, F-actin-interacting protein, is present in aorta, uterus and platelets. FEBS Lett 154: 351–356

Koji-Owada M, Hakura A, Iida K, Yahara I, Sobue K, Kakiuchi S (1984) Occurrence of caldesmon (a calmodulin-binding protein) in cultured cells: Comparison of normal and transformed cells. Proc Natl Acad Sci USA 81: 3133–3137

Laemmli UK (1970) Cleavage of structural proteins during the assembly of the head of bacteriophage T4. Nature 227: 680–685

Lowry OH, Rosebrough NJ, Farr AL, Randall RJ (1951) Protein measurement with the folin phenol reagent. J Biol Chem 193: 265–275

Lynch WP, Riseman VM, Bretscher A (1987) Smooth muscle caldesmon is an extended flexible monomeric protein in solution that can readily undergo reversible intra- and intermolecular sufhydryl crosslinking. J Biol Chem 262: 7429–7437

Marston SB, Smith CWJ (1985) The thin filaments of smooth muscles. J Muscle Res Cell Motil 6: 669–708

Martin RG, Ames BN (1961) A method for determining the sedimentation behaviour of enzymes: Application to protein mixtures. J Biol Chem 236: 1372–1379

Matsumura F, Yamashiro-Matsumura S, Lin JJ-C (1983) Isolation and characterization of tropomyosin-containing microfilaments from cultured cells. J Biol Chem 258: 6636–6644

— — (1985) Purification and characterization of multiple isoforms of tropomyosin from rat cultured cells. J Biol Chem 260: 13851–13859

Montarras D, Fiszman MY, Gros F (1982) Changes in tropomyosin during development of chick embryonic skeletal muscles *in vivo* and during differentiation of chick muscle cells *in vitro*. J Biol Chem 257: 545–548

Ngai PK, Walsh MP (1984) Inhibition of smooth muscle actin-activated myosin $Mg^{++}$-ATPase activity by caldesmon. J Biol Chem 259: 13656–13659

— — (1985) Detection of caldesmon in muscle and non-muscle tissues of the chicken using polyclonal antibodies. Biochem Biophys Res 127: 533–539

O'Farrell PH (1975) High resolution two-dimensional electrophoresis of proteins. J Biol Chem 250: 4007–4021

Olmsted JB (1981) Affinity purification of antibodies from diazotized paper blots of heterogeneous protein samples. J Biol Chem 256: 11955–11957

Riseman VM, Bretscher A (1986) Proteolytic dissection of smooth muscle caldesmon into fragments that retain F-actin, $Ca^{2+}$-calmodulin and tropomyosin binding activities. J Cell Biol 103: 536a

Siegel LM, Monty KL (1966) Determination of molecular weights and frictional ratios of proteins in impure systems by use of gel filtration and density gradient centrifugation. Application to crude preparations of sulfite and hydroxylamine reductases. Biochim Biophys Acta 112: 346–362

Smith CW, Pritchard K, Marston SB (1987) The mechanism of $Ca^{2+}$-regulation of vascular smooth muscle thin filaments by caldesmon and calmodulin. J Biol Chem 262: 116–122

Smith DE, Fisher PA (1984) Identification, developmental regulation, and response to heat shock of two antigenically related forms of a major nuclear envelope protein in Drosophila embryos: Application of an improved method for affinity purification of antibodies using polypeptide immobilized on nitrocellulose blots. J Cell Biol 99: 20–28

Sobue K, Muramoto Y, Fujita M, Kakiuchi S (1981) Purification of a calmodulin-binding protein from chicken gizzard that interacts with F-actin. Proc Natl Acad Sci USA 78: 5652–5655

— Morimoto K, Fujita M, Kakiuchi S (1982) Control of actin-myosin interaction of gizzard smooth muscle by calmodulin- and caldesmon-linked flip-flop mechanism. Biomed Res 3: 188–196

— Takahashi K, Wakabayashi I (1985a) Caldesmon$_{150}$ regulates the tropomyosin-enhanced actin-myosin interaction in gizzard smooth muscle. Biochem Biophys Res Commun 132: 645–651

— Tanaka T, Kanda K, Ashino N, Kakiuchi S (1985b) Purification and characterization of caldesmon$_{77}$: A calmodulin-binding protein that interacts with actin filaments from bovine adrenal medulla. Proc Natl Acad Sci USA 82: 5025–5029

— Takahashi T, Tanaka T, Kanda K, Ashino N, Kakiuchi S, Maruyama K (1985c) Crosslinking of actin filaments is caused by caldesmon aggregates, but not by its dimers. FEBS Lett 182: 201–204

Szpacenko A, Dabrowska R (1986) Functional domain of caldesmon. FEBS Lett 202: 182–186

Towbin H, Staehelin T, Gordon T (1979) Electrophoretic transfer of proteins from polyacrylamide gels to nitrocellulose sheets: Procedure and some applications. Proc Natl Acad Sci USA 76: 4350–4354

Yamashiro-Matsumura S, Matsumura F (1985) Purification and characterization of an F-actin-bundling 55 kilodalton protein from HeLa cells. J Biol Chem 260: 5087–5097

— — (1986) Intracellular localization of the 55-kD Actin-bundling protein in cultured cells: Spatial relationships with actin, alpha-actinin, tropomyosin and fimbrin. J Cell Biol 103: 631–640

— — (1988) Characterization of 83 kDa nonmuscle caldesmon from cultured rat cells: Stimulation of actin binding of nonmuscle tropomyosin and periodic localization along microfilaments like tropomyosin. J Cell Biol 106: 1973–1983

Protoplasma (1988) [Suppl. 2]: 22–26


# Construction and Analysis of *Dictyostelium* Mutants with Defects in Actin-Binding Proteins

M. Schleicher[1,*], E. Wallraff[2], G. Gerisch[2], and G. Isenberg[3]

Max-Planck-Institutes for [1]Psychiatry and [2]Biochemistry, Martinsried,
and [3]Biophysic Department, Technische Universität München, Garching

Received February 19, 1988
Accepted May 14, 1988

Dedicated to Professor Dr. Noburo Kamiya on the occasion of his 75th birthday

## Summary

Monoclonal antibodies against actin-binding proteins from the slime mould *Dictyostelium discoideum* were used to screen mutagenized cells in an attempt to select strains defective in specific components of the cytoskeleton. By a colony blot technique three mutants were detected that were defective in the F-actin crosslinking proteins α-actinin and 120 kDa gelation factor and the F-actin fragmenting protein severin, respectively. Here we focus on the description of the α-actinin mutant HG 1130. Although this protein seems to be highly conserved throughout evolution, its loss did not impair the normal functions as *e.g.* motility, axenic growth, phagocytosis of bacteria, chemotaxis during development and formation of fruiting bodies. These surprising results suggest that evolutionary highly conserved proteins of the nonmuscle cytoskeleton may be replaced in their function by *in vitro* similarly acting proteins.

*Keywords:* Actin-binding protein; Cytoskeleton mutant; *Dictyostelium discoideum*; Development; Motility.

## 1. Introduction

*Dictyostelium discoideum* amoebae are rapidly moving cells (Loomis 1982). The dynamics of the motile system are thought to depend on fast rearrangements of the microfilament network. A number of actin-binding proteins, that might play a key role in the regulation of actin assembly, have been isolated from *D. discoideum*. An overview of the best known actin-binding proteins in *D. discoideum* is shown in Fig. 1. These proteins belong to different subclasses among the actin-binding proteins according to their structure and function. Recently a capping protein was described that caps actin filaments at their fast growing ends in a $Ca^{2+}$-independent manner (Fig. 1 (*1*), Schleicher *et al.*, 1984). Severin, a 40 kDa actin-binding protein, belongs to the subclass of F-actin severing proteins (Fig. 1 (*2*), Andre *et al.* 1988). A number of F-actin crosslinking proteins has been found in *D. discoideum*, the most abundant are α-actinin (Fig. 1 (*3*), Noegel *et al.* 1987), a 120 kDa gelation factor (Fig. 1 (*4*), Condeelis *et al.* 1984), and a 30 kDa crosslinking molecule (Fig. 1 (*5*), Fechheimer 1987). From *in vitro* observations these proteins may function by regulating the assembly state of actin, whereas myosins are believed to produce contractile forces (Fig. 1 (*6, 7*), Cote *et al.* 1985, Warrick *et al.* 1986). The anchoring of the cytoskeleton to membranes might be performed by actin-binding proteins similar to ponticulin, a membrane spanning glycoprotein (Fig. 1 (*8*), Wuestehube and Luna 1987).

The investigation of these cytoskeletal proteins in *D. discoideum* offers a number of advantages. First, most of the actin-binding proteins are functionally and structurally similar to cytoskeletal proteins from higher organisms (for reviews see Stossel *et al.* 1985, Pollard and Cooper 1986). Secondly, the fact that *D. discoideum* cells are haploid, allows easy detection of mutants that can be employed in genetic studies on the cytoskeleton.

---

* Correspondence and Reprints: Max-Planck-Institute for Psychiatry, D-8033 Martinsried, Federal Republic of Germany.

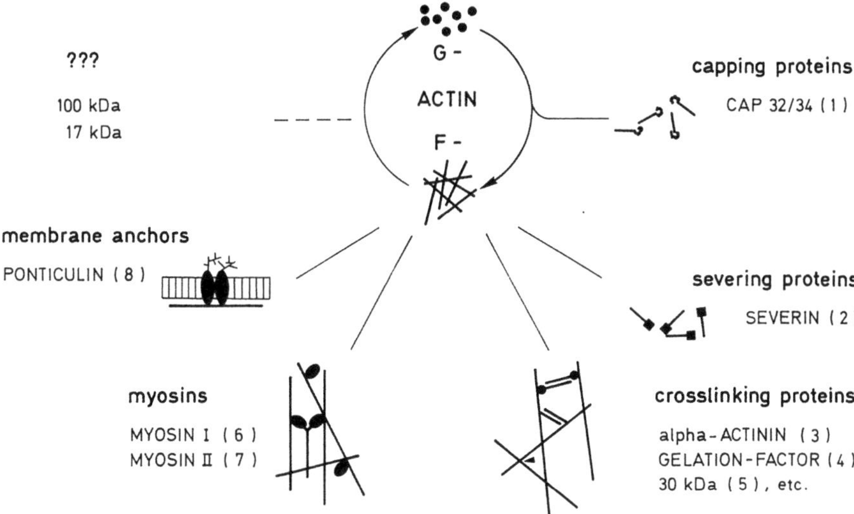



Fig. 1. Actin-binding proteins in *D. discoideum*. Based on the equilibrium between G-actin and F-actin, the actin-binding proteins with capping or severing functions lower the length of actin filaments, thus decreasing the viscosity of the cytoplasm. Crosslinking proteins connect actin filaments and raise the stage of gelation. Myosins in conjunction with filaments produce contractile forces, and putative membrane anchors establish the interaction of the microfilament system with the inner side of the plasma membrane. The overview summarizes recent data and publications: (*1*) Schleicher *et al.* 1984; (*2*) Andre *et al.* 1988; (*3*) Noegel *et al.* 1987; (*4*) Condeelis *et al.* 1984; (*5*) Fechheimer 1987; (*6*) Cote *et al.* 1985; (*7*) Warrick *et al.* 1986; (*8*) Wuestehube and Luna 1987

## 2. Construction of Mutants with Defects in Actin-Binding Proteins

Growth phase cells of AX 2-214 were mutagenized by incubation with 1 mg/ml of 1-methyl-3-nitro-1-nitrosoguanidine for 20 min in phosphate buffer and the cells were plated on SM agar and *Klebsiella aerogenes* (Williams and Newell 1976). Cells from single colonies were transferred in a regular array to agar plates, the colonies replicated using an autoclavable dispenser, blotted after 2–3 days onto nitrocellulose and lysed by freezing and thawing (Fig. 2). Subsequently the nitrocellulose was incubated with iodinated antibodies specific for either α-actinin, severin, the 34 kDa subunit of the capping protein, the membrane associated 17 kDa protein (Scheel *et al.* 1988), or the 120 kDa gelation factor. Figure 3 shows a nitrocellulose filter with 56 *D. discoideum* colonies after incubation with radiolabeled antibody 47-19-2 specific for α-actinin (Schleicher *et al.* 1984). The arrow marks the colony which was not labeled by the antibody. Using this colony blot technique we so far detected three mutants with defects either in 120 kDa gelation factor, severin or α-actinin (HG 1130).

## 3. Biochemical Analysis of the α-Actinin Mutant

Proteins synthesized in the mutant were analyzed using four monoclonal antibodies which in the wild type specifically recognize the 95 kDa polypeptide of the α-actinin homodimer. Cleavage mapping indicated that the binding sites of these antibodies were distributed over a region comprising more than half of the α-actinin

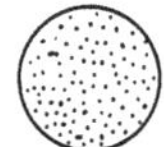

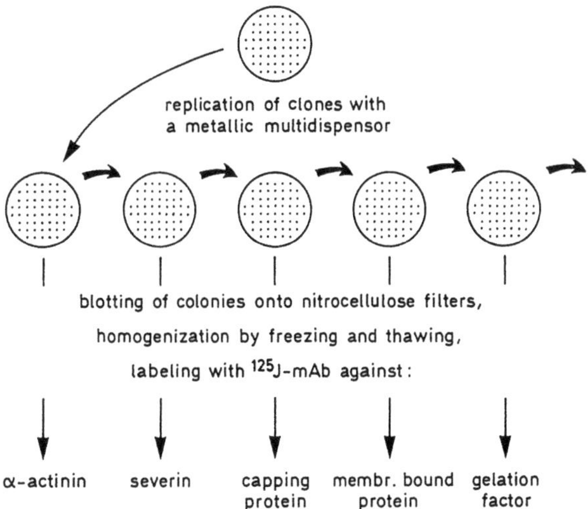



Fig. 2. Diagram of the selection procedure for mutants defective in actin-binding proteins

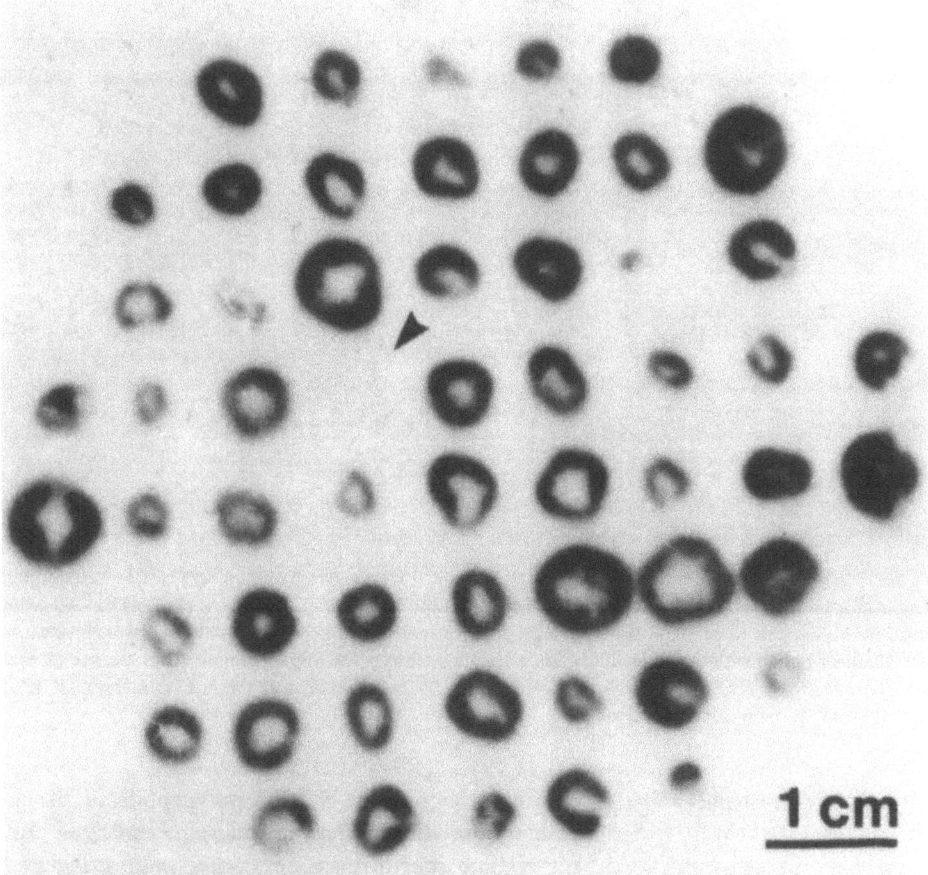


Fig. 3. Autoradiogram of a colony blot of mutagenized AX 2 cells labeled with the monoclonal antibody 47-19-2. One colony (arrow) was of average size and defective in normal antibody binding

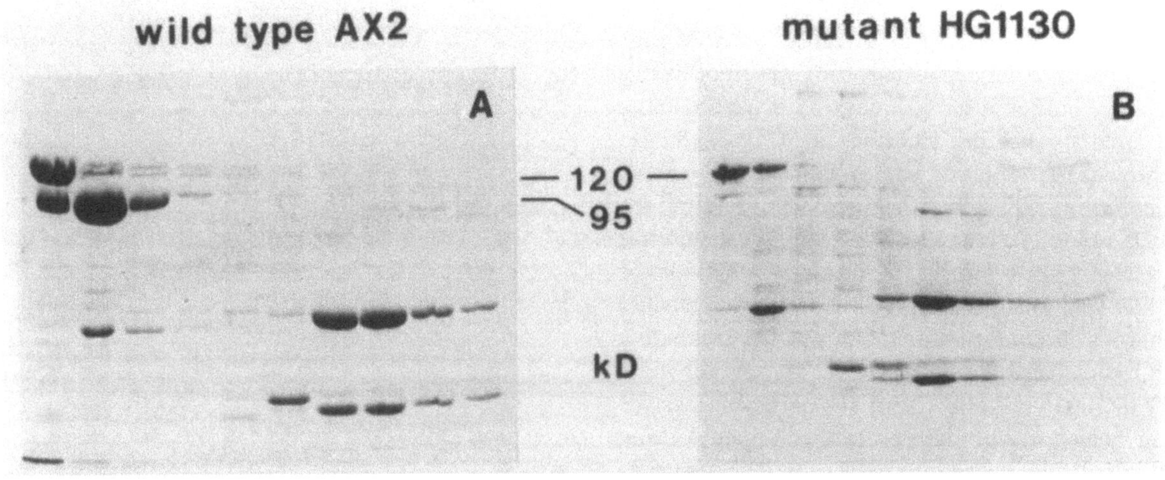


Fig. 4. Comparison of Coomassie blue staining in wild type (*A*) and mutant proteins (*B*). Wild type AX 2 fractions containing 120 kDa gelation factor and α-actinin were obtained by DEAE-cellulose chromatography. Fractions containing these crosslinkers were pooled and subjected to a hydroxylapatite column (Schleicher *et al.* 1988 a). Fractions eluted from this column were analyzed by 7.5–15% gradient SDS-polyacrylamide gels. In parallel, proteins of mutant HG 1130 were fractionated. Both gels were stained with Coomassie blue. In the mutant as in the wild type, the strongly stained band of the 120 kDa gelation factor is seen in fraction 1. No intensely stained band corresponding to wild type α-actinin is recognized in the mutant

polypeptide chain. In the α-actinin mutant HG 1130, three of the antibodies faintly labeled two polypeptides of 95 and 88 kDa; the fourth antibody which binds closest to one end of the polypeptide chain (WALLRAFF *et al.* 1986) faintly labeled the 95 kDa polypeptide only. The 88 kDa polypeptide most likely lacks the C-terminal portion of α-actinin. To quantitate antibody binding activities in wild type and mutant, a competition radioimmuno assay was employed, using an antibody that labeled the two polypeptides. In a mutant cell extract containing soluble proteins, the antibody binding activity was reduced by about 99% when compared to wild type extract.

In order to show that the weak antibody binding to the mutant 95 and 88 kDa bands of the respective polypeptides was due to the extremely low cellular concentrations, we applied the same purification protocol as it was used for the isolation of wild-type α-actinin (SCHLEICHER *et al.* 1988 a). Whereas in the wild type α-actinin was clearly identifiable as a strongly Coomassie blue stained band, no prominent band in the appropriate region was seen after Coomassie blue staining of mutant proteins (Fig. 4). Furthermore, no viscosity increasing activity could be correlated with these fractions of mutant proteins.

## 4. Cell Biological Analysis of the α-Actinin Mutant

Since α-actinin is supposed to be involved in the surface capping of plasma membrane antigens (CARBONI and CONDEELIS 1985), patching and capping of the surface of HG 1130 cells was induced by antibodies directed against the contact site A glycoprotein. Capping of the antigen occurred as efficient in mutant as in the wild type cells.

A sensitive assay for putative alterations of gene expression after mutagenesis is the investigation of the development of *D. discoideum* from amoebae to fruiting bodies. This developmental cycle involves chemotaxis towards cAMP, cell differentiation and motility during all stages of single cells and multicellular aggregates. Most often changes in the genome of mutants can be observed as behavioral changes in the highly coordinated developmental process. Therefore, we deposited the parental strain AX 2 and the mutant HG 1130 on filters and recorded the development over 21 h at 3 h intervals. Both strains developed ripples after 6 h and tight aggregates after 9 h; culmination occurred after 12 h and the formation of fruiting bodies was nearly complete after 21 h of development.

Although the mutant was severely impaired in the expression of α-actinin, the major cell functions seemed to be guaranteed as in the parental strain AX 2. The mutant grew at a normal rate on agar plates with *Escherichia coli* and axenically in nutrient medium with 1.8% maltose according to WATTS and ASHWORTH (1970). This result indicates that phagocytosis of bacteria and pinocytosis during growth in liquid medium may not be affected at all.

However, a significant difference could be detected during longterm experiments in mixtures of wild type and mutant cultures (SCHLEICHER *et al.* 1988 a). Nutrient medium was inoculated with mixed cells and the ratios of wild type and mutant cells were determined at intervals. The initial 1:1 ratio of mutant to wild type cells was shifted to about 3:7 within a period of 11 days, especially if the cells reached repeatedly the stationary phase. This selection advantage of the wild type might be due to a slightly reduced ability for pinocytosis in the α-actinin mutant.

## 5. Is α-Actinin Essential for Major Cell Functions in *D. discoideum*?

An interesting question is how the mutant cells can live with almost no α-actinin, the most prominent F-actin crosslinker in wild type homogenates. The first complete sequence of an α-actinin was obtained from *D. discoideum* (NOEGEL *et al.* 1987) and compared to other proteins in the computer data bank. Striking similarities have been found to proteins from the calmodulin superfamily, because *D. discoideum* α-actinin contains two perfect EF-hand regions that might be the basis of its regulation by $Ca^{2+}$. Homologies to chicken α-actinin (BARON *et al.* 1987) indicate that a putative actin-binding site is highly conserved. Most likely a strong evolutionary pressure kept this region constant. In addition, this stretch of amino acids is very similar to the N-terminal portion of dystrophin, the protein product of the Duchenne muscular dystrophy locus (HAMMOND 1987, HOFFMAN *et al.* 1987, SCHLEICHER *et al.* 1988 b). In view of these data it is puzzling that the lack of the conserved protein in *D. discoideum* does not dramatically impair the normal cell functions.

It is conceivable that the function of α-actinin may not be required under laboratory conditions, because it is guaranteed by more than a single protein. Therefore, we are now working on the construction of double- and triple-mutants in which one actin-binding protein after another is destroyed. The genetic approach to elucidate the function of cytoskeletal proteins might shed light onto the importance of similar proteins in higher organisms where mutant selection can not be performed as easily as in *D. discoideum*.

## Acknowledgements

We thank Dr. A. Noegel for helpful discussions. We are grateful to Daniela Rieger and Barbara Fichtner for their excellent assistance in monoclonal antibody production and cell culture. The work was supported by grant Is 25/4 from the Deutsche Forschungsgemeinschaft to G. Isenberg and G. Gerisch.

## References

Andre E, Lottspeich F, Schleicher M, Noegel A (1988) Severin, gelsolin, and villin share a homologous sequence in regions presumed to contain F-actin severing domains. J Biol Chem 263: 722–727

Baron MD, Davison MD, Jones P, Patel B, Critchley DR (1987) Isolation and characterization of a cDNA encoding a chick α-actinin. J Biol Chem 262: 2558–2561

Carboni JM, Condeelis JS (1985) Ligand-induced changes in the location of actin, myosin, 95 K (α-actinin), and the 120 K protein in amoebae of *Dictyostelium discoideum*. J Cell Biol 100: 1884–1893

Condeelis JS, Vahey M, Carboni JM, DeMey J, Ogihara S (1984) Properties of the 120,000- and 95,000-dalton actin-binding proteins from *Dictyostelium discoideum* and their possible functions in assembling the cytoplasmic matrix. J Cell Biol 99: 119 s–126 s

Cote GP, Albanesi JP, Ueno T, Hammer JA, Korn ED (1985) Purification from *Dictyostelium discoideum* of a low-molecular-weight myosin that resembles myosin I from *Acanthamoeba castellanii*. J Biol Chem 260: 4543–4546

Fechheimer M (1987) The *Dicytostelium discoideum* 30,000-Dalton protein is an actin filament-bundling protein that is selectively present in filopodia. J Cell Biol 104: 1539–1551

Hammonds RG (1987) Protein sequence of DMD gene is related to actin-binding domain of α-actinin. Cell 51: 1

Hoffman EP, Knudson CM, Campbell KP, Kunkel LM (1987) Subcellular fractionation of dystrophin to the triads of skeletal muscle. Nature 330: 754–757

Loomis WF (1982) The development of *Dictyostelium discoideum*. Academic Press, New York

Noegel A, Witke W, Schleicher M (1987) Calcium-sensitive non-muscle α-actinin contains EF-hand structures and highly conserved regions. FEBS Lett 221: 391–396

Pollard TD, Cooper JA (1986) Actin and actin-binding proteins. A critical evaluation of mechanisms and functions. Ann Rev Biochem 55: 987–1035

Scheel J, Ziegelbauer K, Kupke T, Humbel BM, Noegel A, Gerisch G, Schleicher M (1988) Histactin, a histidine-rich actin-binding protein from *Dictyostelium discoideum*; manuscript submitted

Schleicher M, Gerisch G, Isenberg G (1984) New actin-binding proteins from *Dictyostelium discoideum*. EMBO J 3: 2095–2100

— Noegel A, Schwarz T, Wallraff E, Brink M, Faix J, Gerisch G, Isenberg G (1988 a) A *Dictyostelium* mutant with severe defects in α-actinin: its characterization by cDNA probes and monoclonal antibodies. J Cell Sci 90: 59–71

— Andre E, Hartmann H, Noegel AA (1988 b) Actin-binding proteins are conserved from slime molds to man. Dev Gen, in press

Stossel TP, Chaponnier C, Ezzell RM, Hartwig JH, Janmey PA, Kwiatkowski DJ, Lind SE, Smith DB, Southwick FS, Yin HL, Zaner KS (1985) Nonmuscle actin-binding proteins. Ann Rev Cell Biol 1: 353–402

Wallraff E, Schleicher M, Modersitzki M, Rieger D, Isenberg G, Gerisch G (1986) Selection of *Dictyostelium* mutants defective in cytoskeletal proteins: use of an antibody that binds to the ends of α-actinin rods. EMBO J 5: 61–67

Warrick HM, DeLozanne A, Leinwand LA, Spudich JA (1986) Conserved protein domains in a myosin heavy chain gene from *Dictyostelium discoideum*. Proc Natl Acad Sci USA 83: 9433–9437

Watts DJ, Ashworth JM (1970) Growth of myxamoebae of the cellular slime mould *Dictyostelium discoideum* in axenic culture. Biochem J 119: 171–174

Williams KL, Newell PC (1976) A genetic study of aggregation in the cellular slime mould *Dictyostelium discoideum* using complementation analysis. Genetics 82: 287–307

Wuestehube LJ, Luna EJ (1987) F-actin binds to the cytoplasmic surface of ponticulin, a 17-kD integral glyocprotein from *Dictyostelium discoideum* plasma membranes. J Cell Biol 105: 1741–1751

Protoplasma (1988) [Suppl. 2]: 27–36



# Actin-Induced Elongation of Fibers Composed of Cytoplasmic Membrane from *Nitella*

S. Higashi-Fujime*

Department of Molecular Biology, Faculty of Science, Nagoya University Chikusa-ku, Nagoya

Received February 24, 1988
Accepted April 26, 1988

Dediacted to Professor Dr. Noburo Kamiya on the occasion of his 75th birthday

## Summary

As reported previously, cytoplasmic membrane originated from endoplasmic droplets of *Nitella* formed networks of fibrils spread on a glass slide in the activating medium containing Mg-ATP and sucrose, and these fibrils were referred to as the network fibers [Higashi-Fujime S (1980) J Cell Biol 87: 569–578].
After spreading on the glass slide, this network fiber was induced to elongate by muscle actin in ATP dependent manner. Projections were newly formed in the middle of the network fibers and elongated in a straight line on the glass surface. The direction of elongation changed only when it reached on obstacle such as other network fibers or granules, and formed a kink. Elongating fibers suddenly stopped, often retracted and absorbed into the network fiber again. Sometimes they were broken or made bridges between stationary network fibers. Due to these movements, the network pattern was continuously changing. Projections reached about 100 µm long maximally and the maximum speed of elongation was about 30 µm/s. The movement continued for about 5 min after addition of actin. Electron micrographs showed membraneous structures in the fiber.

*Keywords:* Actin; Cytoplasmic membrane; Dark field microscopy; *In vitro* elongation; Network fibers; *Nitella*.

*Abbreviations:* ATP adenosine 5′-triphosphate; EGTA [ethylenebis(oxyethylenenitrilo)]tetraacetic acid; NEM N-ethylmaleimide; DTT dithiothreitol.

## 1. Introduction

In characean cells, the endoplasm streams rotationally along arrays of chloroplasts which are located in the vicinity of the cell wall. Arrays of chloroplasts are connected by subcortical fibrils (Kamitsubo 1972), which are composed of bundles of microfilaments (Palevitz *et al.* 1974, Williamson 1974) and have polarity by decoration with heavy meromyosin (Kersey *et al.* 1976).

In analogy to muscle contraction, the generation of shearing force for streaming (Kamiya and Kuroda 1956) is understood as being due to the interaction between F-actin filaments in stationary subcortical fibrils and myosin in the streaming endoplasm. Actually, rotation of chloroplasts was activated by muscle heavy meromyosin (Kuroda and Kamiya 1975), and myosin coated beads moved on subcortical fibrils (Sheetz and Spudich 1983, Shimmen and Yano 1984). Myosin was isolated from *Nitella* (Kato and Tonomura 1977), and suggested to be located in the endoplasm (Chen and Kamiya 1975) and attached to vesicles (Nagai and Hayama 1979). The wavy motion of endoplasmic filaments anchored on subcortical fibrils (Allen 1974) and the rotational movement of individual F-actin filaments in the subcortical fibrils (Foissner and Jarosch 1981, Jarosch 1976), have also been proposed as possible mechanisms of endoplasmic streaming.

By light microscopy, the behaviour of endoplasmic droplets squeezed out of the cell has been observed. The movements observed in the endoplasmic droplets were rotation and undulation of polygonal or circular fibrils (Jarosch 1956, Kuroda 1964) which were probably derived from the subcortical fibrils and showed repeated formation and disintergration in the droplets (Kuroda 1964, Kamiya and Kuroda 1957). The rotating rings or travelling fibers consisting of F-actin

* Correspondence and Reprints: Department of Molecular Biology, Faculty of Science, Nagoya University, Chikusa-ku, Nagoy 464, Japan.

filaments observed by electron microscopy (Higashi-Fujime 1980) have the same kind of structure as those polygonal fibers.

In addition to *in vitro* movements of these F-actin bundles, rotating rings and travelling fibers (Higashi-Fujime 1980), we found a new type of *in vitro* movement in the same activating medium: elongation of fibrils composed of cytoplasmic membrane derived from endoplasmic droplets isolated from *Nitella*. These fibrils formed a network on a glass surface, and were very sticky and flexible, as reported previously (Higashi-Fujime 1980). Elongation of the network fibers was induced by muscle actin at a low concentration of 0.1 mg/ml in an ATP-dependent manner. This actin-induced elongation was inhibited by muscle tropomyosin.

## 2. Materials and Methods

The internodal cell of *Nitella microcarpa* Braun was used throughout these experiments. The method of *Nitella* cultivation was the same as described previously (Higashi-Fujime 1980).

One edge of the internodal cell was dissected and the whole cell content was squeezed onto a glass slide and the activating medium was added. The composition of the activating medium was the same as that used for the activation of travelling and rotation of cytoplasmic fibers, containing 0.2 M sucrose, 2 mM $MgSO_4$, 1.5 mM ATP, 4 mM [ethylenebis(oxyethylenenitrilo)]tetraacetic acid (EGTA), 0.1 mM $CaCl_2$, and 10 mM imidazole buffer (pH 7.0). The solution was covered with a cover slip (22 mm × 22 mm) using vaseline as a spacer. To wash out chloroplasts whose strong scattering interferred with observation, the medium was perfused from one side of the cover slip to the other by sucking the solution with a small piece of filter paper. Immediately after perfusion to replace the activating medium with the medium containing muscle F-actin at 0.1 mg/ml, the fibers began to elongate. Observations were carried out at room temperature using a dark field microscope as described previously (Higashi-Fujime 1980). Movements were photographed on 16 mm film (Kodak 4 X) at 16 f/s using a Bolex camera.

For electron microscopy, network fibers were attached to a grid with the same procedure as used for observation of moving cytoplasmic fibers: A carbon coated grid was put on a glass slide and the cytoplasm was squeezed onto the grid, then the activating medium was added, and covered with a cover slip. After observing the specimen by dark field microscopy, the cover slip was removed with care. For fixation, the activating medium containing fixative was perfused. The network fibers were fixed with 0.1% $OsO_4$ without changing the structure. The specimen was stained with minimum exposure to a uranylacetate solution of 0.5%.

Muscle proteins were prepared from rabbit skeletal muscle. Actin was prepared by the method of Ebashi (Ebashi and Ebashi 1965) and purified by repeating polymerization and depolymerization (Higashi-Fujime 1983). Myosin was prepared according to Perry (1955) with slight modification (Higashi-Fujime 1985). Tropomyosin was prepared from muscle residues after extraction of myosin by the method of Ebashi *et al.* (1968) and purified with DEAE Sephadex A 25 (Pharmacia Fine Chemicals, Upsalla, Sweden) column in the presence of 6 M urea (van Eerd and Kawasaki 1973).

## 3. Results

### 3.1. Characteristics of the Network Fibers

The network fibers were derived from membrane of endoplasmic droplets as shown in Fig. 1. The endoplasmic droplet was first flattened with a narrow spacer of vaseline, so that all expelled chloroplasts were dispersed in a monolayer on the glass slide. When a droplet was slightly displaced by suction of the cytoplasm with a tip of filter paper, network fibers remained in the same position as before the displacement of the droplet. With addition of the activating medium and application of a cover slip onto the squeezed cytoplasm, the network fibers were spread widely over the glass slide by disruption of endoplasmic droplets.

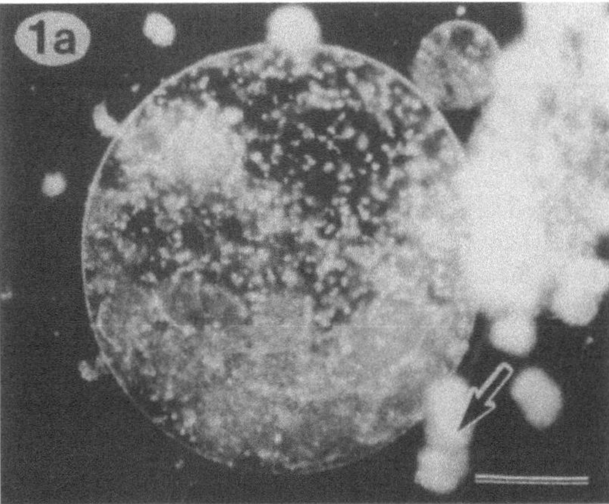
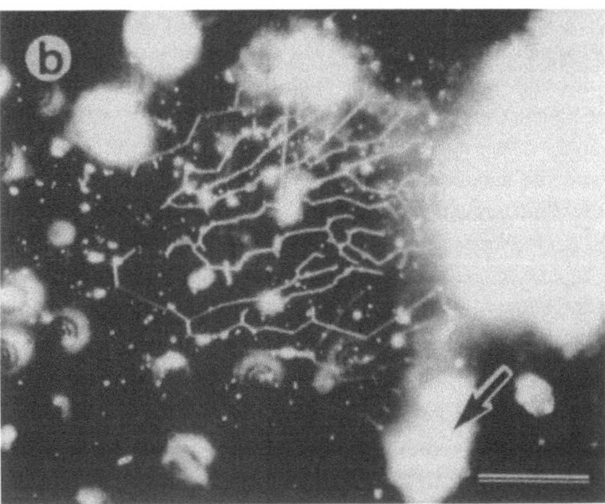

Fig. 1. Origin of the network fibers. *a* An endoplasmic droplet squeezed out of the internodal cell of *Nitella*. The droplet was flattened by narrowing the space between the glass slide and the cover slip. *b* The network fibers formed under the endoplasmic droplet. The same field as in *a*. After the droplet was displaced toward the upper right side of the photograph, network fibers remained where the endoplasmic droplet had been. Arrow in the right lower side of each photograph points to the same chloroplast attached to the glass surface. × 680. Bars 20 µm

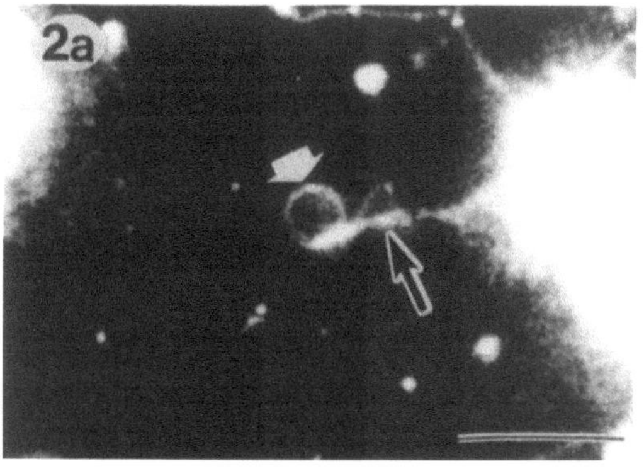

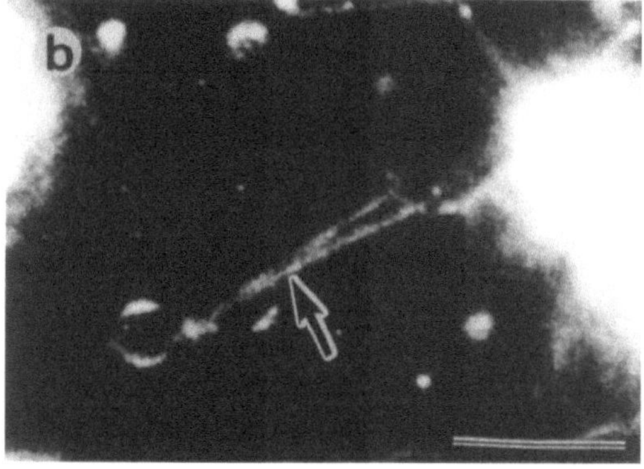

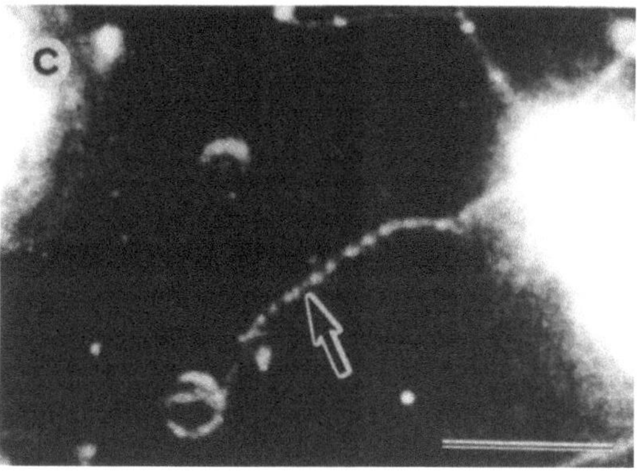

Fig. 2. Passive extension and twisting of the network fibers. *a* A rotating ring (white arrow) attached to the network fiber (edged arrow). *b* By rotation and displacement of the ring toward the lower left corner of the photograph, two network fibers were passively extended. During rotation of the ring, the attached point of the network fibers slipped on the ring, as described previously (Higashi-Fujime 1980). *c* When the point of the fiber attached to the ring rotated with the ring, the fibers twisted and wound around each other. Bright and dark spots on the fibers show such winding. Photographs were taken at time intervals of 6 s. × 1,900. Bars 10 μm

The fibers were attached to the glass surface at some points, but not over their entire lengths. The attached points in the network were readily recognized, since under the flow of the medium fibers were bent markedly but attached points of the fibers remained fixed. Some fibers slacked and swayed in Brownian move-ment, but fibers were usually straight. These straight fibers showed neither Brownian movement nor elongation in the activating medium. The network fiber was very sticky and flexible. As shown in Fig. 2, when the network fibers attached to a rotating ring, they were extended markedly (Fig. 2 *b*), or twisted by many turns with rotation of the ring (Fig. 2 *c*).

The network fibers were stable in the presence of 0.2 M to 0.4 M sucrose. By perfusion of the activating medium containing less than 0.1 M sucrose, network fibers were broken immediately into many pieces and broken fragments were shrunk into small granulary vesicles which remained on the glass surface. The network fibers were disintegrated by 0.5% $OsO_4$ or 0.05% Triton X-100 into many small granular vesicles in the activating medium.

### 3.2. Muscle Actin Induces Elongation of Network Fibers

Washing away chloroplasts on the glass slide by perfusion with a fresh medium enabled us to observe the network fibers more clearly. Immediately after perfusion of the activating medium containing muscle actin at 0.1 mg/ml, network fibers began to elongate.

In the network, some protrusions were newly formed and they elongated greatly (Figs. 3 and 4). The elongating fibers as well as stationary network fibers attached to the glass surface were straight and taut. Fibers changed the direction of elongation when the anterior end of the elongating fiber reached the substratum, granules, or other network fibers (Fig. 3 *d* and *e*, arrowhead), and formed a kink at the attached point. Many protrusions were spontaneously formed at various positions in the network and elongated (Fig. 4). The protrusions were formed even from granular particles as shown in Fig. 3 *b* (edged arrow).

The fibers exhibited various behaviours during elongation. When the fibers stopped elongating, they remained attached to the glass surface, or sometimes they retracted and absorbed into the network. The anterior end of the elongating fiber sometimes attached to another network fiber or formed a bridge between nearby fibers (Fig. 5). Either proximal end or distal end of the elongating fiber or sometimes both ends ran along the attached stationary network fiber(s), and then the elongating fiber was absorbed into the network, as shown in Fig. 5. In other case, the pattern of the network, or polygonal shapes composed of fibers in the network changed frequently. A typical example is shown in Fig. 6. This shape change was brought about by the imbalance of forces between the adjacent network fibers, which was caused by supplying materials from adjacent fibers to the elongating fiber, as will be described later.

Although we can not estimate the fiber width by dark field microscopy, the brightness of the fiber did not change during elongation. Therefore, the width of elongating fiber did not seem to become thinner during elongation. Materials were continuously supplied by flow from the adjacent network fibers to the elongating fiber and were conveyed along the network fibers. This situation is illustrated in Fig. 7: Particular particles or knobs in the network fibers were displaced continuously to the elongating fibers.

The elongating fibers rarely broke in the middle. Broken fragments to the elongating fibers, when this happened, shrank immediately to become small granular vesicles attached to the glass surface. But they never converted into the travelling fibers or rotating rings composed of the bundles of F-actin filaments.

The fibers always elongated along the glass surface and never extended normally to the network plane. The maximum speed of elongation was about 30 µm/s and that of retraction was faster than elongation (about 40 µm/s). Fibers elongated to a maximum length of 100 µm. Elongation continued for about 5 min after addition of actin to the activating medium. After elongation had ceased, additional perfusion of the fresh medium containing actin was not able to induce the elongation any more.

### 3.3. Effect of Varying the Medium Composition: ATP, KCl

Actin-induced elongation of the fibers required ATP. After induction of elongation, perfusion with the activating medium containing actin but deficient in ATP inhibited elongation. A subsequent perfusion with the medium containing ATP and actin induced elongation of the fiber again. Thus, the elongation is ATP-dependent and reversible.

In the absence of KCl, elongation was very active and many protrusions were induced to elongate. At the KCl concentration around 50 mM, few fibers elongated and at 0.1 M KCl, elongation was inhibited. By perfusion with the medium containing N-ethylmaleimide (NEM) at 0.5 mM, the number of elongating fibers decreased with time and stopped finally, and washing out NEM with the medium containing 10 mM dithiothreitol (DTT) induced elongation again.

### 3.4. Effect of Muscle Proteins on Elongation of the Network Fibers

Very active elongation was observed at actin concentrations ranging from 0.02 to 0.4 mg/ml. Tropomyosin of about 0.1 mg/ml reversibly inhibited the elongation induced by actin. To the specimen showing vigorous elongation, the activating medium containing tropomyosin or both actin and tropomyosin was perfused to replace the medium containing actin. Immediately after tropomyosin reached the microscopic field, elongation stopped. Further perfusion with the medium containing actin to wash out tropomyosin, induced elongation again. When muscle myosin at 0.2 mg/ml was applied together with actin, some myosin participated in superprecipitation and the remaining myosin had no effect on the elongation of the network fibers.

### 3.5. Electron Microscopy

Network fibers were crucially damaged not only by uranyl acetate for staining, but also by a fixative of $OsO_4$ (0.5%). Glutaraldehyde 2% in the activating medium did not affect the structure of the network fibers when monitored by dark field microscopy. But glutaraldehyde did not fix the structure sufficiently to prevent damage by staining with uranyl acetate, which broke the structure into vesicular granules, even after 3 hours incubation with glutaraldehyde.

So far as examined, only 0.1% $OsO_4$ resulted in both good retention and fixation of the structure of the network fibers. The minimum exposure to uranyl acetate solution preserved the structure of the fibers fixed with

---

Fig. 3. Elongation of the network fibers. A series of photographs in the same field. Edged arrows in (a), (b) and (d) indicate the elongating fibers. A protrusion for elongation was occasionally formed from a granular aggregate (arrow in b). Bright and dark spots in a line were often recognized in the elongating fibers (d and e), and the straight fiber changed direction of elongation and formed a kink (arrowheads in d and e). × 1,700. Bars 10 µm

Fig. 4. Elongation and retraction of the fibers in a small network. A white arrow in a indicates the position where a branch of the elongating fiber was formed. According to elongation of the newly formed branch (white arrow), the protrusion which extended upwards in the photograph, indicated by an edged arrow in a, suddenly retracted (white arrow in e). This retracted branch elongated again later. With advancing elongation, the proximal end of the elongating fiber displaced a little, and the pattern of the stationary network changes as indicated by edged arrows in c and d. The branch on the other side elongated slowly and continuously. Photographs were taken at time intervals of 1 s. × 2,200. Bars 5 µm

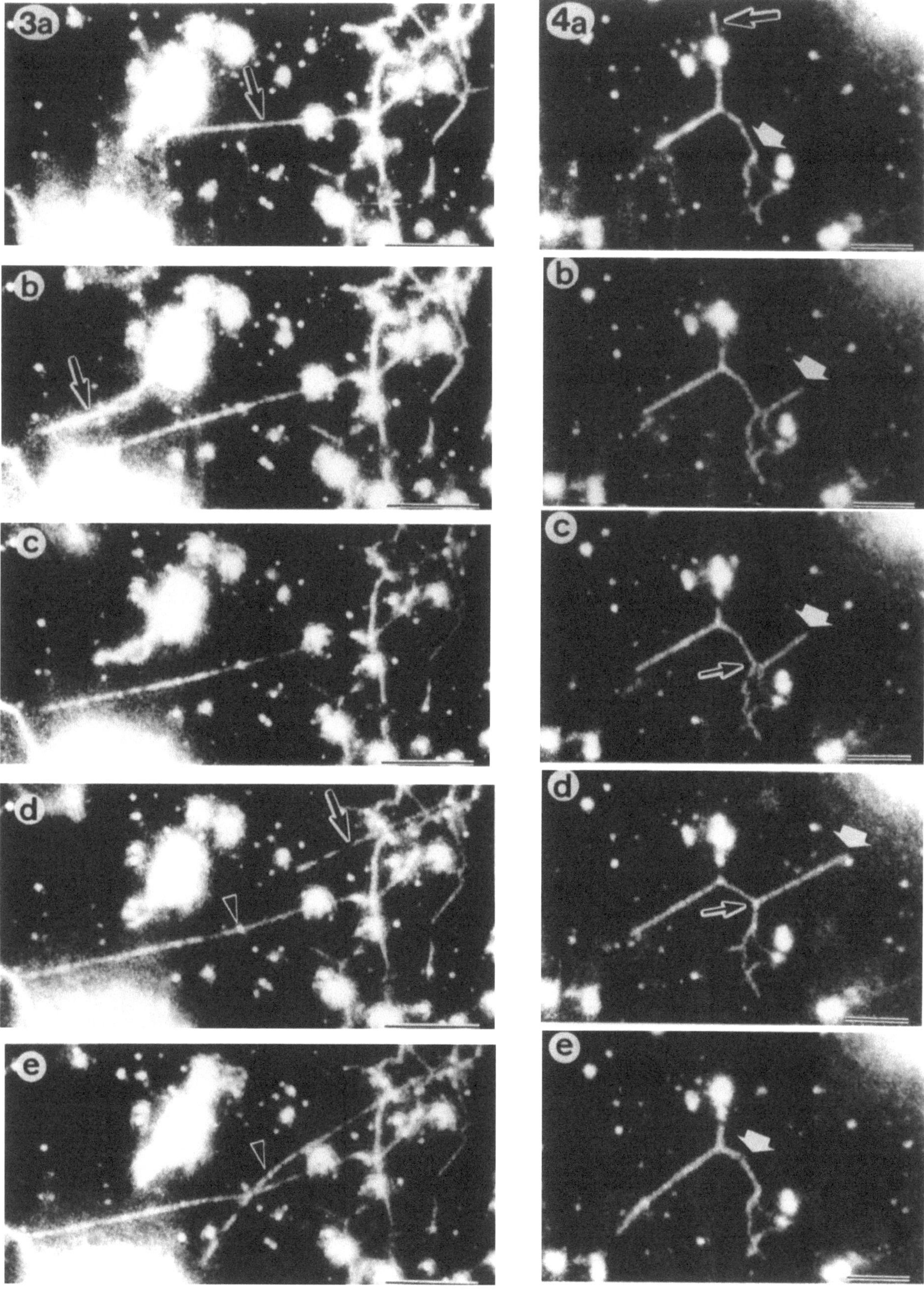

Fig. 3

Fig. 4

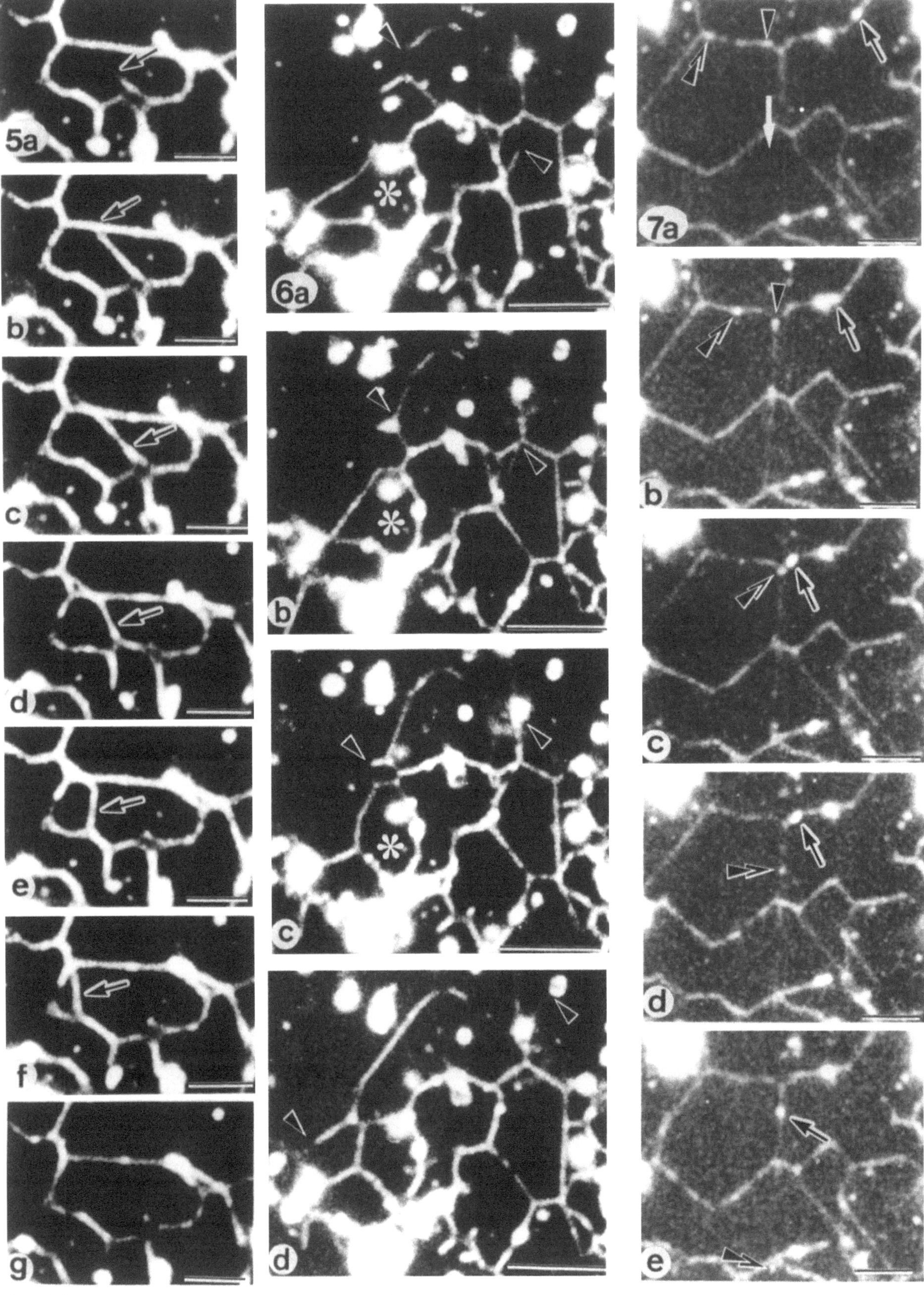

Figs. 5–7

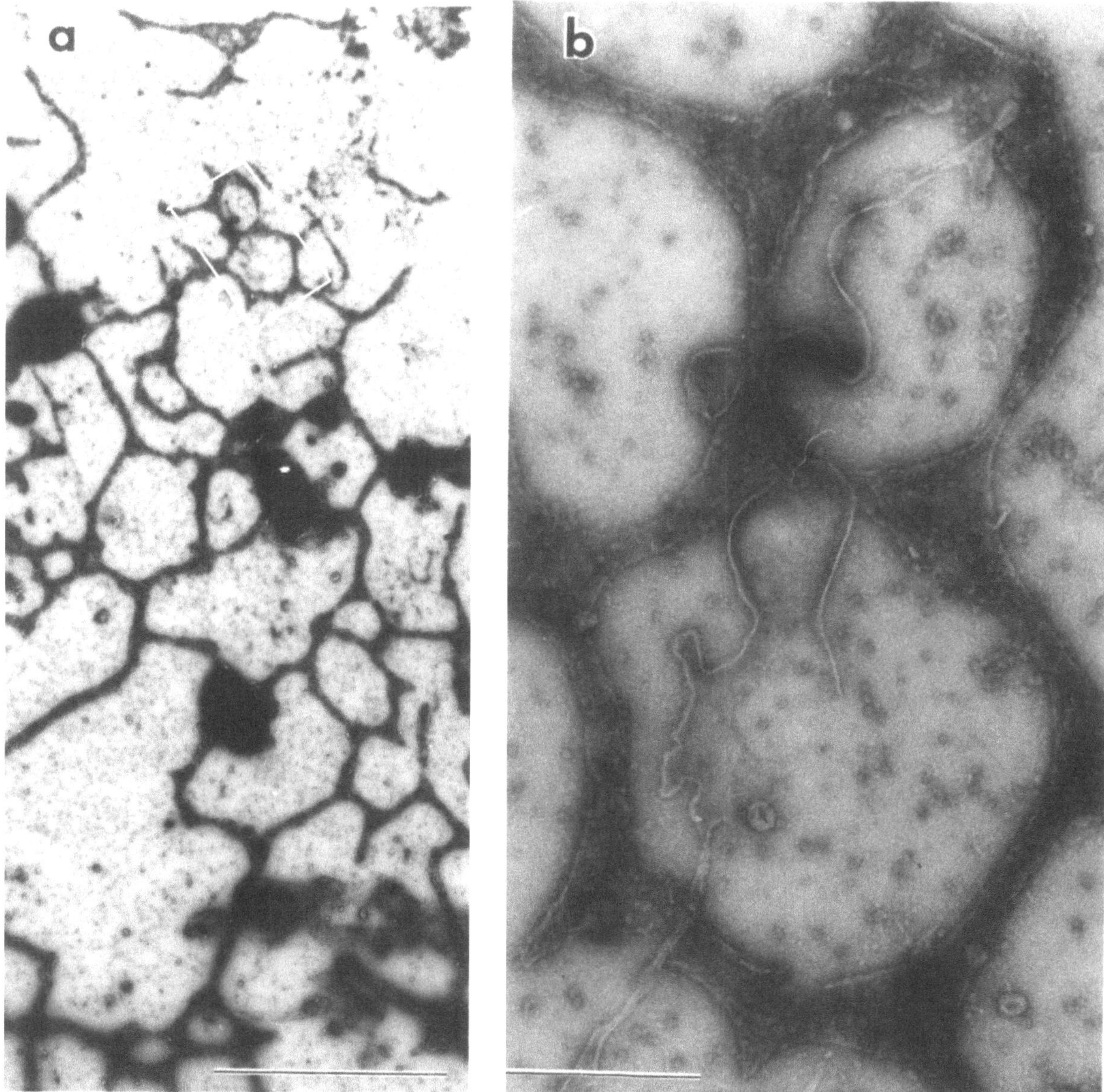

Fig. 8. Electron microscopy of the network fibers. *a* An entire view of a network. × 6,900. Bar indicates 5 µm. *b* Enlargement of the enclosed area shown in *a*. There are densely stained network fibers and membrane fragments stained brightly at the center of the photograph. In the network fibers there is no filamentous structure, but many small granular particles of about 20 nm in diameter are visible. × 58,000. Bar 0.5 µm

---

Fig. 5. Movement of the connecting points of the network fibers. An elongating fiber (an arrow in *a*) attached to the other stationary fibers in the same network (an arrow in *b*), and then both the distal and proximal ends of the elongating fiber began to run in the same direction along the stationary fibers. The arrows in *c–f* point to the translocating fiber. Finally it was absorbed into the network as shown in *g*. Photographs were taken at time intervals of 1/8 s. × 2,200. Bars 5 µm

Fig. 6. Change of the network pattern. Successive photographs taken at time intervals of 0.5 s show the same part of a network. With elongation of many branches of a network, the network pattern changes frequently. Two arrowheads in *a* and in the following pictures indicate tips of elongating fibers. According to elongation of these two fibers, as well as others nearby, polygonal shapes in the network were changed (a typical one is indicated by asterisks). Further elongation markedly changed the pattern. × 1,800. Bars indicate 10 µm

Fig. 7. Conveying materials from the stationary network fibers to the elongating fiber. The direction of elongation of a fiber is indicated by a white arrow in *a*. Single-, double-arrowheads and arrow indicate particular knobs or particles in the network fibers. The subsequent displacement of the particles or knobs clearly shows how the materials are conveyed from the adjacent stationary fibers to the elongating fiber. Photographs were taken at time intervals of 0.75 s. × 2,000. Bars 5 µm.

0.1% $OsO_4$ or occasionally unfixed fibers. On such grids, network fibers were found to be very similar at low magnification, to the network structure observed by dark field microscopy. The fibers were about 0.1 µm wide and looked membranous (Fig. 8 a). At high magnification, many particles about 20 nm in diameter were visible in the fiber, but no filamentous structures such as F-actin filaments were found (Fig. 8 b), even in the medium containing muscle actin.

## 4. Discussion

The network fibers which are induced to elongate by muscle actin, are derived from the cytoplasmic membrane of the endoplasmic droplets. Since the network structures remained at the position where the flattened endoplasmic droplets resided before being washed out, the network fibers must have come from the cytoplasmic membrane, which may contain some membrane-bound components and remaining cytoplasm. In various phenomena of the network fibers exhibit membrane characteristics: (1) They are very flexible, show great passive extension, and are twisted many times, as shown in Fig. 2. (2) Triton X-100 at 0.05% disintegrates the structure immediately into small pieces of granular particles. (3) Uranylacetate for staining, $OsO_4$ at 0.5% for fixation, and lowering the concentration of sucrose also destroy the structure immediately and completely. (4) Excess amount of glutaraldehyde, which links between amino residues, can not fix the network structure at all, which suggests that the main component is not protein(s). (5) Electron micrographs show membraneous structures, including small particles of about 20 nm in diameter.

This result of electron microscopy denies the possibility that (1) the cytoskeletal structure in the endoplasmic droplet contributes to formation of the network fibers, and (2) the mechanism of elongation resembles that in acrosomal reaction, in which the core of an F-actin bundle is involved (Tilney 1975).

The network fibers exhibit quite similar flexibility to the fine fibers found in endoplasmic droplets from *Nitella* (Jarosch 1979), but it is not yet known whether these fibers are identical. Endoplasmic droplets isolated from *Nitella* show myellinoid movement under certain conditions (Jarosch 1956, Yotsuyanagi 1953). Yet, whether or not actin filaments or microtubules are involved in this myellinoid movement has not been confirmed.

From electron microscopic studies of characean cells, endoplasmic reticulum is located in the streaming endoplasm (Williamson 1969) and closely associated with bundles of microfilaments (Bradley 1973). Endoplasmic reticulum shows the network structure in the cytoplasm in plant cells (Honda *et al.* 1964). In cultured cells, endoplasmic reticulum stained with fluorescent dye exhibits a similar structure to that of the network fibers, and is associated with the cytoskeletal structure composed of microtubules (Terasaki *et al.* 1984, 1986). On the other hand, endoplasmic reticulum in onion bulbs also has a similar structure and actin may be involved in movement of endoplasmic reticulum *in vivo* (Quader *et al.* 1987). The structural similarity between the network fibers and endoplasmic reticulum suggests that the network fibers may have derived from endoplasmic reticulum in *Nitella*.

It would be premature to discuss the mechanism of elongation at present, since we have no evidence about the composition of the network fibers. However, some mechanisms which could be postulated for elongation of the network fibers are: (1) Liposomes elongate from a spherical shape to a very long filamentous form depending on the osmolarity at a velocity of about 100 µm/s (Hotani 1984). The network fibers, however, require actin and ATP to elongate. (2) Added actin may elevate the concentration of actin above the critical concentration for polymerization. In the endoplasmic droplets, rotating polygons are thought to have originated from the subcortical fibrils but also they are formed and disintegrated spontaneously and repeatedly (Kuroda 1964), which may be explained by the existence of a large amount of non-polymerizable actin such as the complex of actin with profilin (Carlson *et al.* 1977), fragmin (Hasegawa *et al.* 1980), gelsolin (Yin and Stossel 1979), and so forth. This kind of actin, if any, would polymerize on the membrane with the aid of added F-actin. (3) Added F-actin can attach to the membrane of the network fiber and then react with *Nitella* myosin spread on the glass surface as suggested previously (Higashi-Fujime 1980), and may cause the elongation of the network fiber. As a single F-actin filament is unfortunately invisible by dark field microscopy, it is impossible to confirm whether this is the case or not.

Tropomyosin inhibits elongation of the network fibers. The reason is unknown. Tropomyosin might block the site of F-actin interacting with *Nitella* myosin or some protein which is implicated in elongation. Another possibility is that F-actin changes the physicochemical property when it binds tropomyosin, which may result in the inhibitory effect of tropomyosin.

Endoplasmic streaming requires dephosphorylation of

some protein(s), which makes streaming $Ca^{2+}$-insensitive (Tominaga *et al.* 1985, 1987). Although we have not yet examined the effect of $Ca^{2+}$ on elongation of the network fibers, a similar regulatory mechanism to that in streaming might inactivate a motor protein, if any, responsible for elongation, since elongation continues for about 5 min at room temperature (20–24 °C), and further perfusion of fresh medium containing actin does not activate elongation.

There are some recent reports on myosin-like proteins having an actin activated ATPase: 110 kD-calmodulin complex from microvilli (Collins and Borysenko 1984, Conzelman and Mooseker 1987, Coluccio and Bretscher 1987), and myosin I from *Acanthamoeba* (Pollard and Korn 1973). Particularly, myosin I is a motor protein for vesicular transport (Adams and Pollard 1986) and is preferentially bound to the membrane (Miyata and Korn 1987). It is plausible that a membrane-bound motor protein interacting with actin is involved in elongation of the network fibers and plays an important role in cell motility such as vesicular transport and also endoplasmic streaming in *Nitella*.

## Acknowledgements

We thank Prof. E. Oosawa for his valuable discussions and critical reading of this manuscript.

## References

Adams RJ, Pollard TD (1986) Propulsion of organelles isolated from *Acanthamoeba* along actin filaments by myosin I. Nature 322: 754–756

Allen NS (1974) Endoplasmic filaments generate the motive force for rotational streaming in *Nitella*. J Cell Biol 63: 270–287

Bradley MO (1973) Microfilaments and cytoplasmic streaming: Inhibition of streaming with cytochalasin. J Cell Sci 12: 327–343

Carlson L, Nyström LE, Sundkvist I, Markey F, Lindberg U (1977) Actin polymerizability is influenced by profilin, a low molecular weight protein in nonmuscle cells. J Mol Biol 115: 465–483

Chen JCW, Kamiya N (1975) Localization of myosin in the internodal cell of *Nitella* as suggested by differential treatment with N-ethylmaleimide. Cell Struct Funct 1: 1–9

Collins JH, Borysenko CW (1984) The 110,000-dalton actin- and calmodulin-binding protein from intestinal brush border is a myosin-like ATPase. J Biol Chem 259: 14128–14135

Coluccio LM, Bretscher A (1987) Calcium-regulated cooperative binding of the microvillar 110 K-calmodulin complex to F-actin: Formation of decorated filaments. J Cell Biol 105: 325–333

Conzelman KA, Mooseker MS (1987) The 110 kD protein-calmodulin complex of the intestinal microvillus is an actin-activated Mg-ATPase. J Cell Biol 105: 313–324

Ebashi S, Ebashi F (1965) α-actinin, a new structural protein from striated muscle I. Preparation and action on acto-myosin ATP interaction. J Biochem (Tokyo) 58: 7–12

Kodama A, Ebashi F (1968) Troponin. I. Preparation and physiological function. J Biochem (Tokyo) 64: 465–477

Foissner I, Jarosch R (1981) The motion mechanics of *Nitella* filaments (cytoplasmic streaming): Their imitation in detail by screw-mechanical models. Cell Motil 1: 371–385

Hasegawa T, Takahashi S, Hayashi H, Hatano S (1980) Fragmin: A calcium ion sensitive regulatory factor on the formation of actin filaments. Biochemistry 19: 2677–2683

Higashi-Fujime S (1980) Active movement *in vitro* of bundles of microfilaments isolated from *Nitella* cell. J Cell Biol 87: 569–578

— (1983) Phosphorylation of myosin light chain modulates the *in vitro* movement of fibrils composed of actin and myosin filaments from skeletal muscle. J Biochem (Tokyo) 94: 1539–1545

— (1985) Unidirectional sliding of myosin filaments along the bundle of F-actin filaments spontaneously formed during superprecipitation. J Cell Biol 101: 2335–2344

Honda SL, Hongladarom T, Wildman SG (1964) Characteristic movements of organelles in streaming cytoplasm of plant cells. In: Allen RD, Kamiya N (eds) Primitive motile systems in cell motility. Academic Press, New York, pp 485–502

Hotani H (1984) Transformation pathway of liposomes. J Mol Biol 178: 113–120

Jarosch R (1956) Plasmaströmung und Chloroplastenrotation bei Characeen. Pyton 6: 87—107

— (1976) Dynamic behaviour of the actin fibrils of *Nitella* based on rapid filament rotation. Biochem Physiol Pflanzen 170: 111–131

— (1979) Filament-dynamics of *Nitella*. In: Hatano S, Ishikawa H, Sato H (eds) Cell motility: molecules and organization. Tokyo Univ Press, Tokyo, pp 247–249

Kamitsubo E (1972) Motile protoplasmic fibrils in cells of the Characeae. Protoplasma 74: 53–70

Kamiya N, Kuroda K (1956) Velocity distribution of the protoplasmic streaming in *Nitella* cells. Bot Mag (Tokyo) 69: 544–554

— — (1957) Cell operation in *Nitella*. II Behaviour of isolated endoplasm. Proc Jap Acad 33: 201–205

Kato T, Tonomura Y (1977) Identification of myosin in *Nitella flexilis*. J Biochem (Tokyo) 82: 777–782

Kersey YM, Hepler PK, Palevitz BA, Wessels NK (1976) Polarity of actin filaments in Characean algae. Proc Natl Acad Sci USA 72: 165–167

Kuroda K (1964) Behaviour of naked cytoplasmic drops isolated from plant cells. In: Allen RD, Kamiya N (eds) Primitive motile systems in cell motility. Academic Press, New York, pp 31–41

— Kamiya N (1975) Active movement of *Nitella* chloroplasts *in vitro*. Proc Jpn Acad 51: 774–777

Miyata H, Korn ED (1987) Bound states of *Acanthamoeba* myosin I. J Cell Biol 105: 115 a

Nagai R, Hayama T (1979) Ultrastructure of the endoplasmic factor responsible for cytoplasmic streaming in *Chara* internodal cells. J Cell Sci 36: 121–136

Palevitz BA, Ash JF, Hepler PK (1974) Actin in green alga, *Nitella*. Proc Natl Acad Sci USA 71: 363–366

Perry SV (1955) Myosin adenosinetriphosphatase. Methods Enzymol 2: 582–588

Pollard TD, Korn ED (1973) *Acanthamoeba* myosin I. Isolation form *Acanthamoeba castellanii* of an enzyme similar to myosin. J Biol Chem 248: 4682–4690

Quader H, Hofmann A, Schnepf E (1987) Shape and movement of the endoplasmic reticulum in onion bulb epidermis possible involvement of actin. Eur J Cell Biol 44: 17–26

SHEETZ MP, SPUDICH JA (1983) Movement of myosin-coated fluorescent beads on actin cables *in vitro*. Nature 303: 31–35

SHIMMEN T, YANO M (1984) Active sliding movement of latex beads coated with skeletal muscle myosin on *Chara* actin bundles. Protoplasma 121: 132–137

TERASAKI M, CHEN LB, FUJIWARA K (1986) Microtubules and the endoplasmic reticulum are highly interdependent structure. J Cell Biol 103: 1557–1568

— SONG J, WONG JR, WEISS MJ, CHEN LB (1984) Localization of endoplasmic reticulum in living and glutaraldehyde-fixed cells with fluorescent dyes. Cell 38: 101–108

TILNEY LG (1975) Actin filaments in the acrosomal reaction of *Lymulus* sperm. Motion generated by alterations in the packing of the filaments. J Cell Biol 64: 289–310

TOMINAGA Y, MUTO S, SHIMMEN T, TAZAWA M (1985) Calmodulin and $Ca^{2+}$-controlled cytoplasmic streaming in characean cells. Cell Struct Funct 10: 315–325

— WAYNE R, TUNG HYL, TAZAWA M (1987) Mechanism of $Ca^{2+}$ regulation of cytoplasmic streaming. Muscle Res Cell Motil 8: 285–286

VAN EERD JP, KAWASAKI Y (1973) Effect of calcium (II) on the interaction between the subunits of troponin and tropo-myosin. Biochemistry 12: 4972–4980

WILLIAMSON RE (1974) Actin in the alga, *Chara corallina*. Nature 248: 801–802

WILLIAMSON RE (1979) Filaments associated with the endoplasmic reticulum in the streaming cytoplasm of *Chara corallina*. Eur J Cell Biol 20: 177–183

YIN HL, STOSSEL TP (1979) Control of cytoplasmic actin gel-sol transformation by gelsolin, a calcium-dependent regulatory protein. Nature 281: 583–586

YOTSUYANAGI Y (1953) Recherches sur les phénomènes moteurs dans les fragments de protoplasme isolés II. Mouvements divers déterminés par la condition de milieu. Cytologia 18: 202–217

Protoplasma (1988) [Suppl. 2]: 37–47



# Domain Structure of *Physarum* Myosin Heavy Chain

KAZUHIRO KOHAMA[1,*], MIWA SOHDA[1], KIMIE MURAYAMA[2], and YOH OKAMOTO[3]

[1] Department of Pharmacology, Faculty of Medicine, University of Tokyo, [2] Division of Biochemical Analysis, Central Laboratory of Medical Sciences, and [3] Department of Biochemistry, Juntendo University, School of Medicine, Tokyo

Received February 25, 1988
Accepted April 1, 1988

Dedicated to Professor Dr. NOBURO KAMIYA on the occasion of his 75th birthday

## Summary

*Physarum* myosin is inhibited by $Ca^{2+}$. Specifically its actin-activated ATPase activity (KOHAMA and KENDRICK-JONES 1986) and its movement along actin-cables (KOHAMA and SHIMMEN 1985) are inhibited by physiological levels of $Ca^{2+}$. The involvement of the Ca-inhibition in regulating plasmodial motility is suggested by the Ca-dependent contraction of demembranated plasmodial cells (YOSHIMOTO *et al.* 1981, YOSHIMOTO and KAMIYA 1984, ACHENBACH and WOHLFARTH-BOTTERMANN 1986, 1986 a).

The domain structure of this myosin HC (230 K Mr) was studied by limited proteolysis and SDS PAGE and by preparing HMM and S 1 from proteolysed myosin. The HC of HMM was shown to be comprised of a N-terminal 70 K/60 K fragment, a middle 28 K fragment and a C-terminal 17 K fragment. S 1 is devoid of the 17 K fragment.

These fragments were characterized by (i) a gel overlay technique, which showed that the 28 K fragment contained binding sites for [125I] PLC (phosphorylatable light chain of 18 K Mr, KOHAMA *et al.* 1986) and [125I] CaLC (Ca-binding light chain of 16,131 dalton, KOBAYASHI *et al.* 1988) and that binding sites for [125I] actin was in both the 70 K/60 K fragment and the 28 K fragment. (ii) The digested *Physarum* myosin was subjected to photoaffinity labelling at active site with [3H] UTP, which showed that the 70 K/60 K fragment contained the catalytic site. The reactive SH groups which were rapidly modified with [14C] NEM and fluorescent 1,5,-I-AEDANS were on different fragments.

Unusual amino acid, which is contained in active site peptide of skeletal muscle myosin HC (OKAMOTO and YOUNT 1985) and *Acanthamoeba* myosin II HC (ATKINSON and KORN 1986), was identified as dimethyllysine for *Physarum* myosin.

*Keywords:* Physarum myosin; Myosin heavy chain; Limited proteolysis; Domain structure; Actin-binding sites; Light chain-binding sites; ATPase active site; Sulfhydril residues; Dimethyllysine

* Correspondence and Reprints: Department of Pharmacology, Faculty of Medicine, University of Tokyo, Bunkyo-ku, Tokyo 113, Japan.

*Abbreviations:* ATP adenosine 5′-triphosphate; DTT dithiothreitol; EDTA ethylenediaminetetraacetic acid; EGTA ethyleneglycolbis (β-aminoethyl ether)—N,N,N′,N′-tetraacetic acid; PMSF phenylmethylsulfonyl fluoride; SDS sodium dodecyl sulfate; CBB Commassie Brilliant Blue; SDS PAGE SDS polyacrylamide gel electrophoresis; SH sulfhydryl residue; NEM N-ethylmaleimide; 1,5,-I-AEDANS N-iodo-acetyl-N′-(5-sulfo-1-naphthyl) ethylenediamine; HC heavy chain; HMM heavy meromyosin; S 1 subfragment 1; S2 subfragment 2; LMM light meromyosin; CaLC Calcium binding light chain; PLC phosphorylatable light chain; HPLC high performance liquid chromatography.

## 1. Introduction

Myosin from *Physarum* plasmodium is composed of a pair of HC (230 K Mr) and two pairs of light chains (PLC of 18 K Mr and CaLC of 16,131 dalton) (see KOHAMA 1987, for a review). Electron microscopy reveals that *Physarum* myosin has a structure similar to that of other muscle and nonmuscle myosins. It consists of two globular heads and an extended rod-like tail (HATANO and TAKAHASHI 1971, TAKAHASHI *et al.* 1983, KOHAMA *et al.* 1983, KOHAMA 1987). *Physarum* myosin prepared by the newly developed method (KOHAMA and KENDRICK-JONES 1986) is $Ca^{2+}$-inhibitory, *i.e.*, its actin-activated ATPase activity is inhibited by μM levels of $Ca^{2+}$ (for the physiological significance, see KOHAMA 1987, 1987 a). The $Ca^{2+}$-inhibition is not observed when cysteine residues of myosin HC were modified by NEM (KOHAMA *et al.* 1987). The residues are expected to be analogous to the reactive SH of skeletal muscle myosin.

There is a general agreement that the HC contains the active site for ATPase activity and the binding sites for actin and the light chains. Localisation of these sites in HC of skeletal muscle myosin has been intensively studied with skeletal muscle myosin by the limited proteolysis of skeletal muscle myosin (see HARRINGTON and RODGERS 1984, for a review). Such structural analyses have also been performed with smooth muscle (OKAMOTO *et al.* 1980, MITCHELL *et al.* 1986) and non-muscle (BARYLKO *et al.* 1986, ATKINSON and KORN 1986) myosins. However, no study has been attempted with *Physarum* myosin.

In this study, we identified 3 domains in the HC of *Physarum* myosin head using limited proteolysis and characterized which domain has the reactive SH, the active site, and the binding sites for actin and the light chains. In addition, unusual methylated amino acids, which are supposed to be closely involved in the function of the active site (OKAMOTO and YOUNT 1985, ATKINSON and KORN 1986), were studied.

Part of this study was previously reported in the preliminary form (KOHAMA and SOHDA 1987).

## 2. Materials and Methods

Trypsin (189 u/mg) and chymotrypsin (alpha-chymotrypsin, 53 u/mg) were obtained from Worthington; Soybean trypsin inhibitor (Type II-S) and PMSF from Sigma; 1,5-I-AEDANS from Aldrich; $^{125}$I-labelled Bolton-Hunter reagent and N-ethyl [2,3-$^{14}$C] maleimide from Amersham; and [5,6-$^3$H] UTP from ICN Biochemicals. All other chemicals are of analytical grade.

*Physarum* plasmodia (strain Ng-1) were grown on rolled oats in the dark (KOHAMA and KENDRICK-JONES 1986) by modifying the method described (CAMP 1936, HATANO and OOSAWA 1966). The fresh migrating sheets of plasmodia were harvested and immediately subjected to the preparation procedure used to isolate native actomyosin (KOHAMA and KOHAMA 1984). Myosin* was rapidly purified from native actomyosin by a new method as described previously (KOHAMA and KENDRICK-JONES 1986).

CaLC and PLC were prepared from native actomyosin by modifying the method for preparing CaLC from myosin (KOHAMA *et al.* 1986). The light chains were released from native actomyosin (KOHAMA *et al.* 1985) by incubation at 55 °C for 15 min in 0.4 M NaCl and 0.1 M Na$_2$CO$_3$ (final pH = about 10.5). Native actomyosin thus treated was mixed with 5 vol of cold water and then with 1 N HCl to adjust the pH to 6.5 followed by the centrifugation at 10,000 × **g** for 10 min. The supernate was concentrated by ultrafiltration with an Amicon PM 10 membrane. The concentrate was subjected to ammonium sulfate fractionation; the precipitate between 55% and 85% saturation contained the light chains. The ammonium sulfate precipitate was dialyzed against 4 mM NaPB (Na phosphate buffer at pH 6.3) and applied to the column of DEAE cellulose (DE-52, Whatman) equilibrated with 4 mM NaPB. The column was washed exhaustively with 50 mM NaPB. Then PLC was eluted by 90 mM NaPB. The eluate was concentrated by dialysis against solid sucrose and stored after removing the sucrose by dialysis against 1 mM

NaHCO$_3$. After eluting PLC, 200 mM NaPB was applied to the column. The eluate was fractionated. The fractions containing pure CaLC as judged by SDS PAGE (see below) were pooled and concentrated by dialysis against solid sucrose followed by dialysis against 1 mM NaHCO$_3$. Sometimes the concentrated CaLC was contaminated by PLC as determined by SDS PAGE. In this case, the concentrate was further subjected to the ammonium sulfate fractionation between 63% and 85% saturation. SDS PAGE of CaLC and PLC thus prepared showed single bands (see Fig. 1 in KOHAMA 1987). Actin was prepared from rabbit skeletal muscle by the method of SPUDICH and WATT (1971) with minor modifications (KOHAMA 1980). Rabbit skeletal muscle HMM and S 1 were prepared by a method described previously (MARGOSSIAN and LOWEY 1982).

Myosin was digested by trypsin or chymotrypsin under the conditions specified in each legend. We digested myosin in either 0.15 M or 0.5 M NaCl at pH 7.5. The difference in the ionic strength did not affect the results. The digestion by trypsin or chymotrypsin was stopped by adding 2–3 fold excess (on a weight basis) trypsin inhibitor over trypsin or 0.1–1 mM (final concentration) of PMSF, respectively.

HMM and S 1 of *Physarum* myosin were prepared by subjecting the tryptic or chymotryptic digest of the myosin in 120 mM NaCl, 1 mM EGTA, 20 mM tris-HCl (pH 7.5) and 10 mM 2-mercaptoethanol to HPLC using a size exclusion column of TSK G 3000 SW (Toyo Soda). A typical elution profile was shown in Fig. 2 A. The proteins eluted in the same fractions as those of skeletal muscle HMM and S 1 were concentrated by dialysis against solid sucrose, dialysed against 1 mM NaHCO$_3$ as described previously (KOHAMA *et al.* 1987), and used as the HMM and S 1 preparations, respectively. The HMM and S 1 preparations examined with an electron microscope after rotary shadowing showed homogeneous double heads and single heads, respectively, confirming the previous observations with fractionated trypsin digests using Sepharose 4 B column chromatography (KOHAMA *et al.* 1983, KOHAMA 1987, KOHAMA *et al.* 1987).

SDS PAGE was performed with non-gradient slab gels as described by LAEMMLI (1970) with slight modifications. For gradient slab gels, we used the Phast system equipment (Pharmacia) according to the manufacturer's manuscript. Calibration was usually carried out in each slab by electrophoresing a mixture of molecular weight markers as follows; thyroglobin (330,000 dalton), myosin light chain kinase prepared from chicken gizzard smooth muscle by the method of NAKAMURA and NONOMURA (1984) (140,000 dalton), phosphorylase b (94,000 dalton), albumin (67,000 dalton), ovalbumin (43,000 dalton), carbonic anhydrase (30,000 dalton), trypsin inhibitor (20,100 dalton), myoglobin (17,200 dalton), lactalbumin (14,400 dalton), and myoglobin fragment 1 (8,200 dalton). The molecular mass of these makers are shown in K dalton units in most figures.

Myosin digests were photoaffinity-labelled at the ATPase active site with Mg-[$^3$H] UTP using the method of MARUTA and KORN (1981) with minor modifications (KOHAMA *et al.* 1987). The modified HC fragments were detected by fluorography (see below).

The gel overlay technique was adapted from the procedure initially described by SNABES *et al.* (1981) as modified by MITCHELL *et al.* (1986). SDS PAGE gels (non-gradient) were run and fixed in 50% methanol and 10% acetic acid at about 25 °C for about 30 min, washed for 5 hr with 3 washes of 10% ethanol at 4 °C followed by incubation for 30 min at 4 °C in buffer A (0.2% gelatin, 60 mM NaCl, 2 mM MgCl$_2$, 10 mM K phosphate buffer at pH 7.0, 0.2 mM DTT and 2 mM sodium azide). Gels were incubated at 4 °C for 5 hr in buffer A containing the $^{125}$I-labelled CaLC, PLC or actin (see below). followed by 4 washes with buffer A over 2 nights. The gels were

---

* In this paper, myosin, HMM and S 1 denote myosin, HMM and S 1 from *Physarum*, respectively, unless otherwise specified.

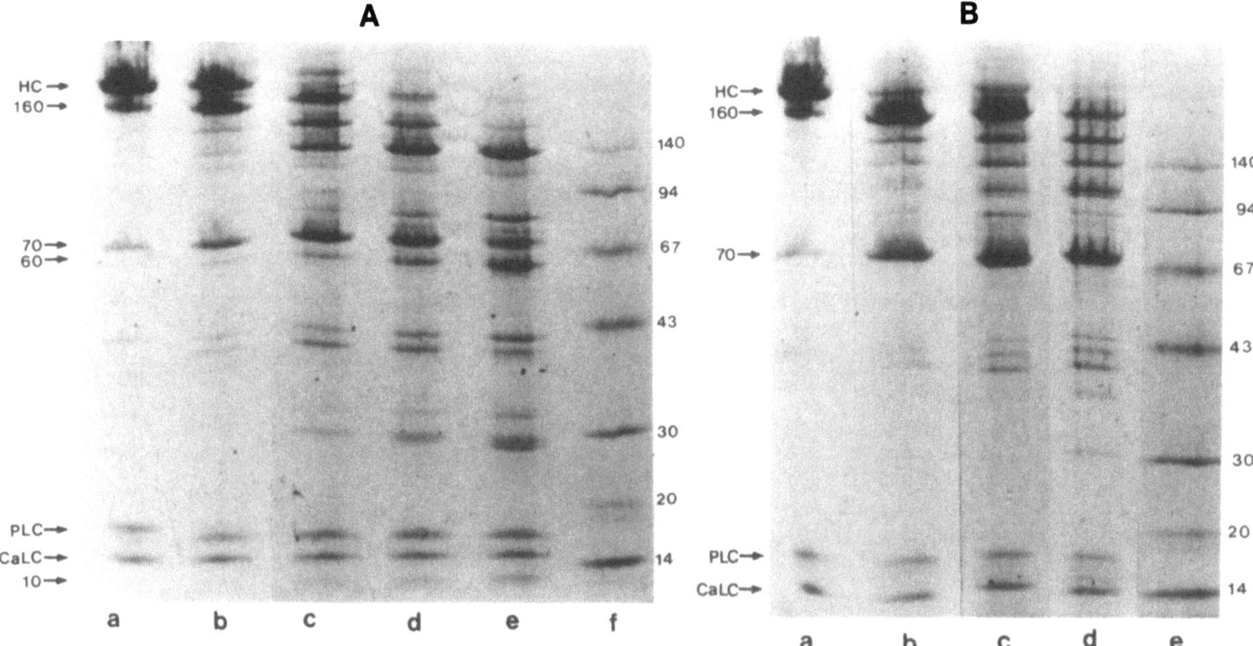

Fig. 1. Time course of myosin digestion with trypsin (*A*) and chymotrypsin (*B*). Myosin (4 mg/ml) was digested with trypsin (trypsin : myosin = 1 : 800, on a weight basis) or chymotrypsin (chymotrypsin : myosin = 1 : 400, on a weight basis) in 0.15 M NaCl and 20 mM tris-HCl (pH 7.5) at 4 °C for specified time. The digests were subjected to SDS PAGE on a 10–15% gradient gel as described in Materials and Methods. The numbers (in K dalton) with arrows were estimated from their mobilities on the gels with molecular weight markers in K dalton (*f* in *A*, *e* in *B*). The estimation of the molecular weights of HC (230 K) and 10 (10 K) were, however, carried out on 8% or 20% gels, respectively, using additional molecular weight markers of thyroglobin (330 K dalton) and myoglobin fraction I (8 K dalton). *A*: *a* undigested control; *b* 5 min; *c* 30 min; *d* 60 min; *e* 120 min. *B*: *a* undigested control; *b* 40 min; *c* 75 min; *d* 270 min. N.B. The time course is schematically summarized in Fig. 7 *B*

rinsed with water to remove the gelatin, fixed with 50% methanol and 7% acetic acid, stained with CBB in 50% methanol and 7% acetic acid, destained with 7% acetic acid, dried and autoradiographed with Kodak X-OMAT AR film at −80 °C in a cassette fitted with Dupont-Cornex intensifying screens. As pointed out by Mitchell *et al.* (1986), light chains are washed out during the gel overlay procedures from the gels separating the digests of skeletal and smooth muscle myosins. However, the HC fragments of these myosins are retained in the gels. The same tendencies were observed with the HC fragments of *Physarum* myosin (see Fig. 5).

Actin and light chains were labelled with $^{125}$I using the Bolton-Hunter reagent as outlined by the manufacturer's (Amersham) manuscript. About 50 μg protein was labelled with about 100 μCi Bolton-Hunter reagent. After stopping the reaction by adding glycine, the proteins thus treated were dialysed against buffer A and used for the gel overlay technique.

Myosin after removing DTT by exhaustive dialysis against 0.5 M NaCl and 2 mM NaHCO$_3$ was used for the modification of reactive SH with [$^{14}$C] NEM or 1,5-I-AEDANS by modifying the method described by Kohama *et al.* (1987) or Reisler (1982), respectively. The modification was performed in 6 mg/ml myosin, 5 μCi/ml [$^{14}$C] NEM 0.5 M NaCl and 100 mM tris-HCl pH 7.5 or in 1 mg/ml myosin, 25 μM 1,5-I-AEDANS, 0.5 M NaCl and 100 mM tris-HCl pH 7.5 at 0 °C for 30 min and stopped by adding 10 mM (final concentration) of DTT. The modified myosins was digested by trypsin (trypsin : myosin = 1 : 30 on a weight basis) at 25 °C for 10 min and sub-

jected to SDS PAGE using 14% gels. The HC fragments modified by [$^{14}$C] NEM or 1,5-I-AEDANS were detected by fluorography (see below) or illumination by an ultraviolet lamp (254 nm), respectively. Fluorography was carried out by immersing the CBB-stained gels in Amplify (Amersham) for 30 min. The gels were subjected to autoradiography (see gel overlay technique).

Unusual amino acid were detected by a high speed amino acid analyzer (Hitachi 835) after hydrolysis of myosin and HMM preparation by 6 N HCl at 105 °C for 24 hr. Commercially purchased monomethyllysine, dimethyllysine, trimethyllysine and 3-methylhistidine were used for the identification of the unusual amino acids. Commercially purchased 6-amino-n-caproic acid was used as an internal standard for the calculation of unusual amino acid content.

## 3. Results

### 3.1. Digestion of Myosin

Myosin was digested with a low concentration of chymotrypsin (enzyme : myosin = 1 : 400, on a weight basis) or trypsin (enzyme : myosin = 1 : 800) at low temperature (4 °C), and the time course of the cleavage of HC (230 Kd) was observed by SDS PAGE (Fig. 1). Initial cleavage by chymotrypsin was found between

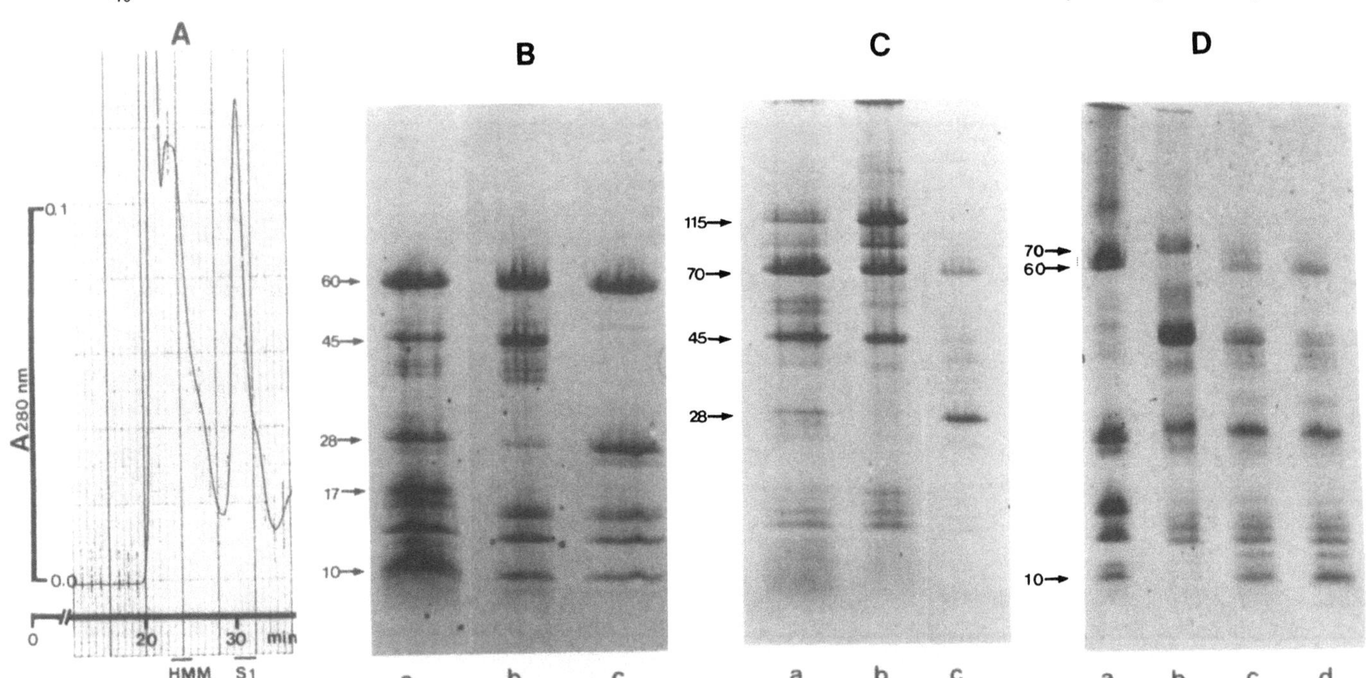

Fig. 2. Preparation of HMM and S 1 and tryptic digestion of HMM. *Physarum* myosin was digested with trypsin (*B*) or chymotrypsin (*C*) and applied to HPLC using TSK G 3000 SW size exclusion column (*A*). The fractions indicated by HMM and S 1 were pooled and concentrated and subjected to SDS PAGE on 8–25% gradient gels (*B, C*) or to further digestion (*D*). The numbers with arrows show the molecular weights in K daltons of the HC fragments. The topography of the HC fragments is schematically shown in Fig. 7 *A*. *A* The absorbance at 280 nm ($A_{280}$ nm) of HPLC of the tryptic digest (lane *a* in *B*). Similar elution profile was obtained with the chymotrytic digest (lane *a* in *C*). *B* SDS PAGE of HMM and S 1 prepared from the tryptic digest. The conditions for limited proteolysis were as follows; in 12 mg/ml myosin, 0.15 M NaCl, 20 mM tris-HCl (pH 7.5), and 0.5 mg/ml trypsin at 25 °C for 13 min. *a* tryptic digest of myosin; *b* tryptic HMM; *c* tryptic S 1. Intramolecular nicking of HMM and S 1 was indicated by the absence of fragments above 60 K in tryptic HMM and S 1 (see Discussion). *C* SDS PAGE of HMM and S 1 prepared from the chymotryptic digest. The conditions for limited proteolysis were as follows; in 10 mg/ml myosin, 0.5 M NaCl, 20 mM tris-HCl (pH 7.5), and 0.3 mg/ml chymotrypsin at 25 °C for 15 min. *a* chymotryptic myosin digest; *b* chymotryptic HMM; *c* chymotryptic S 1. Similar to tryptic HMM and S 1, there are signs of intramolecular nicking in chymotryptic HMM and S 1. The 115 K fragment was intact HMM escaped from the nicking. *D* SDS PAGE of the tryptic digest of chymotryptic HMM. Conditions for the digestion; in 6 µg/ml trypsin, 1 mg/ml chymotryptic HMM, 0.15 M NaCl, 20 mM tris-HCl (pH 7.5) at 25 °C for 0 min (*b*), 5 min (*c*), or 10 min (*d*). *a* Tryptic S 1. The chymotryptic HMM was a different preparation shown in lane *b* in *C*, and did not contain 115 K polypeptides, an intact HC of HMM (see Fig. 7 *A*). Note that the 70 K fragment (70 with an arrow) was digested to th 60 K fragment (60 with an arrow), which supports the idea that the 60 K fragments of tryptic HMM and S 1 are analogous to the 70 K fragments of chymotryptic HMM and S 1

the 160 K and 70 K fragments* (Figs. 1 *A, B*, lanes *b*, and 7 *B*). The 70 K fragment was resistant to prolonged digestion (Fig. 1 *B*, lanes *c* and *d*). The HC was also cleaved rapidly into 160 K and 70 K fragments* by trypsin (Fig. 1 *A*, lane *b*). The 70 K fragment was susceptible to the further digestion with trypsin (Fig. 1 *A*, lanes *c–e*). The decrease in the amount of 70 K fragment was associated with the appearance of 60 K and 10 K fragments (Fig. 1 *A*, lanes *d* and *e*), suggesting that the 70 K fragment cleaved into the 60 K and 10 K frag-

ments (see also Fig. 2 *D*). These 60 K and 10 K fragments were resistant to further proteolysis with a higher concentration of trypsin (enzyme : myosin = 1 : 24) at higher temperature (25 °C) (Figs. 2 *B*, lane *a*, and 7 *B*). Samples from the myosin digest were taken for photoaffinity labelling with [³H] UTP (Fig. 3). The fluorograph of the trypsin or chymotrypsin digests showed that the photoaffinity labelled active site was associated with the 60 K or 70 K fragment, respectively (Fig. 3, lanes *a–d*). This observation indicates that the 70 K/60 K fragment bore active site, which is in accordance with our previous study (Fig. 3 in KOHAMA *et al.* 1987). When the 60 K fragment of tryptic digest was proteolyzed by elevating the trypsin concentration (enzyme : myosin = 1 : 5), [³H] UTP was incorporated into

---

* When weakly digested myosin (Fig. 1 *A* and *B*, lanes *b*) was subjected to size fractionation with HPLC (see Materials and Methods), most of the protein was recovered at the fraction of intact myosin, indicating that this cleavage is a intramolecular "nick" (see also Discussion).

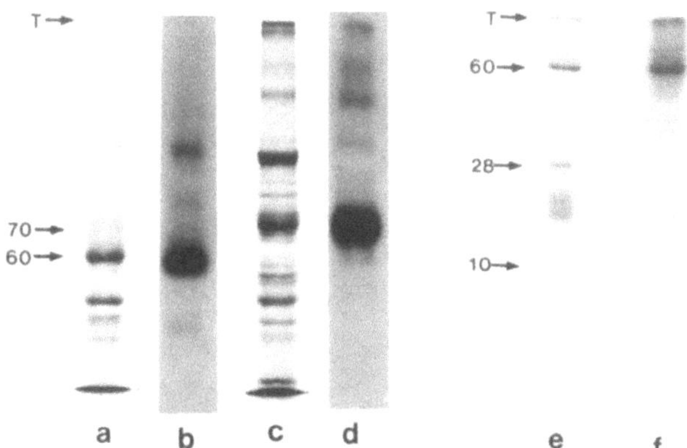



Fig. 3. Identification of the fragment bearing the active site. Tryptic (*a, b*) and chymotryptic (*c, d*) myosin digests and tryptic S 1 (*e, f*) were photoaffinity-labelled with [³H] UTP and subjected to SDS PAGE followed by fluorography as described in the Materials and Methods. The digestion was carried out in 11 mg/ml myosin, 0.03 mg/ml trypsin or chymotrypsin, 0.5 M NaCl, 20 mM tris-HCl (pH 7.5) at 25 °C for 1 hr. Tryptic S 1 was prepared as described in the legend to Fig. 2 *A* and *B. a* and *c* CBB stain of 8% gel. *e* CBB stain of 15% gel. *b, d*, and *f* Fluorogram of *a, c*, and *e*, respectively. The numbers with arrows were HC fragments expressed in K dalton. *T* Gel top. Note that 70 K/60 K fragment bears the active site

both the 60 K and the 43 K fragments (not shown). Thus the active site was localised in the proteolytically stable 43 K core.

### 3.2. Purification of Myosin Heads

The tryptic digests of myosin were subjected to HPLC chromatography using TSK G 3000 SW size exclusion column (Fig. 2 *A*). The proteins eluted in the same fractions as those of skeletal muscle HMM and S 1 were pooled separately, concentrated with solid sucrose and subjected to SDS PAGE. The HC of HMM of *Physarum* myosin was comprised from 60 K, 45 K, and 10 K fragments (Fig. 2 *B*, lane *b*), and the HC of S 1 of *Physarum* myosin was composed of 60 K, 28 K, and 10 K fragments (Fig. 2 *B*, lane *c*). Photoaffinity labelling of S 1 with Mg-[³H] UTP showed that active site was in the 60 K fragment (Fig. 3, lanes *e* and *f*), confirming the results obtained with unfractionated tryptic digests of myosin (Fig. 3, lanes *a* and *b*).

As will be discussed later, the binding sites for the light chains in the HC of HMM and S 1 were the 45 K and 28 K fragments, respectively (Figs. 4 and 5). In the case of vertebrate muscle and nonmuscle myosins, HC binds light chains at the C-terminal region of the myosin heads (Mitchell *et al.* 1986, Barylko *et al.* 1986). Thus we assigned the 45 K and 28 K fragments to the C-terminal part of the HC of HMM and S 1, respec-

tively (Fig. 7 *A*). The 17 K fragment in the tryptic digest (Fig. 2 *B*, lane *a*) was absent in the HC of S 1 (Fig. 2 *B*, lane *c*; Fig. 5, lanes *b* and *d*). This suggests that trypsin cleaves the 45 K fragment into the 28 K and 17 K fragments and that the 17 K fragment is not retained in S 1. Thus, as illustrated in Fig. 7 *A*, the 17 K fragment was assigned to the C-terminal part of the 45 K fragment (see discussion in relation to *Physarum* S 2).

We also purified HMM and S 1 from the chymotryptic digest of *Physarum* myosin using TSK G 3000 SW column (Fig. 2 *C*). The 70 K fragment bearing active site (Fig. 3, lanes *c* and *d*) was contained in both HMM and S 1 (Fig. 2 *C*, lanes *b* and *c*). And the 45 K and 28 K fragments bearing the binding sites for light chains (Fig. 4 *B*, lanes *a–c*) were present in HMM and S 1, respectively. These data are compatible with those of HMM and S 1 purified from the tryptic digests.

The tryptic 60 K fragment bearing the active site (Fig. 3, lanes *a* and *b*) was easily produced from the chymotryptic 70 K fragment by subjecting chymotryptic HMM to tryptic digestion (Fig. 2 *D*). This also suggests that the 10 K fragment produced by the tryptic digestion should be the fragment released from the chymotryptic 70 K fragment (Fig. 2 *D*, lanes *c* and *d*). These observations indicate that the tryptic 60 K fragment is a part of the chymotryptic 70 K fragment, which indication is quite compatible with the observation that the 70 K fragment produced by tryptic digestion of myosin was furthe cleaved into the 60 K and 10 K fragments (Figs. 1 *A* and 7 *B*).

By referring to the results obtained with the tryptic HMM and S 1, we tentatively arranged the these fragments from the N-terminus as follows; 70 K, 28 K, and 17 K (Fig. 7 *A*). However, the cleavage point of 70 K fragment into 60 K and 10 K fragments remains to be determined.

### 3.3. Studies with a Gel Overlay Technique

Radioactively labelled myosin light chains and actin were used as probes with a gel overlay technique to localise their binding regions in the HC. As shown in Fig. 5, both the CaLC and PLC probes bound to the 45 K C-terminal fragment of the HC of HMM (Fig. 5, lanes *e* and *g*) and the 28 K C-terminal fragment of the HC of S 1 (Fig. 5 lanes *f* and *h*). The radioactivity* on the about 40 K fragments of the HC of HMM (Fig. 5,

---

* Relative radioactivities on the 45 K fragment and its nibbled products were different between CaLC and PLC probes (Fig. 5 lanes *e* and *g*), which should be due to the difference in the localization in the 45 K fragment between binding sites for the light chains.

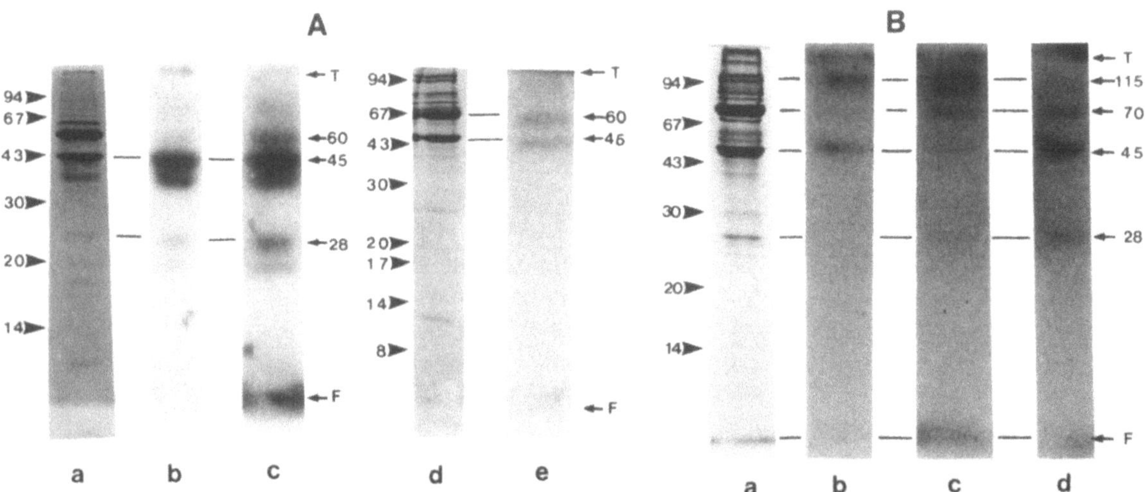

Fig. 4. Binding of [125]I-labelled actin, CaLC and PLC to HC fragments of tryptic and chymotryptic digests. Myosin digests were subjected to SDS PAGE and then to the incubation with [125]I-labelled probes. After exhaustive washing to remove unbound probes, the gels overlaying the probes were stained by CBB, dried and autoradiographed. Numbers (K dalton) with arrowheads are molecular weight markes, and those with arrows are HC fragments. *T* Gel top; *F* Gel front. *A* Experiments with tryptic myosin digests. CBB-stained gels after exhaustive washing for gel overlay procedures (*a* and *d*). Autoradiogram of gels overlaying [125]I-CaLC (*b*), [125]I-PLC (*c*) and [125]I-actin (*e*). Experimental conditions for *a–c*: Digestion in 5 mg/ml myosin, 0.15 M NaCl, 20 mM tris-HCl (pH 7.5), 0.1 mg/ml trypsin at 25 °C for 10 min. SDS PAGE on 14% gel. Experimental conditions for *d* and *e*: Digestion in 11 mg/ml myosin, 0.5 M NaCl, 20 mM tris-HCl (pH 7.5), 0.03 mg/ml trypsin at 4 °C for 2 hr. SDS PAGE on 15% gel. *B* Experiments with chymotryptic myosin digest. CBB-stained gels after exhaustive washing for gel overlay procedures (*a*). Autoradiogram of gels overlaying [125]I-CaLC (*b*), [125]I-PLC (*c*), and [125]I-actin (*d*). Digestion in 10 mg/ml myosin, 0.5 M NaCl, 20 mM tris-HCl (pH 7.5), and 0.2 mg/ml chymotrypsin at 4 °C for 2 hr. SDS PAGE on 15% gels. Note that the 45 K fragment binds CaLC, PLC, and actin. In addition, PLC and actin bind to the 70 K/60 K fragment. The radioactivity of the 115 K fragments (*B*, lanes *b* and *c*) can be explained by assuming that the fragment should be cleaved into the 45 K and the 70 K fragments. In this figure, the radioactivity from CaLC, PLC, and actin probes on 28 K fragment is faint. However, experiment with over-exposed gels definitely detected radioactivity on the 28 K fragment (not shown, see also Fig. 5). The results of this figure are summarized in Fig. 7 *A*

lanes *e* and *g*) can be interpreted as nibbled products of the 45 K fragment of the HC of HMM. There was no sign that the CaLC probe bound to the 70 K/60 K fragments (Fig. 5, lanes *e–h*). However, long exposures of gels overlaying the PLC probe detected weak radioactivity on the 70 K/60 K fragments of the HC of HMM and S1 (not shown, see also Fig. 4 *A*, lane *c*, and *B*, lane *c*), suggesting that a weak interaction between PLC and the 70 K/60 K fragment may occur. Similar experiments were repeated with tryptic digest of myosin (Fig. 4 *A*). The CaLC probe bound to the 45 K (and its nibbled products) and the 28 K fragments, but not to the 70 K/60 K fragment (Fig. 4 *A*, lanes *a* and *b*). The PLC probe also bound to the same fragments as did the CaLC probe in addition to the 70 K/ 60 K fragment (Fig. 4 *A*, lanes *a* and *c*). These data confirm the results obtained with HMM and S1.

The study with the chymotryptic digest (Fig. 4 *B*), which contained the 115 K fragment (70 K/60 K N-terminal fragment + 45 K C-terminal fragment), were quite compatible with the results using the tryptic fragments (Fig. 4 *A*), (i) because CaLC and PLC bound to the 115 K and 45 K fragments and (ii) because PLC also

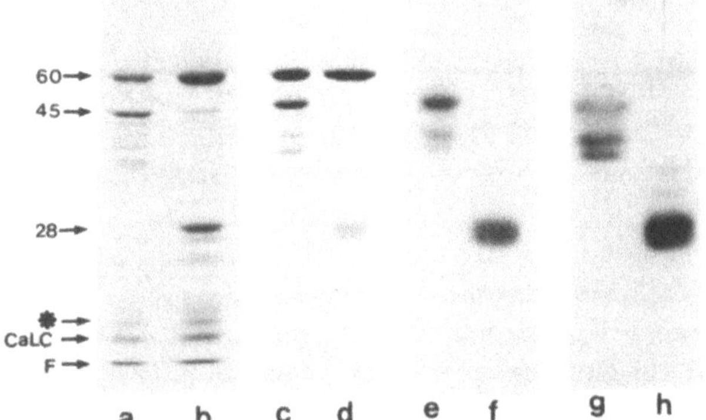

Fig. 5. Binding of [125]I-labelled light chain to tryptic HMM and S1. HMM and S1 were prepared from tryptic myosin digest as described in the legend to Fig. 2 and subjected to SDS PAGE on 13.5% gels followed by gel overlay procedures with [125]I-CaLC and [125]I-PLC. *a*, *c*, *e*, and *g* HMM. *b*, *d*, *f*, and *h* S1. *a*, *b* CBB-stained SDS PAGE of HMM and S1. *c*, *d* CBB-stain of gels after exhaustive washing for gel overlay procedures. *e*, *f* Autoradiogram of gels overlaying [125]I-CaLC. *g*, *h* Autoradiogram of gels overlaying [125]I-PLC. Numbers with arrows are HC fragments in K dalton. An asterisk denotes presumably proteolysed PLC. *F* Gel front. Light chain probes bind to 45 K fragments and their nibbled products in HMM and to 28 K fragments in S1. Comparison between lane *a*/*b* with lane *c*/*d* shows that light chain appear to be washed out during the extensive washing steps in the gel overlay procedures

bound to the 70 K/60 K fragment (Fig. 4 *B*, lanes *a* and *c*).

Deducing from the results with tryptic and chymotryptic fragments of HC using myosin, HMM and S 1, we conclude that the binding site for CaLC and PLC is in the 28 K C-terminal fragment of S 1 and that the 70 K/60 K fragment contains another binding site for PLC (Fig. 7 *A*).

The actin probe bound to the 70 K/60 K fragment as well as to the 45 K fragment of tryptic digest of myosin (Fig. 4 *A*, lane *d* and *e*). In the chymotryptic digest, the 70 K/60 K and 45 K fragments were the major binding sites (Fig. 4 *B*, lane *d*). Accordingly, the 115 K fragment, which was comprised from 70 K/60 K fragment and 45 K fragment at N-terminal and C-terminal, respectively, also showed some signs of actin binding (Fig. 4 *B*, lane *d*). It must be noted that the radioactivity of the actin probe was clearly detected in the 28 K fragment of the chymotryptic digest (Fig. 4 *B*, lane *d*). As the 28 K fragment localised at C-terminal of the 45 K fragments, we conclude that the actin binding sites are localises in both the 70 K/60 K fragment bearing the active site and the 28 K C-terminal fragment of S 1 (Fig. 7 *A*).

Tryptic digest (Fig. 2 *B*, lane *a*) was applied to F-actin affinity column (KOHAMA *et al.* 1985) equilibrated with 30 mM KCl and 20 mM tris-HCl (pH 7.5). After exhaustive washing with the equilibrating buffer, the absorbed materials, *i.e.*, S 1 and HMM, were eluted with 0.5 M KCl containing 1 mM ATP and 20 mM tris-HCl (pH 7.5). SDS PAGE of the eluate contained 60 K, 45 K and 28 K fragments together with light chains (not shown). This experiment supports the conclusion drawn from data obtained with gel overlay technique that the 70 K/60 K fragment and 28 K fragment bear actin-binding site.

Does the 17 K and 10 K fragments of HC bear additional binding sites for actin and light chains? The 10 K fragment, which was often washed out during exhaustive washing of gel overlay procedures, was retained in the gels of Fig. 4 *A* (lanes *a* and *d*). However, the 10 K fragments did not show the radioactivity of the actin (Fig. 4 *A*, lane *e*) or light chains (Fig. 4 *A*, lanes *b* and *c*) probes. We assigned the 17 K fragment to S 2 (see Discussion). As the 17 K fragment was not resistant to proteolysis (see Discussion), it is difficult to neglect the possibility that it has binding sites for actin or light chains. But the possibility is low considering that S 2 of skeletal muscle myosin has neither binding sites for actin nor light chains (HARRINGTON and RODGERS 1984, for a review).

### 3.4. Localisation of the Reactive SH

We previously reported that the Ca-ATPase activity of *Physarum* myosin was enhanced when SH's were modified by NEM and that the activity was modified as the modification proceeded (KOHAMA *et al.* 1987), which suggests that the myosin has two reactive SH such as SH 1 and SH 2 of skeletal muscle myosin (SEKINE and KIELLY 1964). The localisation of SH 1 and SH 2 is at C-terminal 20 K fragment of the HC of skeletal muscle S 1 (HARRINGTON and RODGERS 1984, for a review). However, only one reactive SH has been assigned for the *Dictyostelium* myosin HC (WARRICK *et al.* 1986) and *Acanthamoeba* myosin II HC (HAMMER III *et al.* 1987) by comparing DNA sequence for C-terminal region of the HC of their S 1 with that for the skeletal muscle 20 K fragment (STREHLER *et al.* 1986). *Physarum* myosin was incubated with $^{14}$C-NEM or 1,5-I-AEDANS under similar conditions and digested with trypsin as described in the Materials and Methods. The SDS PAGE of the digest showed that the former was preferentially incorporated into the 70 K/60 K fragment, which confirms the previous results (KOHAMA *et al.* 1987), and that the latter was incorporated into the 45 K fragment (not shown). Therefore, the *Physarum* myosin HC has two reactive SH. There are a few cysteines in the 70 K–60 K fragment of *Dictyostelium* myosin (WARRICK *et al.* 1986) and *Acanthamoeba* myosin II (HAMMER III. *et al.* 1987). Our experiment suggests that the "missing" reactive SH in these myosins is in the 70 K–60 K fragment, which bears the active site (COLLINS *et al.* 1982, ATKINSON and KORN 1986).

### 3.5. Unusual Amino Acids

Trimethylated and dimethylated lysines were found in the active site peptides of skeletal muscle myosin (TONG and ELZINGA 1983) and *Acanthamoeba* myosin II (ATKINSON *et al.* 1986), respectively. *Physarum* myosin contained monomethyllysine (0.86 mol/mol myosin), dimethyllysine (0.94 mol/mol myosin) and trimethyllysine (0.30 mol/mol myosin) (Fig. 6). In *Physarum* HMM, only dimethyllysine was detectable (not shown). We speculate that this dimethyllysine may be in the 70 K/60 K fragment that contains the active site and that post-translational dimethylation, not trimethylation, of lysine in active site is one of the features of myosins from lower enkaryotes.

Methylation of histidine that was also reported for skeletal (JOHNSON *et al.* 1967) and cardiac (MASAKI *et al.* 1986) muscle myosins was not detected in *Physarum* myosin at the stoichiometric level (Fig. 6).

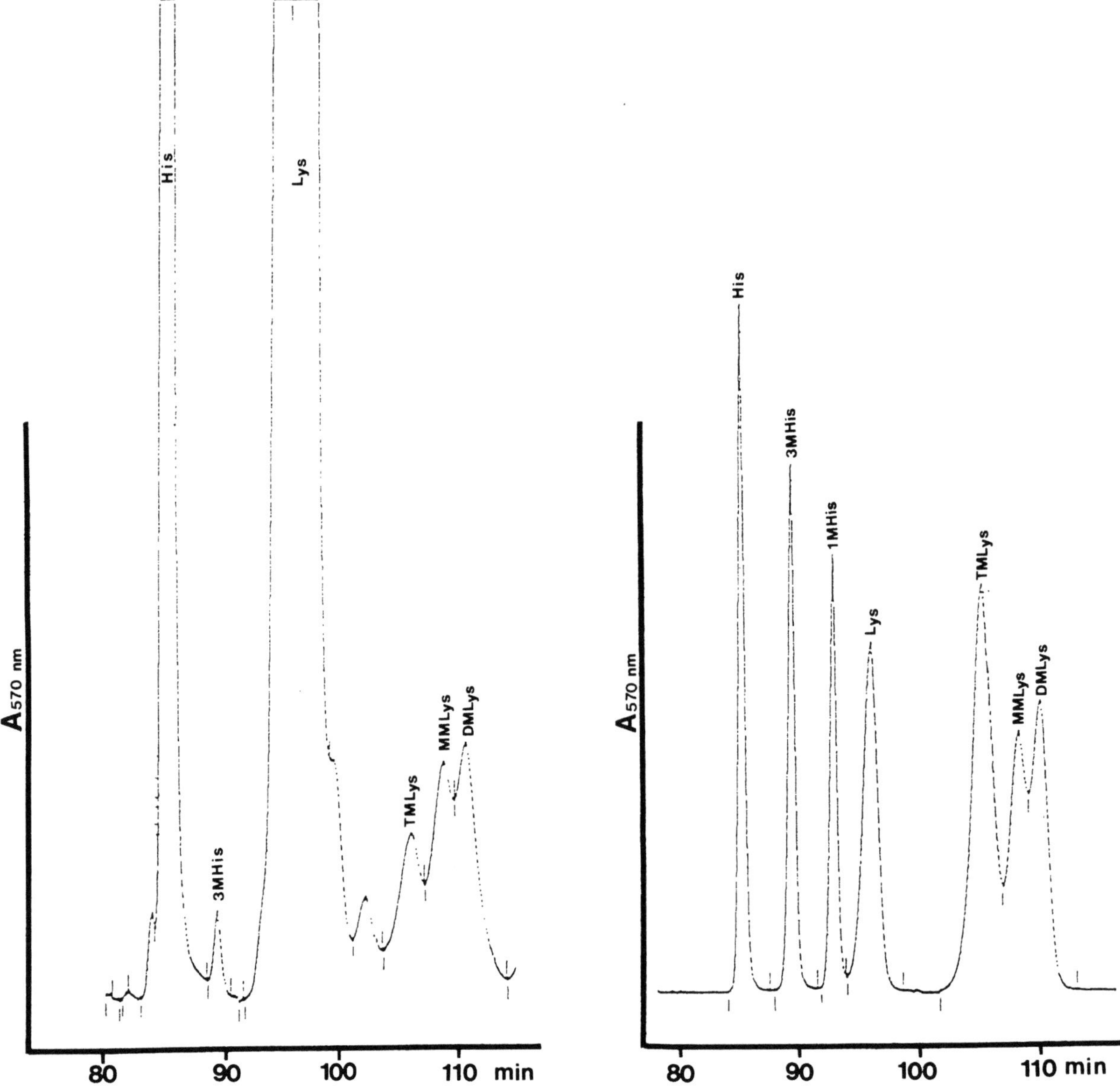



Fig. 6. Detection of methylated lysines in myosin. Hydrolysed myosin was subjected to the amino acid analysis with an amino acid analyzer. The elution pattern of the lysate from the analyzer was shown in left trace. The right trace shows elution pattern (2.5 nmol each) of histidine (*His*), 3-methylhistidine (*3 MHis*), 1-methylhistidine (*1 MHis*), lysine (*Lys*), monomethyllysine (*MMLys*), demethyllysine (*DMLys*), and trimethyllysine (*TMLys*), abscissae, retention time (min); ordinates, absorbance at 570 nm ($A_{570}$ nm, arbitrary unit)

## 4. Discussion

As schematically summarized in Fig. 7 *A*, we tentatively localised the major cleavage sites sensitive to both trypsin and chymotrypsin on *Physarum* myosin HC by following the time course of cleavage (Figs. 1, *A* and *B*, and 7 *B*), and by purifying HMM and S 1 from tryptic or chymotryptic digest of myosin (Fig. 2 *A–C*). The N-terminal 70 K fragment produced by chymotryptic digestion had an additional site for trypsin to cleave the fragment into 60 K and 10 K fragments (Fig. 2 *D*).

We have not yet succeeded in deciding whether the 10 K fragment is at C- or N-terminal in the 70 K fragment.

The N-terminal 70 K/60 K fragment and the C-terminal 28 K fragment of S 1 bear ATPase active site (Fig. 1 in KOHAMA *et al.* 1987 and Fig. 3) and the light chain binding sites (Figs. 4 and 5), respectively. Actin binds to both segments. Skeletal muscle S 1 is comprised of N-terminal 27 K, middle 50 K and C-terminal 20 K fragments (HARRINGTON and RODGERS 1984, for a re-

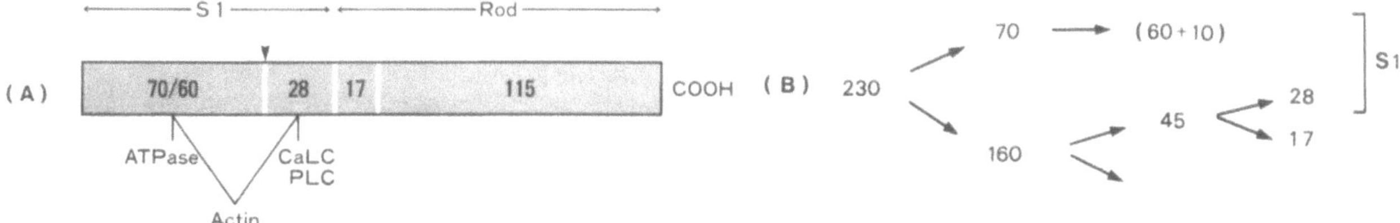

Fig. 7. Tentative model for *Physarum* myosin HC reconstituted from its proteolytic fragments (*A*) and sequence of the HC hydrolysis (*B*). CaLC, PLC and actin show binding sites for the respective proteins. In addition, PLC show weak interaction with 70 K/60 K fragment. ATPase means active site. Reactive SH's were in the 70 K/60 K fragment as determined by incorporation of $^{14}$C-NEM and 1,5,-I-AEDANS, respectively. HC (230 K) is initially cleaved into 70 K and 160 K fragments. The former is further digested to 60 K and 10 K fragments. The 45 K fragment is produced from the latter. The 28 K fragment is produced from the 45 K fragment. The 70 K and 28 K fragments consist HC of S 1

view). The *Physarum* 70 K fragment is homologous with skeletal muscle 27 K + 50 K fragments, because they contain the active site and actin-binding site. The *Physarum* C-terminal 28 K fragment and the skeletal muscle C-terminal 20 K fragment have the ability to bind the light chains and actin. Thus the organization of *Physarum* S 1 functions shows a high degree of similarity with that of sekeletal muscle S 1. However, the localisation of reactive SH was different between S 1's; the *Physarum* reactive SH which was preferentially alkylated with NEM was in the 70 K/60 K fragment (Fig. 1 in Kohama *et al.* 1987), but that of skeletal muscle was apart from the fragment containing the active site (Harrington and Rodgers 1984, for a review).

In the case of skeletal and smooth muscle myosins, the region susceptible to proteolysis is between the N-terminus of S 1 and C-terminus of rod (Fig. 4 in Mitchell *et al.* 1986). However, the most sensitive site of *Physarum* myosin is between the 70 K/60 K fragment (akin to the 27 K + 50 K fragments of skeletal muscle myosin HC) and 28 K fragment (akin to the 20 K C-terminal fragment of skeletal muscle myosin HC) as shown by the arrowhead in Fig. 7 *A*. In fact, fresh myosin often shows a sign of cleavage at this site due to endogenous protease(s) (Fig. 1 *A* and *B*, lanes *a*). Topographically the site is within S 1, because purified S 1 comprises both the 70 K/60 K fragment and the 28 K fragment (Fig. 2 *B* and *C*, lanes *c*). Thus preparation of intact, *i.e.*, nick-free S 1 and HMM were extremely difficult. A similar site which readily releases the N-terminal 70 K–60 K fragment bearing the active site has been reported for brain myosin (Barylko *et al.* 1986) and *Acanthamoeba* myosin II (Collins *et al.* 1982). We speculate that this sensitive site should be specific to nonmuscle myosin and should be conserved among nonmuscle myosins of lower and higher eukaryotes.

We purified S 1 and HMM from myosin digests by size fractionation using skeletal muscle S 1 and HMM as size makers (Fig. 2 *A*). The rationale for the identification are; (i) As reported previously (Kohama *et al.* 1983, Kohama 1987), S 1 and HMM fractions are consist only of single heads and double heads, respectively, as examined with an electron microscope. (ii) S 1 and HMM have ATPase activities (not shown) and [$^3$H] UTP binding activity in their HC (Fig. 3). SDS PAGE of S 1 and HMM shows that they are fragmented (Figs. 2, *B* and *C*, and 5, lanes *a* and *b*). The globular shape of their heads indicates that they are nicked inside the molecule. There is also nibbling of the 45 K fragment of HMM, which is indicated by the gel overlay technique (Fig. 5). The cleavage site between S 2 and LMM of skeletal muscle myosin is known to be variable according to conditions for the digestion (Sutoh *et al.* 1978). Therefore, we speculate that the nibbling site of 45 K is at its C-terminal, which is S 2 and LMM junction of *Physarum* myosin (Fig. 7 *A*).

Myosin digests using low concentrations of protease contain enormous numbers of fragments above 70 K dalton in SDS PAGE (Fig. 1 *A* and *B*). In myosin digests with higher concentration of protease, these fragments are almost absent (Fig. 2 *B* and *C*, lanes *a*), suggesting that the myosin rod has many cleavage sites. This is different from the cleavage of rod of skeletal muscle myosin, which has two distinct cleavage sites at the S 1/S 2 junction and the S 2/LMM junction (Harrington and Rodgers 1984, for a review).

Comparing SDS PAGE between S 1 and HMM, the 17 K fragment at C-terminal of S 1 appears to be *Physarum* S 2 (Fig. 7). However, the size is much smaller than that of skeletal muscle S 2, *i.e.* 45–70 K dalton (Sutoh *et al.* 1978). Thus the molecular weight of *Physarum* HMM preparation is expected to be smaller than that of skeletal muscle HMM preparation. The HPLC gel filtration column used in this study (see Materials and Methods), however, could not detect difference in the molecular weights due to poor separation of marker proteins above 300 K dalton (not shown).

When myosin was exposed to low concentration of trypsin, PLC was quickly digested (compare lane *b* with lane *a* in Fig. 1 *A*). The light chain is fairly resistant to proteolysis with low concentrations of chymotrypsin (Fig. 1 *B*). Chymotrypsin partially cleaved PLC (Fig. 2 *C*, lane *b*). On the other hand, there was no clear indication that CaLC was digested by trypsin or chymotrypsin (Figs. 1 and 2). Although we have not yet carried out an intensive study to examine whether the light chains of S 1 and HMM preparation are intact or not, these preparations appear to have partially proteolysed PLC and intact CaLC (Fig. 5, lanes *a* and *b*).

## Acknowledgements

We thank H. KUROKAWA for the help in HPLC. Our thanks are also due to Professors Y. NONOMURA and T. SEKINE for their discussions, interest, and support and to Dr. J. KENDRICK-JONES for suggesting a gel overlay technique. This work was partly supported by grants from the Ichiro Kanehara Foundation (to Y. O.), the Naito Foundation (to K. K.), the Fugaku Trust for Medical Research (to K. K.), Sankyo Foundation of Life Science (to K. K.), the Yamada Science Foundation (to K. K.), Takeda Science Foundation (to K. K.), and the Aino Hospital Foundation (to K. K.), and Grant-in-Aids for Scientific Research from the Ministry of Education, Science and Culture of Japan (to K. K. and Y. O.).

## References

ACHENBACH F, WOHLFARTH-BOTTERMANN K-E (1986) Reactivation of cell-free models of endoplasmic drops from *Physarum polycephalum* after glycerol extraction at low ionic strength. Eur J Cell Biol 40: 135–138

— — (1986 a) Successive contraction-relaxation cycles experimentally induced in cell-free models of *Physarum Polycephalum*. Eur J Cell Biol 42: 111–117

ATKINSON MA, KORN ED (1986) The purification and characterization of a globular subfragment of *Acanthamoeba* myosin II that is fully active when cross-linked to F-actin. J Biol Chem 261: 3382–3386

— ROBINSON EA, APPELLA E, KORN ED (1986) Amino acid sequence of the active site of *Acanthamoeba* myosin II. J Biol Chem 261: 1844–1848

BARYLKO B, TOOTH P, KENDRICK-JONES J (1986) Proteolytic fragmentation of brain myosin and localisation of the heavy-chain phosphorylation site. Eur J Biochem 158: 271–282

CAMP WG (1936) A method of cultivating myxomycete plasmodium. Bull Torrey Bot Club 63: 205–210

COLLINS JH, CÔTÉ GP, KORN ED (1982) Localization of the three phosphorylation sites on each heavy chain of *Acanthamoeba* myosin II to a segment at the end of the tail. J Biol Chem 257: 4529–4534

HAMMER III JA, BOWERS B, PATERSON BM, KORN ED (1987) Complete nucleotide sequence and deduced polypeptide sequence of a nonmuscle myosin heavy chain gene from *Acanthamoeba*: evidence of a hinge in the rodlike tail. J Cell Biol 105: 913–925

HARRINGTON WF, RODGERS ME (1984) Myosin. Ann Rev Biochem 53: 35–73

HATANO S, OOSAWA F (1966) Isolation and characterization of plasmodium actin. Biochim Biophys Acta 127: 488–498

— TAKAHASHI K (1971) Structure of myosin A from the myxomycete plasmodium and its aggregation at low salt concentrations. J Mechnaochem Cell Motil 1: 7–14

JOHNSON P, HARRIS CI, PERRY SV (1967) 3-Methylhistidine in actin and other muscle proteins. Biochem J 105: 361–370

KOBAYASHI T, TAKAGI T, KONISHI K, HAMADA Y, KAWAGUCHI M, KOHAMA K (1988) Amino acid sequence of the calcium-binding light chain of myosin from the lower enkaryote, *Physarum polycephalum*. J Biol Chem 263: 305–313

KOHAMA K (1980) Heterogeneity of amino acid incorporation rate in adult skeletal muscle. J Biochem 87: 997–999

— (1987) Ca-inhibitory myosin: their structure and function. Adv Biophys 23: 149–182

— (1987 a) Inhibitory Ca-control of actin-myosin-ATP interaction in *Physarum polycephalum*: Myosin-linked and actin-linked natures. Fortsch Zool 34: 97–107

— CRAIG K, KOHAMA T, KENDRICK-JONES J (1983) Characterization of Ca-sensitive *Physarum* myosin (Abstr). Eur J Cell Biol [Suppl] 1: 25

— KENDRICK-JONES J (1986) The inhibitory Ca$^{2+}$-regulation of the actin-activated Mg-ATPase activity of myosin from *Physarum polycephalum* plasmodia. J Biochem 99: 1433–1446

— KOHAMA T (1984) Myosin confers inhibitory Ca$^{2+}$-sensitivity on actin-myosin-ATP interaction of *Physarum polycephalum* under physiological conditions. Proc Japan Acad 60 B: 435–439

— — KENDRICK-JONES J (1987) Effect of N-ethylmaleimide on Ca-inhibition of *Physarum* myosin. J Biochem 102: 17–23

— SHIMMEN T (1985) Inhibitory Ca$^{2+}$-control of movement of beads coated with *Physarum* myosin along actin-cables in *Chara* internodal cells. Protoplasma 129: 88–91

— SOHDA M (1987) Characterization of proteolytic fragment of *Physarum* myosin heavy chain. Cell Struct Funct 12: 681 (Abstr)

— TAKANO-OHMURO H, YAMAGUCHI Y, KOHAMA T (1986) Isolation and characterization of myosin from amoebae of *Physarum polycephalum*. J Biol Chem 261: 8022–8027

— UYEDA TQP, TAKANO-OHMURO H, TANAKA T, YAMAGUCHI T, MARUYAMA K, KOHAMA T (1985) Ca$^{2+}$-binding light chain of *Physarum* myosin confers inhibitory Ca$^{2+}$-sensitivity on actin-myosin-ATP interaction via actin. Proc Japan Acad 61 B: 501–505

LAEMMLI UK (1970) Cleavage of structural proteins during assembly of the head of bacteriophage T. Nature 227: 680–685

LOWRY OH, ROSEBROUGH NJ, FARR AL, RANDALL RJ (1951) Protein measurement with the folin phenol reagent. J Biol Chem 193: 265–275

MASAKI T, TAKANO-OHMURO H, IIZUKA H, OKAMOTO J, OBINATA T (1986) 3-Methylhistidine content and pH dependency of ATPase activity of adult and embryonic chicken cardiac ventricular myosins. J Biochem 100: 1091–1094

MARUTA H, KORN ED (1981) Direct photoaffinity labeling by nucleotides of the apparent catalytic site on the heavy chains of smooth muscle and *Acanthamoeba* myosins. J Biol Chem 256: 499–502

MARGOSSIAN SS, LOWEY S (1982) Preparation of myosin and its

subfragments from rabbit skeletal muscle. Methods Enzymol 85 B: 55–71

Mitchell EJ, Jakes R, Kendrick-Jones J (1986) Localisation of light chain and actin binding sites on myosin. Eur J Biochem 161: 25–35

Nakamura S, Nonomura Y (1984) A simple and rapid method to remove light chain phosphatase from chicken gizzard myosin. J Biochem 96: 575–578

Okamoto Y, Okamoto M, Sekine T (1980) Two opposite effects of ATP on the chymotryptic cleavages in smooth muscle myosin head. J Biochem 88: 361–371

— Yount R (1985) Identification of an active site peptide of skeletal muscle myosin after photoaffinity labelings with N-(4-azido-2-nitrophenyl-2-amionoethyl) diphosphate. Proc Natl Acad Sci USA 82: 1575–1579

Reisler E (1982) Sulfhydryl modification and labeling of Myosin. Methods Enzymol 85 B: 84–93

Sekine T, Kielly WW (1964) The enzymatic properties of N-ethyl-maleimide modified myosin. Biochim Biophys Acta 81: 336–345

Snabes MC, Boyd III AE, Bryan J (1981) Detection of actin-binding proteins in human platelets by $^{125}$I-actin overlay of polyacrylamide gels. J Cell Biol 90: 809–812

Spudich JA, Watt S (1971) The regulation of rabbit skeletal muscle contraction. I. Biochemical studies of the interaction of the tropomyosin-troponin complex with actin and the proteolytic fragments of myosin. J Biol Chem 246: 4866–4871

Strehler EE, Strehler-Page M-A, Perriard J-C, Periasamy M, Nadal-Ginard B (1986) Complete nucleotide and encoded amino acid sequence of a mammalian myosin heavy chain gene. Evidence against intron-dependent evolution of the rod. J Mol Biol 190: 291–317

Sutoh K, Sutoh K, Karr T, Harrington WF (1978) Isolation and physico-chemical properties of a high molecular weight subfragment-2 of myosin. J Mol Biol 126: 1–22

Takahashi K, Ogihara S, Ikebe M, Tonomura Y (1983) Morphological aspects of thiophosphorylated and dephosphorylated myosin molecules from the plasmodium of *physarum polycephalum*. J Biochem 93: 1175–1183

Tong SW, Elzinga M (1983) The sequence of the $NH_2$-terminal 204-fragment of the heavy chain of rabbit skeletal muscle myosin. J Biol Chem 258: 13100–13110

Warrick HM, De Lonanne A, Leinwand LA, Spudich JA (1986) Conserved protein domains in a myosin heavy chain gene from *Dictyostelium discoideum*. Proc Natl Acad Sci USA 83: 9433–9437

Yoshimoto Y, Kamiya N (1984) ATP and calcium-controlled contraction in a saponin model of *Physarum polycephalum*. Cell Struct Funct 9: 135–141

— Matsumura F, Kamiya N (1981) Simultaneous oscillations of $Ca^{2+}$ efflux and tension generation in the permealized plasmodial strand of *Physarum*. Cell Motil 1: 433–443

Protoplasma (1988) [Suppl. 2]: 48–56


# Immunoelectron Microscopic Localization of the *Physarum* 36,000-Dalton Actin Binding Protein on the Surface of Vesicular Structures in the Plasmodium

S. OGIHARA

Department of Biology, College of General Education, Osaka University

Received February 26, 1988
Accepted July 5, 1988

Dedicated to Professor Dr. NOBURO KAMIYA on the occasion of his 75th birthday

## Summary

In an attempt to assign the *in vivo* function of the 36 kDalton actin binding protein (ABP-36) purified from *Physarum* plasmodia, it was localized in the Triton insoluble fraction prepared from the endoplasm at the electron microscopic resolution using immunogold reagents. The ABP-36 was specifically localized on the surfaces of vesicles of 0.5 to 1.0 µm in diameter, and the vesicles are often clustered. This Triton insoluble fraction binds exogenous added actin filaments and the binding is inhibited by the antibody against the ABP-36. Treatment with Mn-ATP enhanced the binding by a factor of 4, and then the myosin heavy chain and a 28 kDalton protein, but not the ABP-36, were phosphorylated. Out of the cytoplasmic structures *in situ* that were searched, vesicular structures closely resembling those stained with the immunogold were found in the endoplasm of intact plasmodia. Those vesicles fuse with each other and the plasma membrane when the endoplasm is mechanically isolated. The fusion phenomenon results in the formation of the plasmalemma invagination systems, which are reported to be connected morphologically with each other by acto-myosin fibrils. These results suggest that the ABP-36 is located on the exocytotic vesicles and that it may play a crucial role in linking the cytoskeleton to the plasma membrane.

*Keywords:* 36 kDalton actin binding protein; Exocytosis; Immunoelectron microscopy; *Physarum* plasmodium; Vesicle fusion.

*Abbreviations:* ABP-36 36 Dalton actin binding protein; SDS-PAGE sodium dodecyl sulfate polyacrylamide gel electrophoresis; CBB Coomasie Brilliant Blue; BSA bovine serum albumin; PBS phosphate-buffered saline; PMSF phenylmethylsulfonyl fluoride; DTT dithiothreitol; ATP adenosine triphosphate; IgG immunoglobulin G.

---

* Correspondence and Reprints: Department of Biology, College of General Education, Osaka University, Toyonaka, Osaka 560, Japan.

## 1. Introduction

An actin binding protein was isolated by OGIHARA and TONOMURA (1982) from a Triton insoluble actin-containing cytoskeleton of the plasmodium. It has a molecular weight mass of 36,000 Dalton (ABP-36) and bound 7 mol of actin per mol. Addition of stoichiometric amount of the ABP-36 to F-actin enhanced the sedimentability of actin and the complex formed was electron microscopically found to be large aggregates of F-actin, suggesting a crosslinking property of the ABP-36. Also the ABP-36 inhibited the actin-myosin interaction.

Using immunoelectron microscopy, I here show that the ABP-36 is localized on the surfaces of vesicular structures in the Triton insoluble fraction prepared form the endoplasm of the plasmodium. Those vesicles are likely to be identical with the exocytotic vesicles that play a crucial role in the formation of the plasmalemma invagination systems. The ABP-36 may be located on the exocytotic vesicles and have a property of linking the cytoskeleton to the plasma membrane.

## 2. Materials and Methods

### 2.1. Preparation of the Triton Insoluble Fraction

To prepare a Triton insoluble fraction of morphological and biochemical homogeneity, the endoplasmic veins, the method of which was originally developed by BARANOWSKI and WOHLFARTH-BOTTERMANN (1982), were employed as a starting material, since homogenizing the plasmodium, which should have some uneven distribution in its protein and lipid constituents depending on where to harvest, would result in a heterogeneous preparation. In brief, plasmodial strands were punctured by cutting with a razor blade, and the pure endoplasm that poured out was sucked up into a 10 μl glass micropipet. When it was immediately extruded into water, it formed a long cylindrical cytoplasmic mass, *i.e.*, an endoplasmic vein.

The endoplasmic veins aged precisely for 5 min in water at room temperature (20–22 °C) were pelleted at 15,600 × **g** for 10 sec, and homogenized for 1 min on ice in 1 ml of an ice-cold homogenizing solution contained in a 1.5 ml Eppendorf tube, with a glass pestle made exclusively to fit the tube. The homogenizing solution contained 0.2% Triton X-100, 2 mM $MgCl_2$, 5 mM Tris-Cl, 50–100 μM $CaCl_2$, 0.2 mM PMSF, 150 μg/ml aprotinin, 10 μg/ml chymostatin, 1 μg/ml pepstatin A, pH 7.6. The homogenate was centrifuged at 15,600 × **g** for 5 min, and washed three times in 1 ml of ice cold $PH_4$ Buffer containing 50 mM KCl, 2 mM $MgCl_2$, 50–100 μM $CaCl_2$, 20 mM Tris-Cl, pH 7.6. The final pellet was designated as Triton insoluble fraction.

### 2.2. Protein Purification

The ABP-36 was isolated from *Physarum* plasmodia by a method modified from the original procedure (OGIHARA and TONOMURA 1982). In brief, the ABP-36 purified following the original method which included salting-out and Sephacryl S-300 gel filtration was further purified with DEAE Sepharose CL-6B ion-exchange chromatography and a hydroxylapatite column.

### 2.3. Electrophoretic Procedure and Western Blot Analysis

SDS-PAGE was carried out according to the procedure of LAEMMLI (1970) in a microslab apparatus using 5–15% polyacrylamide gels. Electrophoretic transfer of proteins form unstained gels to nitrocellulose paper was done according to the method of TOWBIN *et al.* (1979). Nitrocellulose paper was then blocked, dried, and incubated with affinity-purified antibodies (1–2 μg/ml) overnight at 37 °C. The procedure essentially followed the method of CARBONI *et al.* (1985).

### 2.4. Anti-ABP-36 Antibody and Affinity Purification

Purified ABP-36 was run on a curtain of SDS-PAGE, and about 5 mm strips on both the right and the left edges were cut out and stained with CBB. The stained strips were lined up with the unstained gel, and a band corresponding with the ABP-36 was cut out and dialyzed against PBS overnight. The gel in the dialysis bag was crushed thoroughly, and emulsified well with an equal volume of Freund Complete Adjunvant. It was injected to guinea pigs, and the sera were occasionally screened on Western blots of the total plasmodial proteins using an avidin-biotin-peroxidase method. The sera were affinity-purified by the method of TALIAN *et al.* (1983). In brief, the purified ABP-36 band prepared as above was cut out, and the gel was electro-blotted onto nitrocellulose paper. The paper was blocked in 25 mg/ml BSA/PBS at pH 7.2. The nitrocellulose paper was incubated with the antiserum diluted 1 : 200 in 1 mg/ml BSA/PBS 16 hr at room temperature on a rotary shaker. Bound antibody was released from the paper with 0.2 M glycine-Cl buffer at pH 2.8, and immediately neutralized with 1 N NaOH to pH 6–8. The purified antibody was dialyzed against PBS overnight, supplemented with 1 mg/ml BSA or gelatin, and stored at − 80 °C until use.

### 2.5. Actin Binding Assay

The Triton insoluble fraction starting from the endoplasmic vein of 2.8 μl was resuspended in 100 μl of $PH_4$ Buffer plus 0.2 mM PMSF. It was mixed with 0.17 mg/ml rabbit skeletal muscle G-actin, and incubated at 20 °C for 40–60 min. The mixture was then pelleted at 15,600 × **g** for 5 min at 4 °C. The amount of actin was quantified by densitometry of SDS-PAGE of the pellets. Bound actin was determined by subtracting actin content in the Triton insoluble fraction from those of samples mixed with G-actin. In anti-ABP-36 inhibition experiments, prior to mixing with actin, the Triton insoluble fraction, starting from 2.8 μl of the original endoplasm per tube, was first incubated with various dilutions of the anti-ABP-36 antiserum in 200 μl PBS containing 1 mg/ml BSA, 2 mM $MgCl_2$ and 0.2 mM PMSF for 30 min at 4 °C with gentle mixing.

### 2.6. Phosphorylation Analysis

The Triton insoluble fraction was prepared as above starting from 2.8 μl endoplasm. It was suspended in 50 μl of $PH_4$ Buffer supplemented with 10 mM $MnCl_2$ and 50 μM $CaCl_2$. After 5 minute preincubation at 25 °C, phosphorylation reactions were started by adding 50 μM [$^{32}$P]-ATP and incubating the mixture for 60 sec at 25 °C, and stopped by addition of 1 ml of ice-cold IP Buffer (0.2% Nonidet P-40, 1 mM ATP·2Na, 100 mM potassium phosphate buffer, 20 mM Tris-Cl buffer, 5 mM EDTA, 1 mM DTT, 25 mM sodium pyrophosphate, 100 mM NaF, 10 mM $NaHSO_3$, pH 7.5). It was then centrifuged at 15,600 × **g** for 5 min and the whole precipitates were subjected to SDS-PAGE. The specific activity of radioactive ATP was chosen so that the final pellets would have about 30,000 c.p.m. per tube. SDS-PAGE gels were stained with CBB, dried on No. 2 filter paper, and exposed to Kodak XAR-5 film with intensifying screens at − 80 °C.

### 2.7. Densitometry

Gels were scanned at 600 nm with a Shimazu CS-9000 scanning densitometer equipped with an automatic peak detection and peak area computation device.

### 2.8. Electron Microscopy and Immunoelectron Microscopy

For conventional electron microscopy, samples were fixed with 4% glutaraldehyde in 80 mM potassium phosphate buffer, pH 6.0, containing 0.4% tannic acid for 1 hr at room temperature. They were washed twice in 40 mM potassium phosphate buffer, pH 6.0, and fixed with 1% $OsO_4$ in 40 mM potassium phosphate buffer, pH 6.0, for 1 hr at room temperature. They were washed twice with water, stained *en bloc* for 15 min in 1% uranyl acetate, carried through a series of dehydration steps in ethanol and propylene oxide. The samples were embedded in a 50 : 50 Epon-araldite mixture. Ultrathin sections were stained with uranyl acetate and lead citrate. For immunoelectron microscopy, 17 nm colloidal gold was synthesized according to the method of SLOT and GEUZE (1985) using tannic acid as reducing agent. It was conjugated with affinity-purified goat anti-guinea pig IgG, mixed with 1% BSA, washed twice in 1% BSA/TBS at pH 8.2, resuspended in 50% glycerol, 1% BSA, 0.02% $NaN_3$ in TBS, and stored at − 20 °C until use. The Triton insolube fractions were incubated in 10 mM Mn-ATP in $PH_4$ Buffer containing 50 μM $CaCl_2$. They were fixed in an excess volume of PLP fixative (MCLEAN and NAKANE 1974) containing 2% paraformaldehyde, 75 mM DL-lysine, 37.5 mM potassium phosphate, pH 7.4, and 10 mM $NaIO_4$ for 3 hr at 4 °C with gentle shaking. They were pelleted, washed three times in PBS over a period of 3 hr, and mixed with 1% BSA in PBS for 30 min on ice for blocking nonspecific staining. Blocked samples were incubated with 27 µg/ml affinity-purified anti-ABP-36 at 4 °C overnight, washed three times in ice-cold 0.1% BSA in PBS, for 30 min each time, and stained overnight with colloidal gold-anti guinea pig IgG conjugates at optical density of 2.0 at 520 nm. They were washed for five times in 0.1% BSA in PBS over a period of 5 hr and then once in PBS. They were fixed in 2% glutaraldehyde containing 80 mM potassium phosphate, pH 7.0 for 30 min at room temperature, washed in 40 mM potassium phosphate, pH 6.0, at room temperature, and post-fixed in 2% OsO$_4$, 40 mM potassium phosphate, pH 6.0, for 30 min at room temperature. All steps hereafter were the same as for the conventional electron microscopy described above.

## 3. Results

### 3.1. Antibody Specificity and Occurrence of the ABP-36 in the Triton Insoluble Fraction

The affinity-purified antibody was used in the immunoelectron microscopy described in this paper. In the immunoblot analysis of antibody specificity, anti-actin antibody was used to stain nitrocellulose paper strips immediately adjacent to those stained with anti-ABP-36 antibody to make sure that the stained band really had the same mobility as the band stained with CBB

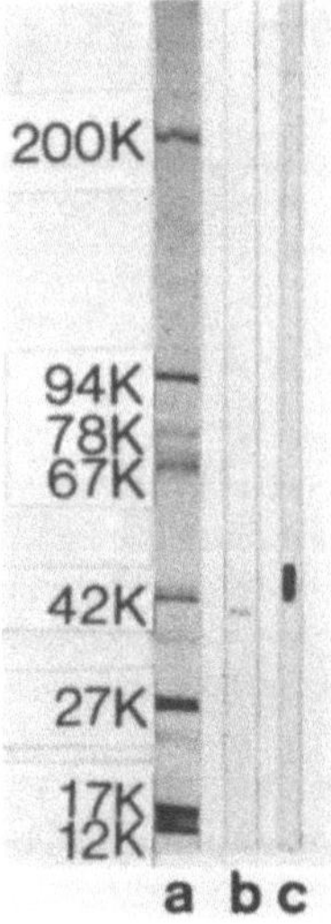

Fig. 1. The specificity of the anti-ABP-36 antibody. *a* Molecular weight standards. Nitrocellulose blots of the total homogenate of *Physarum* plasmodium were stained with the affinity-purified antibody against the ABP-36 (*b*) or with the affinity-purified anti-actin (*c*). Molecular masses of the standards are: 200,000 (rabbit skeletal muscle myosin heavy chain), 94,000 (phosphorylase b), 78,000 (transferrin), 67,000 (bovine serum albumin), 42,000 (rabbit sekeletal muscle actin), 27,000 (concanavalin A), 17,000 (myoglobin), and 12,000 (cytochrome c). The concentrations of the primary antibodies used were: 2.3 µg/ml (*b*) and 1 µg/ml (*c*)

(Fig. 1). This would be very difficult to attain without an immunostained standard, actin, because of smiling of SDS-PAGE gels, shrinkage of gels in the electrotransfer buffer. The affinity-purified anti-ABP-36 stained a single band of molecular weight of 36,000 daltons on the blot of total homogenate of plasmodial proteins, demonstrating that it was monospecific (Fig. 1 *b*).

The immunogold particles detecting the ABP-36 in the Triton insoluble fraction are found on surfaces of some vesicular structures of about 0.5–1.0 µm in diameter (Fig. 2). The gold is not present inside the vesicles. The vesicle membrane does not appear to be bilayers by the PLP fixation method. The vesicles are closely associated with each other, and the gold probes are also found in the narrow gaps between vesicles (Fig. 2 *a*). In Fig. 2 *b* are shown larger vesicles stained with gold particles (arrows). The contour of the vesicle is not smooth compared with vesicles just closely clustered as shown in Fig. 2 *a*.

Those vesicles are often found in close association with a sheet-like structure as shown in Fig. 2 *c*. Except for rare cases as shown in Fig. 2 *f*, the ABP-36 is not present on the sheet-like structure. Figure 2, *d* and *e*, shows magnified views of the stained vesicles. The ABP-36 is very close to the vesicle membrane, and filamentous actin is not observed. Therefore, the binding of the ABP-36 to the vesicles is not likely to be mediated by filamentous actin. An oblique section shown in Fig. 2 *e* suggests rather random distribution of the gold particles on the membrane surface than any trace of periodic pattern.

### 3.2. Binding of Actin Filaments to the Triton Insoluble Fraction and Its Enhancement by Mn-ATP

The occurrence of the ABP-36 on the vesicular structures in the Triton insoluble fraction and the high accessibility of high molecular weight proteins such as IgG as shown in Fig. 2 denoted that exogenous added actin might bind to the Triton insoluble fraction *in vitro*. The results are shown in Fig. 3. When rabbit skeletal muscle G-actin was mixed with the Triton insoluble fraction prepared from 2.8 µl of the endoplasmic vein, let to polymerize in PH$_4$ buffer, and pelleted by centrifugation which sedimented virtually no actin filaments, about 2 µg of actin was resolved on SDS-PAGE. This shows that 1.6 µg of exogenous actin, about 10% of the total added actin, co-precipitated with the fraction, taking the actin content in the fraction (0.4 µg) into consideration (Fig. 3, compare lanes

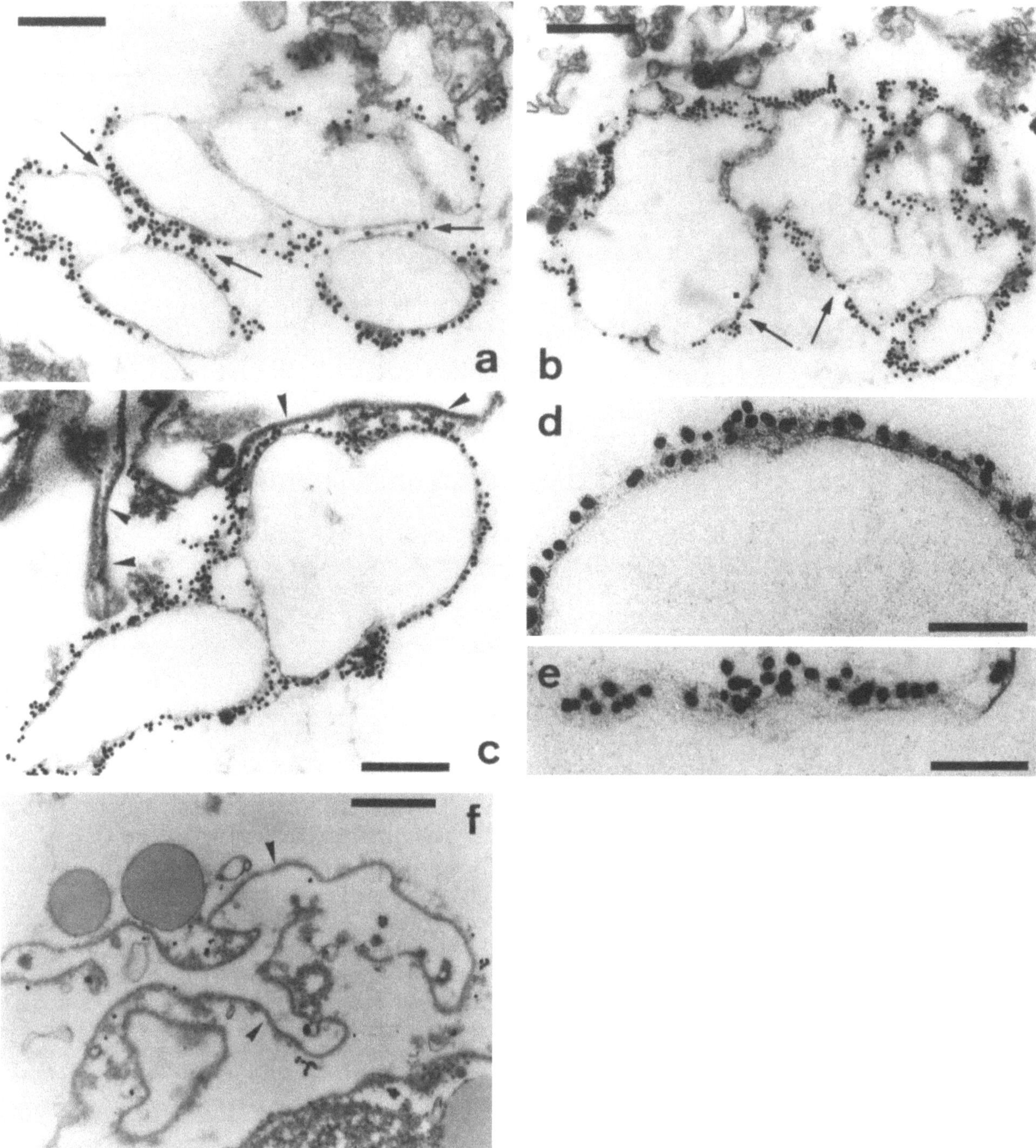

Fig. 2. Vesicular structures contained in the Triton insoluble fraction are stained with anti-ABP-36. The immunogold particles conjugated with the secondary antibody are predominantly found on the surfaces of vesicular structures (*a–e*). The signals are low on the sheet-like structure (*f*, arrowheads). The stained vesicles are closely associated (*a–c*) and the signals are also found in the narrow gaps between vesicles (*a*, arrows). Occasionally there are larger vesicles which exhibit irregular contour and the contour lines are discontinuous in some places (*b*, arrows). The vesicles are often found in close proximity to the sheet-like structure (*c*, arrowheads). *d* and *e* Magnified views of the vesicle surface, the cytoplasmic side of the vesicles being upward in the panels. Scale bars are 0.5 µm (*a, b, c*, and *f*) and 0.2 µm (*d, e*)

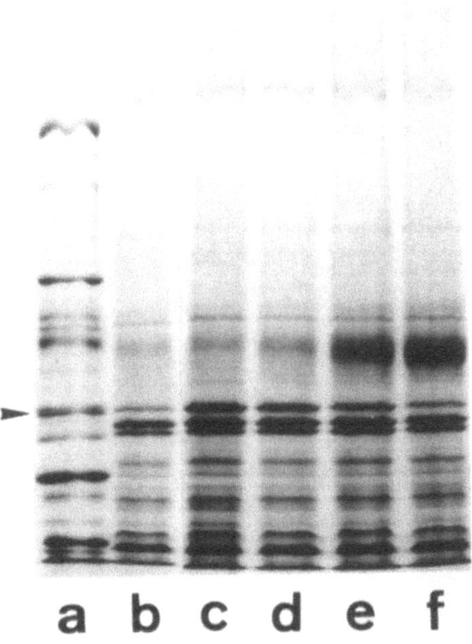
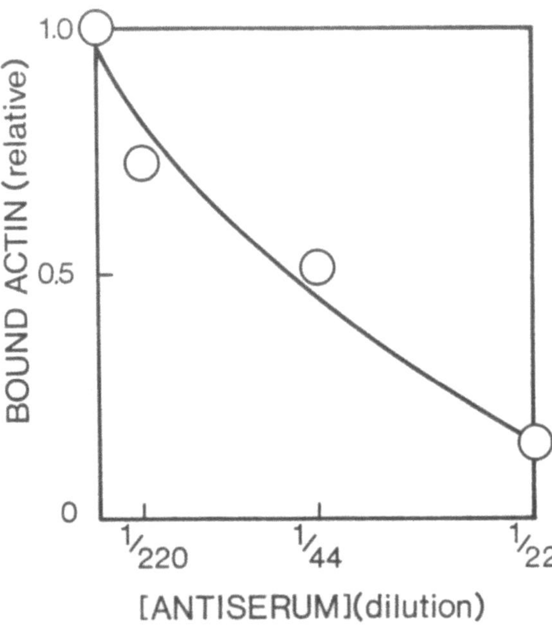


Fig. 3. Actin binding to the Triton insoluble fraction and its inhibition by anti-ABP-36. In the left panel, the Triton insoluble fraction (*b*) is mixed with muscle G-actin and pelleted at speeds at which most F-actin alone remains in the supernatant (*c–f*). Preincubation of the fraction with the antiserum against the ABP-36 at dilutions of 1 : 220 (*d*), 1 : 44 (*e*) and 1 : 22 (*f*) inhibits co-sedimentation of actin with the fractions in a concentration-dependent manner as shown in the right panel. Actin concentration added was 0.17 mg/ml. *a* Molceluar weight standards identical to those in Fig. 1. The mobility of actin is indicated by the arrow

*b* and *c*). Although the content of the ABP-36 in the Triton insoluble fraction can not be determined very accurately even by densitometry because of its scarcity, it falls in a range of 0.1 to 0.2 μg per lane. This would bind approximately 1 to 2 μg of F-actin since purified ABP-36 binds 7 mol of actin per mol ABP-36 (Ogihara and Tonomura 1982), and it is found to be the case. To test whether or not actin binding to the fraction is specifically mediated by the ABP-36, the fraction was pretreated with anti-ABP-36 antiserum. The results are shown in Fig. 3. Those incubated with the antiserum contained about 50,000 Dalton IgG heavy chain and the amounts increased as higher concentrations of the antiserum were used (Fig. 3, compare lanes *d*, *e*, and *f*). As seen in the SDS-PAGE gel, the higher the concentrations of the added antiserum were, the lower the amounts of the bound actin were. This is more clearly shown in the right figure obtained by densitometric scanning of the gel. With the antiserum at 1 : 22 dilution, 85% of the actin binding activity of the fraction was blocked. A possibility of participation of some proteolytic activity likely contained in the serum, and hence of the decrease in the actin binding activity is excluded for the following reasons. First, 0.2 mM PMSF and 1 mg/ml of BSA were added in both the pretreatment and the actin binding assay to block pro-

teases. Second, the antiserum was pre-heated at 56 °C for 30 min to inactivate proteases. Finally, the treated fractions had identical SDS-PAGE patterns to that of the untreated (Fig. 3, compare lanes *b*, *c*, *d*, *e*, and *f*) indicating no proteolytic digestion.

Table 1 shows the enhancement of the actin binding activity of the Triton insoluble fraction by treatment with 10 mM Mn-ATP. This treatment caused the amount of bound actin to increase both in the presence and absence of phalloidin. Also note the increased binding by the addition of phalloidin both in the treated

Table 1. *Enhancement of actin binding to Triton insoluble fraction by Mn-ATP treatment*

| Mn-ATP treatment[a] | G-actin (0.17 mg/ml) | Phalloidin (15 μM) | Relative amount of bound actin[b] |
|---|---|---|---|
| — | + | — | 1.00 |
| — | + | + | 1.19 |
| + | + | — | 3.87 |
| + | + | + | 4.48 |

[a] See Materials and Methods for details
[b] The amount of actin was quantified by densitometry of the SDS-PAGE gels as shown in Fig. 3. Actual amount of bound actin was calculated by subtracting actin content in Triton insoluble fraction from those of samples mixed with G-actin

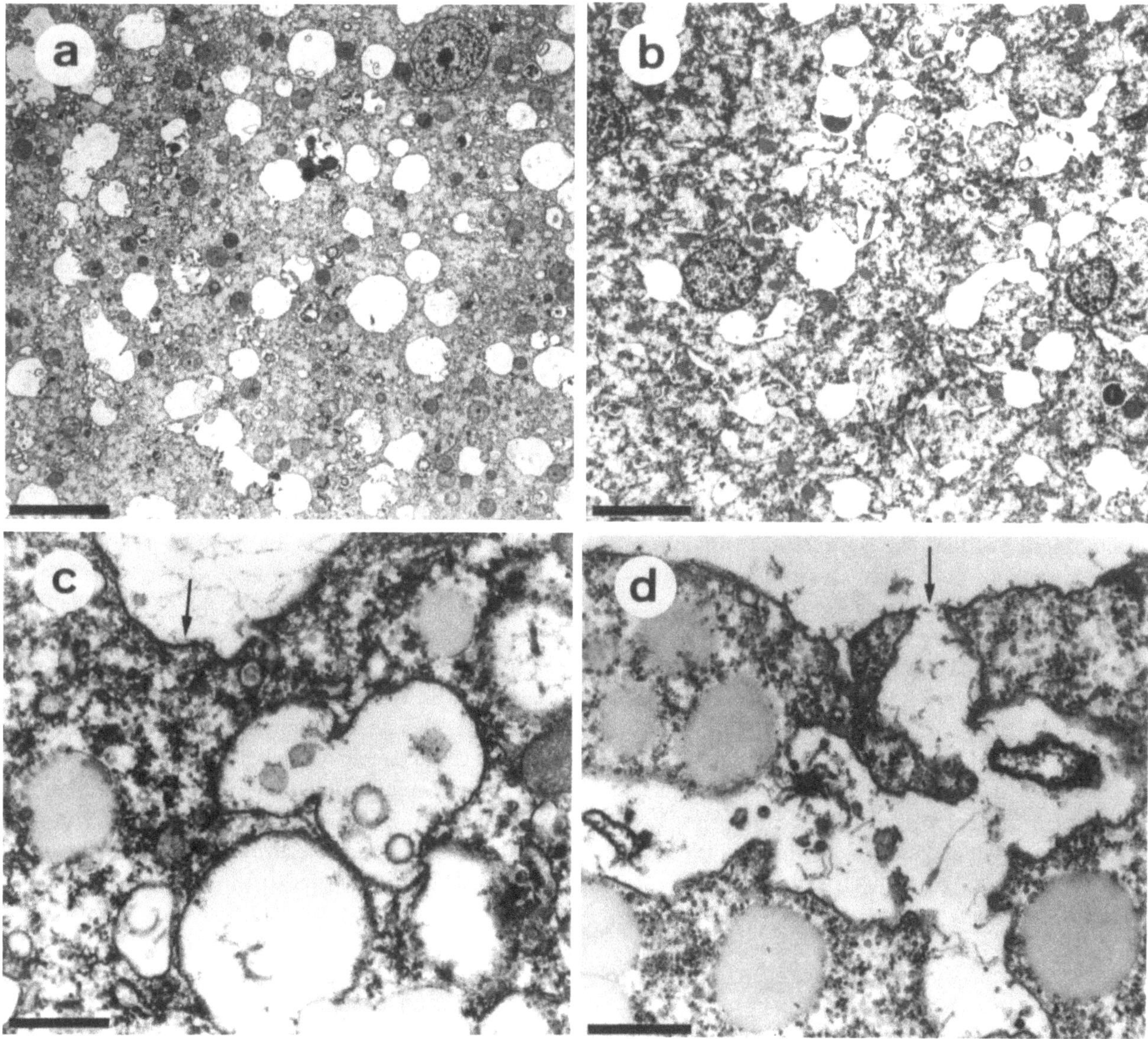

Fig. 4. Electron micrographs of the vesicles in the endoplasm. Before extrusion of the endoplasm, there are numerous vesicular structures in the endoplasm region of the plasmodium (*a*). There are vesicles of about 0.5 μm in diameter. Those vesicles change their morphology within 5 sec after the endoplasm is punctured out to make the endoplasmic veins (*b*, see Materials and Methods for details). *c* and *d* Vesicles found in close proximity to the plasma membrane (arrow) in the endoplasmic veins aged for 5 min in water. In *d*, a cluster of the vesicles is open to the outer medium at the place indicated by the arrow. Scale bars are 1.0 μm (*a, b*) and 0.5 μm (*c, d*)

and in the untreated fractions. This phalloidin effect is consistent with the previous report (Ogihara and Tonomura 1982) in a sense that the bound actin is likely to be in F-form. Neither MnCl$_2$ nor ATP when used solely enhanced the activity (data not shown). To characterize the effect of the Mn-ATP treatment (Nakamura *et al.* 1985, Brunati *et al.* 1985), phosphorylated proteins were analyzed in autoradiography. The prominently phosphorylated bands in the Triton insoluble fraction are myosin heavy chain and a protein which has a molecular weight of 28,000 Daltons. The

ABP-36 was not phosphorylated under the conditions employed.

### 3.3. Vesicular Structure in situ

Intact plasmodia and endoplasmic veins were fixed for conventional electron microscopy to see whether vesicular structures morphologically similar to those in the Triton insoluble fraction exist or not, because there is a possibility that the immunostained vesicles are an artifact induced by the fixation with PLP and the lysis with Triton X-100, especially with a relatively high concentration of $Ca^{2+}$ (50 to 100 µM) contained in the Triton solution. There were vesicles in the intact plasmodia which fulfills the following criteria; round, empty and about 0.5 to 1.0 µm, characteristic of the immunostained vesicles (Fig. 4). These vesicles are of relatively homogeneity in morphology, and spaced almost evenly in the intact endoplasm (Fig. 4 *a*). However, they change their morphology and the spatial arrangement in the endoplasmic veins. In Fig. 4 *b* are shown the vesicles in an endoplasmic vein that was fixed within 5 sec after puncturing a plasmodial strand. They have lost their original morphology and appear to have been elongated and connected with each other, and hence begin to take various sizes and morphologies. A remarkable aspect in this change is that they seem to have begun fusing with each other. This phenomenon of fusion of the vesicles was most clearly observed in regions close to the plasma membrane. Figure 4, *c* and *d*, shows magnified views of such region of an endoplasmic vein 5 min after the puncture. Several vesicles are clustered closely and two of them seem to have fused already (Fig. 4 *c*). The fusion phenomenon appears to have proceeded further in Fig. 4 *d*, in which the vesicles seem to have lost their original morphology except that the tubular hollow body has slight constriction in some places, retaining the vestiges of fusion. The most striking feature of this fusion phenomenon is that such tubular bodies eventually fuse with the plasma membrane as shown in Fig. 4 *d* (arrow).

## 4. Discussion

The ABP-36 was first found in and purified from the actin-containing cytoskeleton of *Physarum* plasmodium prepared with Triton X-100 (Ogihara and Tonomura 1982). It binds stoichiometrically 7 mol of actin per mol. Morphological examination and co-sedimentation assays of the complex of the ABP-36 and F-actin suggested its actin-crosslinking activity. However, no information was obtained concerning its localization *in situ* and function *in vivo*. That the immunogold staining pattern shown in Fig. 2 is specific for the ABP-36 is certified by several reasons. First, the anti-ABP-36 was affinity-purified and the antibody gave only one band on the immunoblot of the total plasmodial proteins (Fig. 1). Second, an affinity-purified secondary antibody was used to conjugate with colloidal gold. Also the staining, washing and blocking procedure was exhaustive to achieve specific staining. A previously reported method for immunoelectron microscopy was

of full use (Condeelis *et al.* 1987). Relatively low concentrations of the primary antibody were used, and the colloidal gold-antibody conjugates were routinely centrifuged to remove micro-aggregates before use, both helping obtain specific staining patterns. The fixation method employed, the PLP fixative, which fixes glycoproteins, helping paraformaldehyde and hence enabling to lower the paraformaldehyde concentrations seems to have worked. An artifactual rearrangement of the antigen is unlikely for two reasons; (1) an unfixed specimen was far less structurally preserved than those fixed with PLP and there was unspecific dispersion of the gold particles, and (2) a specimen fixed with 4% paraformaldehyde showed preservation comparable to those fixed with PLP while it had much fewer signals than PLP suggesting an antigenicity loss (data not shown).

The Triton insoluble fraction binds exogenous added actin and the amount of bound actin is close to the amount calculated basing upon the *in vitro* binding stoichiometry (Ogihara and Tonomura 1982). At least 85% of the actin binding to the Triton insoluble fraction was blocked by the anti-ABP-36 (Fig. 3). Therefore, the actin binding to the Triton insoluble fraction could largely be mediated by the ABP-36. The rest of the binding activity appears to be attributable to some other actin binding proteins contained in the fraction. The immunoelectron microscopy shown in Fig. 2 clearly demonstrated that the ABP-36 is localized on the surface of the vesicles. Then, it is highly likely that the ABP-36 possessed of the actin binding activity has distinct location on the vesicles in the Triton insoluble fraction. Although the Triton insoluble fraction is relatively crude and the ratio of the vesicles to the sheet-like structure which seems to have been derived from the plasma membrane has not been determined, the sheet-like structure may have a low, if any, actin binding activity.

It might be noted that it has become possible to visualize the ABP-36 bound to the surface of the vesicles by Ca ions contained in solutions used to prepare the Triton insoluble fraction when compared with the solutions used to stabilize the cytoskeleton (Ogihara and Tonomura 1982). The extremely low ionic strength and relatively higher pH might have helped as well. This procedure yielded less actin in the Triton-resistant fraction. A highly likely explanation for the less actin is that fragmin (Hasegawa *et al.* 1980) seems to have worked at $Ca^{2+}$ concentrations of 50 to 100 µM to sever actin filaments.

Table 1 indicates that Mn-ATP enhanced the actin

binding activity of the Triton insoluble fraction. The ABP-36 was not detected to be phosphorylated under the same conditions as used for the actin binding assay, and this offers a contrast to some other actin binding proteins of molecular weight of about 36 kDalton such as calpactins (Gerke and Weber 1984, Fava and Cohen 1984, Glenney and Glenney 1985, Shadle *et al.* 1985, Glenney *et al.* 1987, Glenney 1986 a, b, for review see Brugge 1986), chromobindins (Creuts *et al.* 1983, Burgoyni 1984, Summers and Creuts 1985) and lipocortins (Hirata *et al.* 1984, Giesow *et al.* 1986, Pepinsky and Sincelair 1986). Phosphorylation of some other proteins may be involved in the regulation of the activity. The phosphorylated proteins analyzed in the autoradiography were prominently myosin heavy chain and a 28 kDalton protein. The plasmodial myosin acquires the actin activatable Mg-ATPase activity and the thick filament forming activity only when the heavy chain is phosphorylated (Ogihara *et al.* 1983). It is not known whether or not the heavy chain phosphorylation affects its affinity for actin. As to the 28 kDalton protein, little is known, except that it becomes less phosphorylated in the endoplasmic vein with time (data not shown). Identification of this protein is in progress.

The property of the ABP-36 that may be important for its *in vivo* function is that it is predominantly found on the surface of the vesicles (Fig. 2). Out of cytoplasmic structures that were searched, vesicular structures closely resembling those stained with the anti-ABP-36 antibody were found in the endoplasm of intact plasmodia (Fig. 4 *a*). Those endoplasmic vesicles begin to fuse with each other within 5 sec when the endoplasm is extruded to make endoplasmic veins (Fig. 4 *b*). Fusion continues furthermore, resulting in the formation of larger vesicles especially in regions near the plasma membrane (Fig. 4 *c*). Larger vesicles, often taking tubular morphology, eventually fuse with the plasma membrane, forming plasmalemma invagination systems (Fig. 4 *d*). These sequential processes could be collectively exocytotic. The behaviours of the exocytotic vesicles may explain the occurrence of larger vesicles with irregular contour (Fig. 2 *b*) and vesicles in close proximity to the sheet-like structure in the Triton insoluble fraction (Fig. 2 *c*). Achenbach *et al.* (1979) and Wohlfarth-Bottermann (1983) reported that the exocytotic vesicles play a crucial role in the formation of the plasmalemma invagination systems in the plasmodium. The invagination systems are known to occupy large surface area (Wohlfarth-Bottermann 1983, Baranowski and Wohlfarth-Bottermann

1982) and importantly shown first by Wohlfarth-Bottermann (1962, 1964) to be connected with each other and with the plasma membrane by the actomyosin fibrils. The ABP-36 could be located on the exocytotic vesicles, and if so, should have a property of linking the cytoskeleton to the plasma membrane. Several actin-binding proteins which are categorized in the crosslinker such as α-actinin or fodrin have been shown to be localized on the plasma membrane or the secretory vesicles (see Niggli and Burgess 1987 for review). The possibility of such two properties for the ABP-36 will be explored. The Triton insoluble fraction, a lysed plasmodium, has been used in this study. However, with suitable postembedding immunocytochemical techniques, it should be possible to localize the ABP-36 in fixed whole plasmodium.

## Acknowledgements

I am grateful to Dr. J. Condeelis of Albert Einstein College of Medicine for actin antibody, and Dr. Makoto Miyata of Osake University for skeletal muscle actin. I thank Hiromo Sesaki for help in preparing the manuscript. This study is supported by a grant-in-aid from the Ministry of Education, Science, and Culture of Japan. Part of this study was carried out under the NIBB Cooperative Program (86–123).

## References

Achenbach F, Achenbach U, Wohlfarth-Bottermann KE (1979) Plasmalemma invaginations, contraction and locomotion in normal and caffeine-treated protoplasmic drops of *Physarum*. Eur J Cell Biol 20: 12–23

Baranowski Z, Wohlfarth-Bottermann KE (1982) Endoplasmic veins from plasmodia of *Physarum polycephalum*: A new strand model with defined age, structure and behaviour. Eur J Cell Biol 27: 1–9

Brugge JS (1986) The p 35/p 36 substrates of protein-tyrosine kinases as inhibitors of phospolipase $A_2$. Cell 46: 149–150

Burgoyne RD (1984) Mechanisms of secretion from adrenal chromaffin cells. Biochim Biophys Acta 779: 201–216

Brunati AM, Marchiori F, Pinna LA (1985) Isolation and partial characterization of distinct forms of tyrosine protein kinases from rat spleen. FEBS Lett 188: 321–325

Carboni JM, Condeelis JS (1985) Ligand-induced changes in the location of actin, myosin, 95 K (α-actinin), and 120 K protein in amoebae of *Dictyostelium discoideum*. J Cell Biol 100: 1884–1893

Condeelis J, Ogihara S, Bennett H, Carboni J, Hall A (1987) Ultrastructural localization of cytoskeletal proteins in *Dictyostelium* amoebae. In: Spudich JA (ed) Methods in cell biology, vol 28, *Dictyostelium discoideum*: molecular approaches to cell biology. Academic Press, Orlando, pp 191–207

Creuts CE, Dowling LG, Sando JJ, Villar-Palasi C, Whipple JH, Zaks WJ (1983) Characterization of the chromobindins: Soluble proteins that bind to the chromaffin granule membrane in the presence of $Ca^{2+}$. J Biol Chem 258: 14664–14674

Fava RA, Cohen S (1984) Isolation of a calcium-dependent-35-kilodalton substrate for the epidermal growth factor receptor/kinase from A-431 cells. J Biol Chem 259: 2636–2645

Gerke V, Weber K (1984) Identity of p36K phosphorylated upon Rous sarcoma virus transformation with a protein purified from brush borders: calcium-dependent binding to non-erythroid spectrin and F-actin. EMBO J 3: 227–233

Giesow MJ, Fritsche U, Hexham JM, Dash B, Johnson T (1986) A consensus amino-acid sequence repeat in *Torpedo* and mammalian $Ca^{2+}$-dependent membrane binding proteins. Nature 320: 636–638

Glenney Jr JR (1986 a) Phospholipid-dependent $Ca^{2+}$ binding by the 36-kDa tyrosine kinase substrate (calpactin) and its 33-kDa core. J Biol Chem 261: 7247–7252

— (1986 b) Two related but distinct forms of the $M_r$36,000 tyrosine kinase substrate (calpactin) that interact with phospholipid and actin in a $Ca^{2+}$-dependent manner. Proc Natl Acad Sci USA 83: 4258–4262

— Brian T, Powell MA (1987) Calpactins: Two distinct $Ca^{2+}$-regulated phospholipid- and actin-binding proteins isolated from lung and placenta. J Cell Biol 104: 503–511

— Glenney P (1985) Comparison of $Ca^{++}$-regulated events in the intestinal brush border. J Cell Biol 100: 754–763

Hasegawa T, Takahashi S, Hayashi H, Hatano S (1980) Fragmin: A calcium ion sensitive regulatory factor on the formation of actin filaments. Biochemistry 19: 2677–2683

Hirata F, Matsuda K, Notsu Y, Hattori T, DelCarmine R (1984) Phoshporylation at a tyrosine residue of lipomodulin in mitogen-stimulated murine thymocytes. Proc Natl Acad Sci USA 81: 4717–4721

Laemmli UK (1970) Cleavage of structural proteins during the assembly of the head of bacteriophage T4. Nature 227: 680–685

McLean IW, Nakane PK (1974) Periodate-lysine-paraformaldehyde fixative: A new fixative for immunoelectron microscopy. J Histochem Cytochem 22: 1077–1083

Nakamura S, Takeuchi F, Tomizawa T, Takasaki N, Kondo H, Yamamura H (1985) Two separate tyrosine protein kinases in human platelets. FEBS Lett 184: 56–59

Niggli V, Burgess MM (1987) Interaction of the cytoskeleton with the plasma membrane. J Membr Biol 100: 97–121

Ogihara S, Tonomura Y (1982) A novel 36,000-dalton actin-binding protein purified from microfilaments in *Physarum* plasmodia which aggregates actin filaments and blocks actin-myosin interaction. J Cell Biol 93: 604–614

Ogihara S, Ikebe M, Takahashi K, Tonomura Y (1983) Requirement of phosphorylation of *Physarum* myosin heavy chain for thick filament formation, actin activation of $Mg^{2+}$-ATPase activity, and $Ca^{2+}$-inhibitory superprecipitation. J Biochem 93: 205–223

Pepinsky RB, Sinclair LK (1986) Epidermal growth factor-dependent phosphorylation of lipocortin. Nature 321: 81–84

Shadle PJ, Gerke V, Weber K (1985) Three $Ca^{2+}$-binding proteins from porcine liver and intestine differ immunologically and physicochemically and are distinct in $Ca^{2+1}$ affinities. J Biol Chem 260: 16354–16360

Slot JW, Geuze HJ (1985) A new method of preparing gold probes for multiple-labeling cytochemistry. Eur J Cell Biol 38: 87–93

Summer TA, Creutz CE (1985) Phosphorylation of a chromaffin granule-binding protein by protein kinase C. J Biol Chem 260: 2437–2443

Talian JC, Olmsted JB, Goldman RD (1983) A rapid procedure for preparing fluorescein-labeled specific antibodies from whole antiserum: Its use in analyzing cytoskeletal architecture. J Cell Biol 97: 1277–1282

Towbin H, Staehelin T, Gordon J (1979) Electrophoretic transfer of proteins from polyacrylamide gels to nitrocellulose sheets: Procedure and some applications. Proc Natl Acad Sci USA 76: 4350–4354

Wohlfarth-Bottermann KE (1962) Weitreichende, fibrillare Protoplasmadifferenzierungen und ihre Bedeutung für die Protoplasmaströmung. II. Lichtmikroskopische Darstellung. Protoplasma 54: 514—539

— (1964) Differentiation of the ground cytoplasm and their significance for the generation of the motive force of amoeboid movement. In: Allen RD, Kamiya N (eds) Primitive motile systems in cell biology. Academic Press, New York, pp 79–109

— (1983) Dynamic cellular phenomena in *Physarum* possibly accessible to laser techniques. In: Earnshaw JC, Steer MW (eds) The application of laser light scattering to the study of biological motion. Plenum Press, New York (NATO ASI Series, Series A: Life Sci, vol 59, pp 501–507)

Protoplasma (1988) [Suppl. 2]: 57–62



# Manipulating Single Microtubules

S. Inoué*

Marine Biological Laboratory, Woods Hole, Massachusetts

Received May 7, 1988
Accepted June 14, 1988

Dedicated to Professor Dr. Noburo Kamiya on the occasion of his 75th birthday

## Summary

Microtubules which make up the spindle fibers in living cells are highly dynamic. This paper first reviews some history of visualizing individual microtubules (which are only 25 nm in width) and other submicroscopic filaments with the light microscope. We then describe the asymmetric growth and shortening of freshly exposed microtubule ends observed with video-enhanced DIC microscopy. Plus and minus ends were exposed by cutting the microtubule with a UV microbeam. Our results indicate that a simple GTP cap is inadequate to explain the stability of a particular microtubule end. Instead, a propagating (conformational) change is likely to be involved in the rapid disassembly of microtubules whenever the cap at the plus (but not minus) end has been removed. The observations suggest a potentially important mechanism that can selectively govern the growth and shortening of individual microtubules in living cells.

*Keywords:* Microscopy; Microtubule polarity; Polymerization; UV-microbeam; Video microscopy.

Snaring a slippery eel is no mean task. But snaring slime mold, and measuring its tortional force and streaming pressure! My hat off to Professor Noburo Kamiya who, for nearly five decades, has spearheaded biophysical research and revealed much that lies at the basis of cytoplasmic streaming by managing these and other fabulous fetes.

In celebrating the accomplishments of such a master and in commemorating his 75th birthday, I would like to take the occasion to recall the role that another biophysical research tool, the light microscope, has played in visualizing the dynamic activities of microtubules.

As widely recognized now, microtubules are often la-

bile, slender tubular polymers of tubulin that establish or modulate cell shape, govern or bring about directed transport of organelles including cell centers and chromosomes, and provide structural framework or positional information crucial to many activities of living cells.

## 1. Visualizing Individual Microtubules and Other Submicroscopic Filaments

Although individual microtubules measure a scant 25 nm in diameter, or only a twentisecond of the wavelength of green light, Summers and Gibbons (1971) already visualized the gliding of axonemal doublet microtubules directly under the light microscope nearly two decades ago. Likewise, using darkfield illumination, Macnab and Koshland (1974) observed individual 14 nm filaments in splayed bacterial flagella a few years later. (See Reichert 1909, especially Fig. 5, p. 31, and Figs. 9 and 10, p. 49, for earlier darkfield microscopical studies on bacterial flagellar bundles and their splaying.)

Thus, visualizing an object far smaller than the limit of resolution with the light microscope is not a new event. (Additionally, see *e.g.*, Hotani 1976, Kamiya and Asakura 1976, for paper describing the conformational changes in bacterial flagellin, Nagashima and Asakura 1980, for flexibility of actin filaments, and Kamiya 1982, for the rotation of microtubules, all observed with darkfield microscopy; Allen *et al.* 1985, and Schnapp *et al.* 1985, for gliding and transport behavior of microtubules observed with video-enhanced DIC; Higashi-Fujime 1986, Yanagida *et al.*

* Correspondence and Reprints: Marine Biological Laboratory, Woods Hole, MA 02543, U.S.A.

1984, and Toyoshima *et al.* 1987, for gliding motions of actin and myosin observed with video-enhanced darkfield and fluorescence microscopy.)

Despite this history, many still ask today: "How is it possible to use video enhancement to visualize objects that are unresolvable with the light microscope?" The answer is that the criteria for *visualizing* and *resolving* an object are not the same (see, *e.g.*, Gage 1932, p. 124–125).

We can visualize a small or thin object, and even determine its exact position to a very small fraction of a wavelength of light, so long as the object provides a discrete enough diffraction image to signal its presence and location (*e.g.*, Gelles *et al.* 1987, Kamimura 1987). It is even possible to estimate with a light microscope how many unresolved filaments or microtubules make up a single diffraction image (Inoué 1988 a).

On the other hand, resolution depends on our ability to tell the twoness of the diffraction images of two adjacent objects, so that the resolution of a light microscope normally cannot exceed the Abbe limit. (For a discussion of some exceptions, see Inoué 1988 a).

## 2. Birefringence of Microtubules

As I have described elsewhere, the fluctuation in birefringence, seen with a sensitive polarizing microscope, revealed the lability and dynamic nature of the fibrils that make up the spindle fibers (reviews: Inoué 1981, 1988 b). The parallel behavior of these birefringent fibrils and microtubules led us to conclude that the latter were responsible for the birefringence of the spindle fibers in living cells (Inoué and Sato 1967). But our good friend Dan Mazia was not that easily convinced and challenged us to provide quantitative proof that this was actually the case.

Our Biophysical Cytology group at the University of Pennsylvania was able to respond to Dan, albeit nearly a decade after his challenge. We were able to establish that the form birefringence of the microtubules (seen in isolated spindles whose birefringence was unchanged from the living) in fact precisely accounts for the spindle birefringence in living cells (Sato *et al.* 1975).

The birefringence of the spindle fibers allowed us to postulate several dynamic properties of their fibrils, including the fact that they were in a colchicine, temperature, and pressure sensitive equilibrium with their subunits. Although such intepretation was challenged for over a decade, these properties became ascribable to microtubules once methods for isolating labile mi-

crotubules were discovered by Weisenberg (see Inoué 1981, for summary reference).

## 3. Polarity of Microtubules

Microtubules, like actin, have been shown to be polarized filaments, *i.e.*, they tend to grow predominantly at one end (the plus end) or grow faster at that end relative to the other end (the minus end) (Allen and Borisy 1974, Summers and Kirschner 1979). For example, in the spindle and asters the plus end of their microtubules generally point away from the centriole (Telzer and Haimo 1981, McIntosh and Euteneuer 1984).

The polar ends of spindle fibers, cut with a UV microbeam in dividing plant cells, rapidly disappear. But in a few minutes they grow back from the severed fiber, still intact between the irradiated area and the kinetochore or phragmoplast (Inoué 1964). In animal cells the fibers again persist between the kinetochore and irradiated area after an area of reduced birefringence (arb) is induced by an UV microbeam. The kinetochore side of the fiber again grows back soon, but the exact behavior of the arb and the poleward fiber appears to depend on the particular type of cell irradiated (Forer 1965, Gordon and Inoué 1979).

## 4. Dynamic Instability

Recently Mitchison and Kirschner (1984) concluded that microtubules in a population, even *in vitro*, do not all behave uniformly. At "equilibrium", some microtubules grow steadily while others undergo a catastrophic change and then rapidly depolymerize for some length, then resume their growth again. These striking events, or the dynamic instability of microtubules, have now been recorded with the light microscope under video-enhanced darkfield illumination by Horio and Hotani (1986), and with video-enhanced DIC by Walker *et al.* (1988).

Under appropriate solution conditions, microtubules *in vitro* grow about two to three times faster on the plus end than the minus end. The plus and minus ends both spontaneously undergo catastrophe with about the same frequency and shorten rapidly, but the minus end is "rescued" more frequently so that each shortening event is less extensive on the minus end (Walker *et al.* 1988). Specialized regions containing GTP-tubulin rather than GDP-tubulin, or "caps" on the ends are postulated to prevent their catastrophe and spontaneous shortening (Kirschner and Mitchison 1986). Richard Walker, Ted Salmon and I decided to try

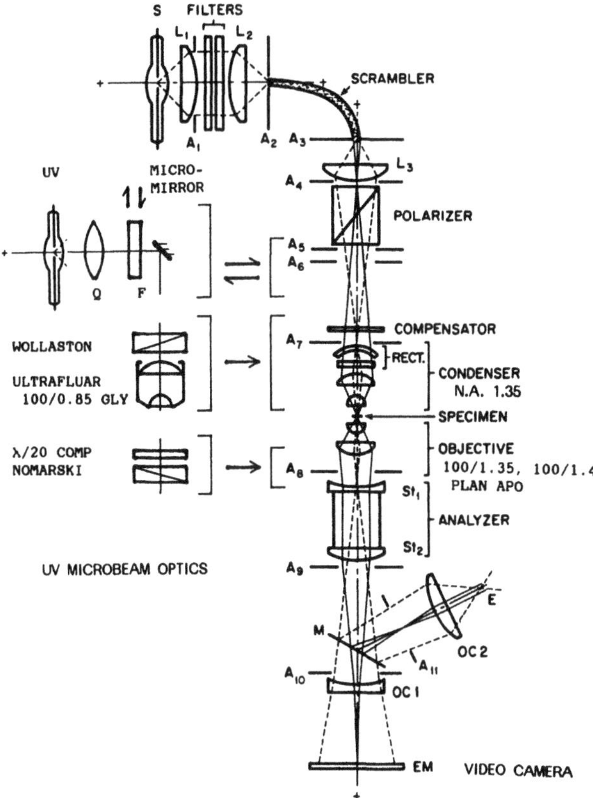



Fig. 1. UV-microbeam arrangement incorporated into the video, high extinction polarizing microscope. See text

cutting off the tip of a microtubule and induce a freshly cut end to observe how its growth behavior would be affected.

## 5. Cutting a Microtubule

In order to cut a microtubule at a known location and observe its behavior before and after the cut, we modified our high extinction, inverted polarizing microscope (INOUÉ 1986, Figs. III-21, 22) in the following way. A first surface micromirror (ca. 0.2 mm wide × 0.7 mm long) was placed in front of the field stop, ca. 150 mm from the condenser aperture, at 45 degrees to the axis of the microscope (Fig. 1). The micromirror, illuminated with a concentrated mercury arc lamp, acted as the small source of UV for irradiating the microtubule.

The rectified condenser on the microscope was replaced with an × 100/0.85 N.A. Zeiss Ultrafluar objective lens and focussed to produce a sharp, reduced image of the micromirror into the specimen plane. A quartz cover slip, glycerol immersed to the Ultrafluar lens, supported the microtubules on the condenser side of the

specimen space. On the other side, the preparation was covered with a glass coverslip oil contacted with the objective lens.

A Nomarsky prism slider from a Zeiss objective lens (× 63/1.4), placed in a slot above the Ultrafluar, provided a reasonable match with the Nikon Nomarsky prism present below the 100/1.4 and 100/1.35 Plan Apo Nikon objective lenses. Without special modification, the Nomarsky prism above the condenser transmitted ca. 20% of the UV energy effective for cutting the microtubules.

To visualize the low contrast DIC image of the individual microtubules, the video signal, enhanced by the analog circuit in the Newvicon camera (Dage/MTI, Michigan City, IN, model 65), was exponentially averaged every two frames and displayed with background subtraction and contrast enhancement using an Image-1/AT digital image processor (Universal Imaging Corp. Media, PA). Averaging of more than two frames reduces the camera noise further but blurs the image of the microtubule which is undergoing Brownian motion. We used procedures developed by WALKER et al. (1987) for observing the assembly of individual microtubules nucleated from the plus and minus ends of isolated axonemes.

With this setup, we could clearly visualize the individual microtubules in DIC, with the image of the UV micromirror superimposed at the location to be irradiated (Fig. 2). All that was needed for irradiation was to remove the UV-cut filter placed in front of the UV source.

## 6. Response of the Cut Microtubule Ends

The MAPs-free microtubules were grown in a synthetic medium containing 16 micromolar (1.6 mg/ml), phosphocellulose-column-purified porcine brain tubulin in a solution containing: 100 mM PIPES, 2 mM EGTA, 1 mM MgSO₄, 1 mM GTP, at pH 6.9. At room temperature, the tubulin concentration must have been slightly below the critical concentration since microtubules grew only on the two ends of the purified axonemal fragments which were mixed together with the tubulin solution to act as seeds for microtubule growth. Flagellar axoneme fragments, free from sperm head and dynein outer arms, were prepared from osmotically shocked sea urchin sperm (*Lytechinus pictus*) following the method described by BELL et al. (1982). The short fragments of axoneme adhered to the quartz coverslip, while most of the microtubules growing on their ends did not; these microtubules waved laterally exhibiting Brownian motion.

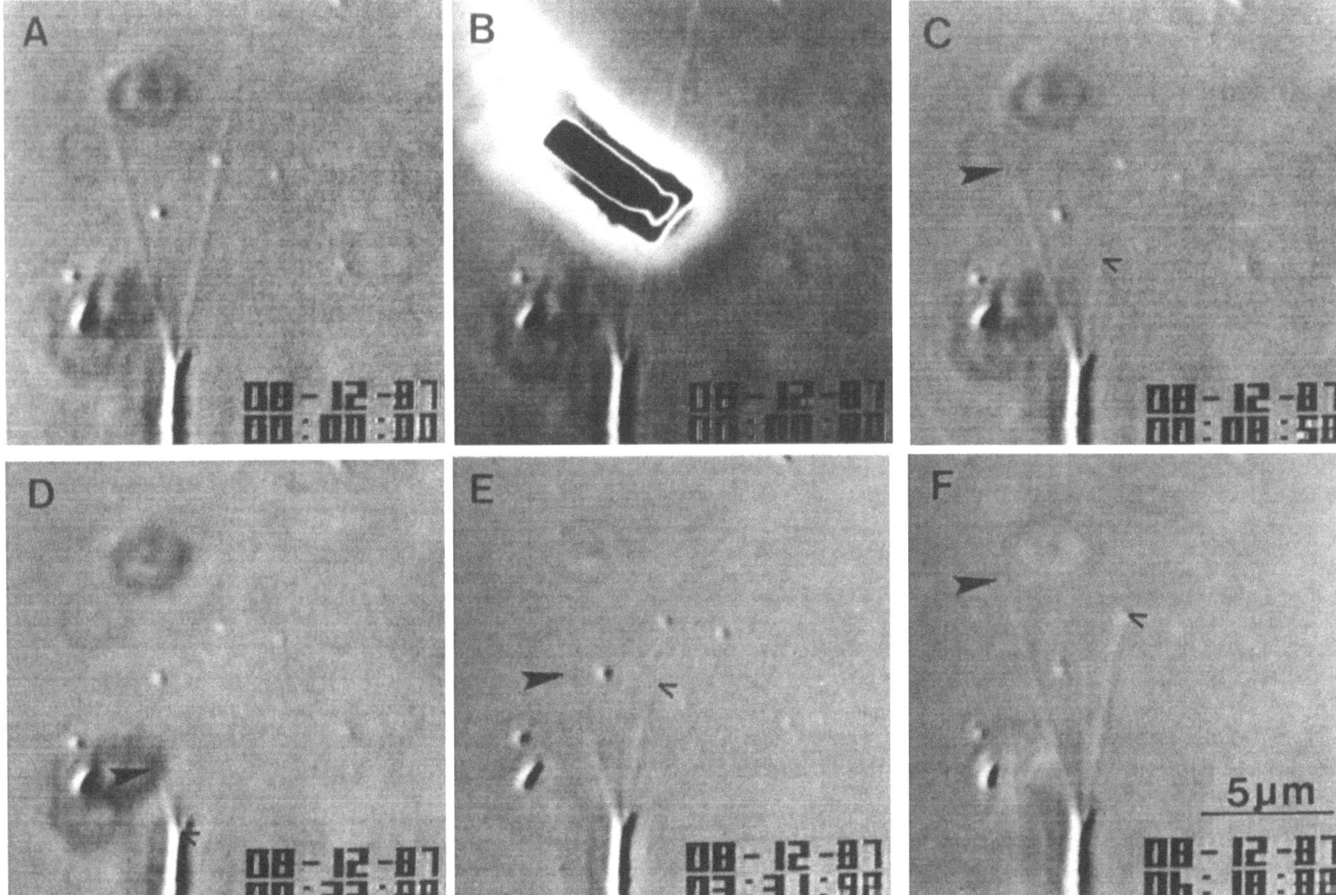

Fig. 2. Rapid shortening and regrowth of cut, plus-end microtubules. *A* Before cutting, 2 microtubules (V-shaped, reaching the top of the figure) grown on the plus-end of an axoneme fragment (high contrast, vertical object to the left of the date-time marker). *B* Image of UV-micromirror superimposed on the microtubules, just before removing the UV-barrier filter. *C, D* Rapid shortening of the 2 microtubules captured at 8 and 32 seconds after start of irradiation. *E, F* Regrowth after shortening. ▼, ∇ Tips of the shortening and re-growing plus-end microtubules. Time in minutes: seconds: second/100 after start of 3-second long, heterochromatic UV irradiation. Nikon 100/1.4 Plan Apo objective in DIC contrast. Newvicon video image (Dage-MTI model 65) 2-frame jumping averaged with background subtraction (Universal Imaging Image-1/AT) and recorded on a laser disk recorder (Panasonic OMDR TQ-2021FBC). ×3,600

Microtubules which were not stuck to the coverslip were completely severed with a 3-second irradiation with the UV microbeam. Those that were stuck could not be severed even with prolonged irradiation.

The severed tips of either plus- or minus-end growing microtubule showed vigorous Brownian motion and drifted away. When a severed minus end was created on a microtubule attached to the axoneme, it immediately started to regrow at the characteristic minus-end growth rate. Severed plus ends showed totally different behavior.

After cutting, a microtubule with a severed plus end suddenly began to shorten, and shortened rapidly all the way back to the axonemal seed. Once the microtubule had shortened down to the seed, it would start growing back again at the characteristic plus-end growth rate.

(Fig. 2. For quantitative data on the shortening and growth rates observed, see Walker *et al.* 1988).

What we found then, is a totally asymmetric behavior of freshly-cut plus and minus microtubule ends (Fig. 3). Rather than simply exhibiting different growth rates, the minus end shows no sign of rapid shortening while the plus end shortens rapidly all the way to the seed. Could such a drastic disparity in behavior be explained on the basis of somewhat different on and off constants at the two ends? Or does it imply even a different capping or stabilizing mechanism for the two ends?

At this point, it seems reasonable to conclude: (a) that the act of cutting with the UV microbeam does not denature or otherwise drastically modify the tubulin exposed at the fresh end of the cut; and (b) that perhaps it is not the simple presence of a (GTP-tubulin, or

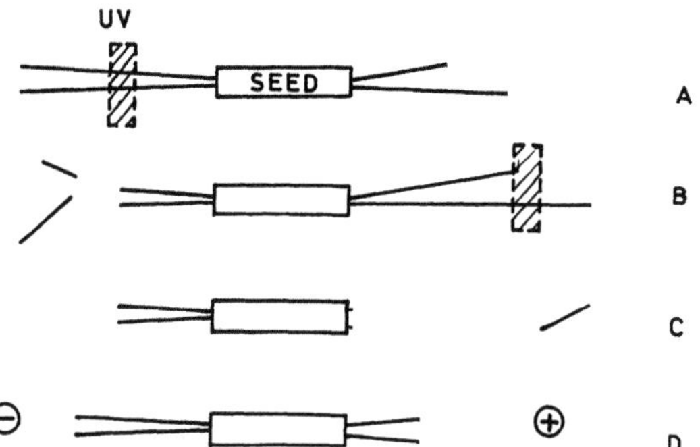


Fig. 3. Schematic behaviour of freshly cut microtubule ends. *A* UV microbeam irradiation of microtubules growing on minus end of axonemal seed. Subsequent behaviour of microtubules are shown in *B–D*. *B* Irradiation of microtubule or its tip, growing on the plus end of the seed. Subsequent rapid shortening and relatively rapid regrowth are shown in *C* and *D* (see also photographs in Fig. 2 and description in text)

whatever) cap that prevents the catastrophe and rapid shortening, but that a propagated conformational change or some chain reaction along the microtubule (that results from the loss of a cap at the plus end) induces or promotes rapid shortening.

Whatever the explanation turns out to be, we have, by direct microscopic observation and manipulation of individual microtubules, found another intriguing, polarized dynamic behavior of labile microtubules. These observations provide a better foundation for interpreting the behavior of microtubules which were irradiated with UV microbeam in living cells. Furthermore, they suggest that the growth and shortening of a particular microtubule in a living cell could be regulated by agents which nick the plus-end tip, or otherwise modify the conformation or arrangement of its tubulin molecule.

## Acknowledgements

This article was written to commemorate Dr. Noburo Kamiya's 75th birthday. The author is grateful to his co-workers, Richard Walker and Ted Salmon for allowing him to use some data on UV severing of microtubules before the full report has appeared in print. This work was supported by grant, NIH R37 GM31617-06 and NSG DCB 8518672 to S. I. and NSF DCB 8616621 and NIH GM 24364 to E.D.S.

## References

Allen C, Borisy GG (1974) Structural polarity and directional growth of microtubules of *Chlamydomonas* flagella. J Mol Biol 90: 381–402

Allen RD, Weiss DG, Hayden JH, Brown DT, Fujiwake H, Simpson M (1985) Gliding movement of and bidirectional transport along single native microtubules from squid axoplasm: Evidence for an active role of microtubules in cytoplasmic transport. J Cell Biol 100: 1736–1752

Bell CW, Fraser C, Sale WS, Tang W-J, Gibbons IR (1982) Preparation and purification of dynein. In: Wilson L (ed) The cytoskeleton, part A, cytoskeletal proteins, isolation and characterization. Academic Press, New York (Methods in cell biology, vol 24, pp 373–379)

Forer A (1965) Local reduction of spindle fiber birefringence in living *Nephrotoma suturalis* (Loew) spermatocytes induced by ultraviolet microbeam irradiation. J Cell Biol 25: 95–117

Gage SH (1932) The microscope. Comstock, Ithaca, New York

Gelles J, Schnapp BJ, Sheetz MP (1987) Tracking kinesin-driven movements with nanometre-scale precision. Nature 331: 450–453

Gordon GW, Inoué S (1979) Unexpected increase in poleward velocities of mitotic chromosomes after UV irradiation of their kinetochore fibers. J Cell Biol 83: 376 a

Higashi-Fujime S (1986) *In vitro* movements of actin and myosin filaments from muscle. Cell Motil Cytoskel 6: 159–162

Horio T, Hotani H (1986) Visualization of the dynamic instability of individual microtubules by dark-field microscopy. Nature 321: 605–607

Hotani H (1976) Light microscopic study of mixed helices in reconstituted *Salmonella* flagella. J Mol Biol 106: 151–166

Inoué S (1964) Organization and function of the mitotic spindle. In: Allen RD, Kamiya N (eds) Primitive motile systems in cell biology. Academic Press, New York, pp 549–598

— (1981) Cell division and the mitotic spindle. J Cell Biol 91: 131 s–147 s

— (1986) Video microscopy. Plenum Press, New York

— (1988 a) Imaging of unresolved objects, superresolution, and precision of distance measurement, with video microscopy. In: Taylor DL, Wang Y-L (eds) Fluorescence microscopy of living cells in culture: quantitative fluorescence microscopy: imaging and spectroscopy. Academic Press, New York (Methods in cell biology, vol 30, pp 529–538)

— (1988 b) The living spindle. In: K Dan Festschrift: Advances in cell division research. Zool Sci 5: 529–538

— Sato H (1967) Cell motility by labile association of molecules. J Gen Physiol 50: 259–292

Kamimura S (1987) Direct measurement of nanometric displacement under an optical microscope. Appl Optics 26: 3425–3427

Kamiya R (1982) Extrusion and rotation of the central-pair microtubules in detergent-treated *Chlamydomonas* flagella. Cell Motil 1: 169–173

— Asakura S (1976) Helical transformations of *Salmonella* flagella *in vitro*. J Mol Biol 106: 167–186

Kirschner MW, Mitchison T (1986) Microtubule dynamics. Nature 324: 621

Macnab R, Koshland Jr DE (1974) Bacterial motility and chemotaxis: Light-induced tumbling response and visualization of individual flagella. J Mol Biol 84: 399–406

McIntosh JR, Euteneuer U (1984) Tubulin hooks as probes for microtubule polarity: an analysis of the method and an evaluation on microtubule polarity in the mitotic spindle. J Cell Biol 98: 535–533

Mitchison T, Kirschner M (1984) Dynamic instability of microtubule growth. Nature 312: 237–242

Nagashima H, Asakura S (1980) Dark-field light microscopic study of the flexibility of F-actin complexes. J Mol Biol 136: 169–182

Reichert K (1909) Ueber die Sichtbarmachung der Geisseln und die Geisselbewegung der Bakterien. Centralbl Bakteriol Parasiteuk Infektionskrankh 1 Abt Orig [A]51: 14–94

Sato H, Ellis GW, Inoué S (1975) Microtubular origin of mitotic spindle form birefringence. J Cell Biol 67: 501–517

Schnapp BJ, Vale RD, Sheetz MP, Reese TS (1985) Single microtubules from squid axoplasm support bidirectional movement of organelles. Cell 40: 455–462

Summers KE, Gibbons IR (1971) Adenosine triphosphate-induced sliding of tubules in trypsin-treated flagella of sea-urchin sperm. Proc Natl Acad Sci USA 68: 3092–3096

Summers K, Kirschner MW (1979) Characteristics of polar assembly and disassembly of microtubules observed *in vitro* by darkfield microscopy. J Cell Biol 83: 205–217

Telzer BR, Haimo LT (1981) Decoration of spindle microtubules with dynein: Evidence for uniform polarity. J Cell Biol 89: 373–378

Toyoshima Y, Krone S, McNally E, Niebling K, Toyoshima C, Spudich JA (1987) Myosin subfragment-1 is sufficient to move actin filaments *in vitro*. Nature 328: 536–539

Walker RA, O'Brien ET, Pryer NK, Soboeiro M, Voter WA, Erickson HP, Salmon ED (1987) Calculation of transition frequencies for individual microtubules exhibiting dynamic instability behavior. J Cell Biol 105: 29a

Walker RA, Inoué S, Salmon ED (1988) Analysis of the mechanism of microtubule dynamic instability using a UV microbeam to sever elongating microtubules. Publication of 46th Annual Meeting of the Electron Microscopic Society of America, August 7–12, 1988, in press

Yanagida T, Nakase M, Nishiyama K, Oosawa F (1984) Direct observation of motion of single F-actin filaments in the presence of myoin. Nature 307: 58–60

# Cytoskeleton in Cellular Structure and Activity

Protoplasma (1988) [Suppl. 2]: 65–75



# Cytochalasin-Induced Ultrastructural Alterations in *Nicotiana* Pollen Tubes

S. A. LANCELLE and P. K. HEPLER*

Botany Department, University of Massachusetts, Amherst, Massachusetts

Received December 23, 1987
Accepted February 16, 1988

Dedicated to Professor Dr. NOBURO KAMIYA on the occasion of his 75th birthday

## Summary

Cytochalasin causes rapid inhibition of cytoplasmic streaming and growth in pollen tubes. In the present study we describe the ultrastructural changes that occur in pollen tubes of *Nicotiana alata* following treatment with cytochalasins. We utilize rapid freeze fixation and freeze substitution as a method superior to conventional chemical fixation for the preservation of cytoskeletal elements and membrane systems. The results show that either cytochalasin B or cytochalasin D causes microfilaments to form massive bundles throughout the cytoplasm, including the tipmost region of the pollen tube. Other effects include a loss of organelle zonation and vesicle aggregation in the tube tip, the accumulation of large stacks and coils of endoplasmic reticulum, and a crystalline rearrangement of a membrane component of some of the vacuoles. Following removal of cytochalasin the pollen tube ultrastructure returns to its normal configuration.

*Keywords:* Actin; Cytochalasin; Cytoskeleton; Freeze substitution; Microfilaments; Pollen tubes; Rapid freeze fixation.

*Abbreviations:* CB cytochalasin B; CD cytochalasin D; CF conventional chemical fixation; DMSO dimethylsulfoxide; MES 2(N-morpholino)ethanesulfonic acid; MF microfilament; MT microtubule; PEG polyethylene glycol; PM plasma membrane; RER rough endoplasmic reticulum; RF-FS rapid freeze fixation-freeze substitution; RP rhodamine-labelled phalloidin.

## 1. Introduction

Actin filaments play a crucial role in the generation of cytoplasmic streaming in plants (KAMIYA 1981). In the pollen tube, a rapidly growing cell that exhibits vigorous cytoplasmic streaming, the presence of an extensive actin containing microfilament (MF) system has been established, primarily by fluorescence microscopy after staining with rhodamine-labelled phalloidin (RP) (PERDUE and PARTHASARATHY 1985, PIERSON *et al.* 1986). Although MFs have been reported at the electron microscope level in conventionally fixed cells and in glycerinated preparations (CONDEELIS 1974, FRANKE *et al.* 1972), the cellular ultrastructural morphology of these elements as well as other cytoskeletal components, including microtubules (MTs), has been most convincingly demonstrated in pollen tubes of *Nicotiana alata* preserved by rapid freeze fixation followed by freeze substitution (RF-FS) (LANCELLE *et al.* 1987). In brief, the observations show an extensive system of long, thin MF cables running parallel to the long axis of the pollen tube. They are closely associated with various organelles, but none exclusively. In the tube tip, where vesicles containing wall precursor material aggregate, the MFs are present as single elements or groups of only a few rather than as bundles. Other structures that are present in the *N. alata* pollen tube and that may be composed of actin include MT-associated MFs and inclusions of filamentous semicrystalline material (CRESTI *et al.* 1985, 1986; LANCELLE *et al.* 1987).

In an effort to resolve more clearly the structural and functional relationship of these MFs, we have undertaken an investigation of pollen tube ultrastructure under conditions in which MF function has been blocked. It is well established that cytochalasins quickly inhibit cytoplasmic streaming and tip growth in pollen tubes (FRANKE *et al.* 1982, HERTH *et al.* 1972, MASCARENHAS

---

* Correspondence and Reprints: Botany Department, University of Massachusetts, Amherst, MA 01003, U.S.A.

and LAFOUNTAIN 1972, PICTON and STEER 1981). The implication from fluorescence studies has been that cytochalasin causes the MFs to disperse or depolymerize and thus renders them nonfunctional (PERDUE and PARTHASARATHY 1985). We find, in marked contrast, that MFs are actually arranged in much more massive and randomly oriented bundles after cytochalasin treatment. In addition, there is a disappearance of the fine MFs in the tube tip, a loss of cytoplasmic zonation, and a crystalline rearrangement of a component of the tonoplast. These effects are seen after treatment with either cytochalasin B (CB) or cytochalasin D (CD) and are completely reversible upon removal of the drug from the growth medium.

## 2. Materials and Methods

Pollen from greenhouse-grown *Nicotiana alata* was sown on the surface of liquid culture medium containing 50 mM PEG-400 (ZHANG and CROES 1982), 1 mM $CaCl_2$, 0.01% $H_3BO_3$, and 15 mM MES buffer, pH 5.5. After $2^1/_2$–3 hr the culture medium was drawn off and replaced with one of the following: medium + 1 µM CB and 0.01% DMSO; medium + 5 µM CD and 0.25% DMSO; medium + 0.01% DMSO (CB control); medium + 0.25% DMSO (CD control); fresh medium only (overall control).

For staining with RP, pollen tubes were grown and treated as described above, then harvested by centrifugation and fixed in 3% formaldehyde, 50 mM PEG-400, and 15 mM MES, pH 6.8 for $^1/_2$ hr. Staining was carried out in the dark in 0.33 µM RP (Molecular Probes, Inc.) and 15 mM MES, pH 6.8, for 1 hr in microcentrifuge tubes. Cells were then rinsed 3 times in buffer, and mounted in buffer for examination on a Reichert fluorescence microscope equipped with a 546 exciter filter, a 560 dichroic, and no barrier filter. Photographs were taken on Tri-X film pushed to ASA 1600 with Diafine developer.

For preparation by CF, cells were fixed in 2% glutaraldehyde and 2% formaldehyde in 50 mM cacodylate buffer plus 5 mM $CaCl_2$, pH 7.2., for 2 hr. After several rinses in fresh buffer, the cells were postfixed in osmium ferricyanide as described by HEPLER (1981). After staining 2 hr *en bloc* in 2% aqueous uranyl acetate, dehydration was carried out in acetone and embedment in Epon-Araldite resin. Sections were stained 3–5 min in Reynold's lead citrate before examination on a JEOL 100CX electron microscope.

For rapid freeze fixation and freeze substitution, cells were treated as previously described (LANCELLE *et al.* 1986) at recorded times. After substitution, the cells were stained *en bloc* in 5% uranyl acetate in methanol for 2 hr before embedment in Epon-Araldite resin. Sections were handled as described above.

## 3. Results

### 3.1. Control Cells

Cytoskeletal structure after RF-FS in *N. alata* pollen tubes grown under control conditions has been described in detail in a previous publication (LANCELLE *et al.* 1987). Therefore we include here only a few micrographs and a brief description of the structure of pollen tubes grown under control conditions, emphasizing those details not covered in the earlier publication. We have observed no difference between DMSO-treated cells and the overall controls.

After RF-FS, we observe many bundles of up to approximately 20 MFs throughout the cytoplasm (Figs. 1 and 2) except at the very tip, where the MFs occur singly or in bundles of only a few. The bundles run parallel to the long axis of the pollen tube. Various organelles, but none exclusively, are found in close association with them, but no crossbridges are evident. Staining of chemically fixed cells with RP (Fig. 3) or with fluorescently labelled anti-actin (TANG and HEPLER, in preparation) confirms that the MFs form an extensive network and are composed of actin.

In addition to the cytoplasmic MF bundles, cortical MTs have one or more closely associated parallel MFs that are crossbridged to them at regular intervals (this is also seen in cytochalasin-treated samples and is illustrated in Fig. 18). Immediately subjacent to many cortical MTs are elements of tubular ER that are oriented parallel to the MTs.

Also present throughout the pollen tube cytoplasm are inclusions that appear to be composed of tightly packed filaments in a semicrystalline array. These inclusions

---

Figs. 1–5. Micrographs of *N. alata* pollen tubes grown under control conditions. All except Fig. 3 are from material prepared by rapid freeze fixation and freeze substitution (RF-FS)

Fig. 1. MF bundle typical of those found throughout the cytoplasm. Bar, 0.25 µm

Fig. 2. Low magnification view of pollen tube cytoplasm shows organelles and elements of RER scattered throughout the cytoplasm. Arrowheads indicate MF bundles. Bar, 1 µm

Fig. 3. Pollen tube fixed in formaldehyde and stained with rhodamine-labelled phalloidin shows a large number of actin bundles and their predominantly axial orientation. Bar, 5 µm

Fig. 4. A vesicle bearing wall precursor material (asterisk) is in the process of fusing with the plasma membrane in the tip region of the pollen tube. Serial sections of many pollen tube tips revealed at most one such vesicle caught in the process of fusing. Bar, 0.25 µm

Fig. 5. A near median section through a pollen tube tip reveals the large number of vesicles present in the "clear zone" at the tip, as well as a large number of coated areas (arrowheads) in the plasma membrane. Bar, 0.5 µm

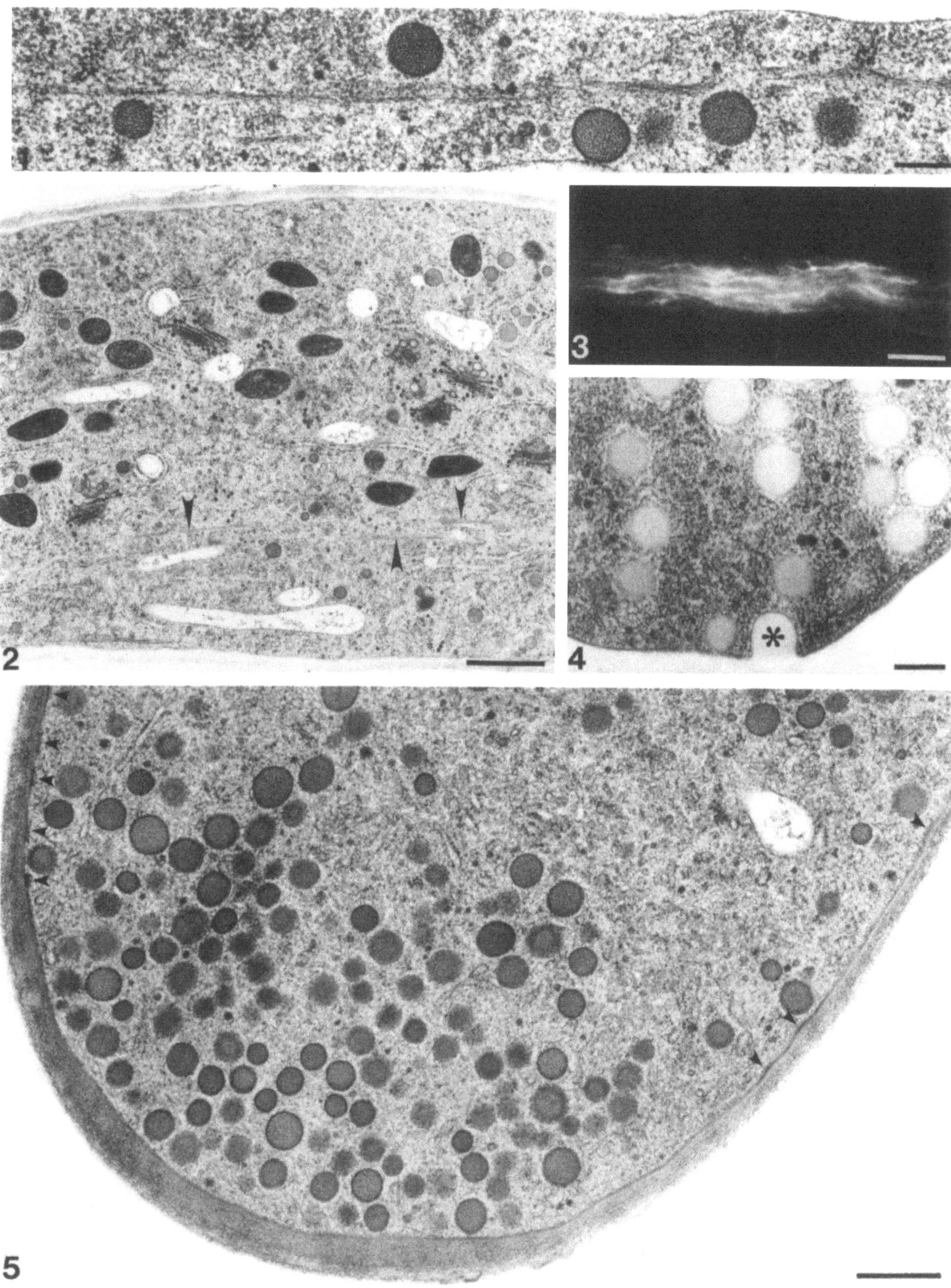

Figs. 1–5

are associated with dictyosomes and their chemical composition and function are unknown.

The organelle zonation typical of pollen tube tips, as well as the accumulation of large vesicles (Figs. 4 and 5), approximately 0.25 µm in diameter and bearing wall precursor material, are visible in near-median sections. Serial sections reveal only an occasional vesicle in the process of fusing with the plasma membrane (PM) (Fig. 4). The PM also contains abundant coated areas and nearby coated vesicles (Fig. 5); these are present in the very tip and along the length of the tube, but no attempt has yet been made to quantify them or to compare their numbers in different areas of the tube.

### 3.2. Effect of Cytochalasin Treatment

The results described for cytochalasin treatment apply to both CB and CD: No differences in structure between the two have been observed. Curiously, optimal stoppage of cytoplasmic streaming was obtained with 5 µM CD and 1 µM CB, a result opposite general findings, since CD is generally considered to be 10 × more potent than CB (see Cooper 1987). This discrepancy is unexplained.

Upon addition of cytochalasin at the optimal concentration, most of the pollen tube cytoplasm stops streaming almost instantaneously, but it takes 3–5 min for streaming within the pollen grain to stop completely. Pollen tubes frozen 3–5 min after addition of cytochalasin (the shortest possible time between addition of the drug and preparation of a sample for freezing) already show alterations in ultrastructure. There are some fine bundles of MFs present, but most have disappeared. Instead, massive bundles with no particular orientation in relation to the long axis of the pollen tube have emerged (Figs. 6–9, 11) and they are often seen in close association with various organelles (Fig. 9). These MF bundles can occasionally be seen after CF, where they are much more tightly packed (Fig. 6) than their general appearance after RF-FS (Figs. 7–9). The loosely fibrillar appearance of the bundles after RF-FS is most likely an artifact; it is possible

that they are especially prone to freezing damage and therefore show evidence of such damage while nearby structures do not. The semicrystalline inclusions associated with dictyosomes (see Cresti *et al.* 1985, 1986; Lancelle *et al.* 1987) are still present throughout the tube and seem unaffected by cytochalasin treatment. Staining of cytochalasin-treated pollen tubes with RP results in little localized fluorescence above a diffuse background glow.

At the tip of the pollen tube, few, if any, of the 0.25 µm vesicles are present, presumably having fused to the PM (Figs. 9–12). Replenishment of vesicles to the tip is prevented by the lack of cytoplasmic streaming, and organelle zonation is lost (Fig. 10). Coated vesicles and coated areas of the PM are still present in abundance (Fig. 11), even after 45 min in cytochalasin (the longest period of drug treatment). The massive MF bundles may extend well into the tip and in some cases appear directly attached to the PM (Figs. 9 and 11). No other ultrastructural effects on the PM have been observed. The general organization of the cytoplasm changes, and this is also apparent within 3–5 min. In ungerminated *N. alata* pollen grains, RER is "packaged" in tightly coiled whorls and layered stacks that gradually unwind and spread apart during pollen tube emission and growth, so that in the pollen tube itself, elements of RER are not so closely associated (Cresti *et al.* 1985). Upon addition of cytochalasin, however, the RER once again becomes stacked, often in large masses throughout the cytoplasm (Figs. 12 and 13; *cf.* Fig. 2). RER is not usually found in the pollen tube tip under control conditions, but it is often found there in abundance after cytochalasin treatments (Figs. 10 and 12).

Under control conditions, the many small vacuoles present in the vegetative cytoplasm are long and extended, and often can be seen in close association with fine MF bundles, but not apparently crossbridged to them. After cytochalasin treatment, many of the vacuoles are generally more round in shape, while others become branched and distorted. These show, in tangential section, a crystalline rearrangement of an unknown membrane component (Figs. 14 and 15). In fa-

---

Figs. 6–9. Massive MF bundles in pollen tubes treated with cytochalasin. Fig. 6 is from material prepared by conventional chemical fixation (CF); Figs. 7–9 are from material prepared by RF-FS. Bar, 0.25 µm

Fig. 6. After CF, the MF bundles are seen only occasionally, but they are large and densely packed

Figs. 7 and 8. After RF-FS, the large MF bundles are seen frequently throughout the cytoplasm, but take on a loosely fibrillar appearance that may be due to local freezing damage

Fig. 9. Large MF bundles (arrowheads) without any particular orientation near the tip of the pollen tube. Small arrowheads indicate an apparent attachment site of the MFs to the plasma membrane

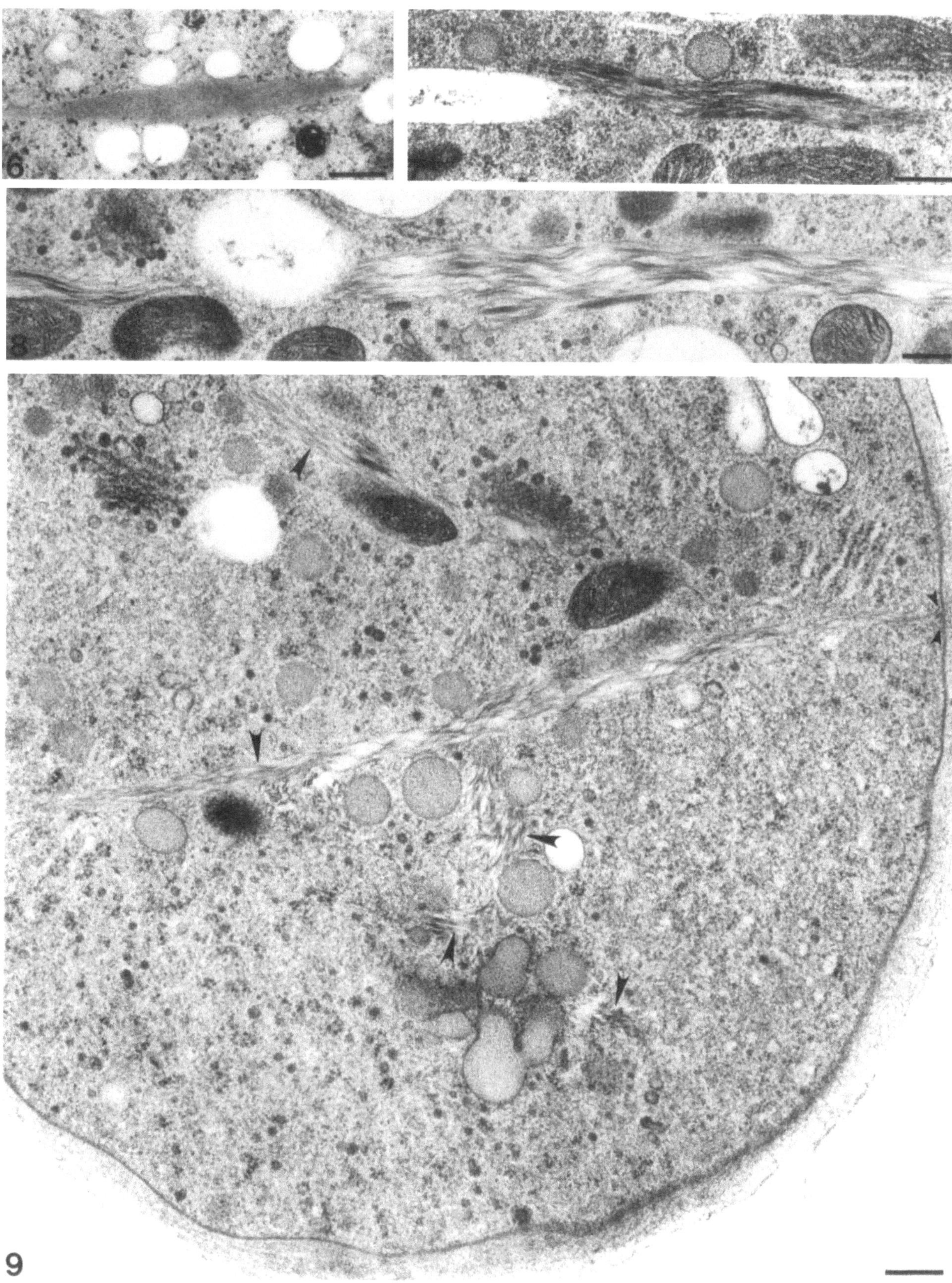

**9**

vourable transverse sections of the membrane, this component appears to extend outward from the membrane and even attach to other areas of the tonoplast or to nearby MFs in a regular pattern (Figs. 14–17). The crystalline pattern and regular attachments are not present under control conditions, and are never observed with CF.

The MT-associated MFs that run the length of cortical MTs appear unaffected by cytochalasin treatment (Fig. 18), and are still present 45 min after addition of the drug. Cortical MT-ER associations are maintained as well.

We observed no discernible ultrastructural effects on the generative cell. The effect of cytochalasin on generative cell motility within the pollen tube is undetermined.

Removal of cytochalasin from the medium restores control-type ultrastructure within approximately 10 min.

## 4. Discussion

Many of the cytochalasin-induced ultrastructural changes described here are previously unreported for pollen tubes. Franke *et al.* (1972) report that in *Lilium* and *Clivia* pollen tubes prepared by CF, cytochalasin stops cytoplasmic streaming and growth but has no effect on the ultrastructure. Picton and Steer (1981) did observe a marked decrease in vesicle numbers at the tip of *Tradescantia* pollen tubes after CD treatment, but state that the remainder of the tube appears unaffected. There may be differences among species, but it is also possible that the discrepancies between the present study and those published previously are due to the inability of CF to adequately preserve membrane systems (Mersey and McCully 1978) and cytoskeletal elements. In pollen tubes in particular, MTs and MFs have been difficult to fix and observe at the EM level with CF (Lancelle *et al.* 1987).

The extensive MF system present in the pollen tubes grown under control conditions apparently functions not only in cytoplasmic streaming but also in maintaining the general structure and organization of the cytoplasm. Almost immediately after addition of cytochalasin and concomitant with the appearance of the large MF bundles is an equally dramatic loss of cytoplasmic organization. The organelle zonation normally found in the tip region becomes severely disrupted. Elements of RER once again become closely associated into many-layered stacks and coils, reminiscent of their appearance in the pollen grain prior to germination (Cresti *et al.* 1985). These stacks often extend into the pollen tube tip, where RER is normally excluded.

It is unclear whether the many fine bundles of MFs seen before cytochalasin treatment aggregate into the massive bundles seen after treatment, or whether they disaggregate or depolymerize, and the large bundles arise by some other mechanism. The former seems more likely based on observations showing MFs at all times after cytochalasin treatment. It has been suggested that the semicrystalline filamentous inclusions that are present in *Nicotiana* pollen grains and tubes may be precursors for filamentous actin (Cresti *et al.* 1986, Heslop-Harrison *et al.* 1986). Certain inclusions in pollen from other species have been shown to contain actin based on RP staining (Heslop-Harrison *et al.* 1986), but direct evidence for *Nicotiana* is lacking. Nonetheless, the massive MF bundles resemble the shorter, tightly packed inclusions; whether or not they are related, however, is unknown.

In cultured animal cells, treatment with cytochalasin results in reorganization of actin MFs into a variety of forms, including contraction into large felt-like masses (Miranda *et al.* 1974, Weber *et al.* 1976, Rathke *et al.* 1977, Yahara *et al.* 1982). Although these cells are not directly comparable to plant cells, "rodlike" or "ribbonlike" structures composed of MFs and reminiscent of the large bundles we have described here have also been reported (Miranda *et al.* 1974, Rathke *et al.* 1977, Yahara *et al.* 1982). Cande *et al.* (1973) also observed large, hypercondensed MF bun-

---

Figs. 10 and 11. Micrographs of the pollen tube tip area after cytochalasin treatment. Figs. 10 and 11 are from material prepared by RF-FS; Fig. 12 from material prepared by CF

Fig. 10. Low magnification micrograph shows the complete loss of organelle zonation after cytochalasin treatment. Elements of RER, normally excluded from the tip area, are present in abundance after cytochalasin treatment. Bar, 1 µm

Fig. 11. MF bundles, normally not present in the tip area, are seen frequently after cytochalasin treatment. Small arrowhead indicates point of apparent attachment to the plasma membrane. Large arrowheads indicate coated areas of the PM. These coated pits seem unaffected by cytochalasin treatment, at least up until the 45 min maximum treatment time used here. Bar, 0.25 µm

Fig. 12. Elements of RER become stacked and coiled, and often accumulate in the tip region. Bar, 0.5 µm

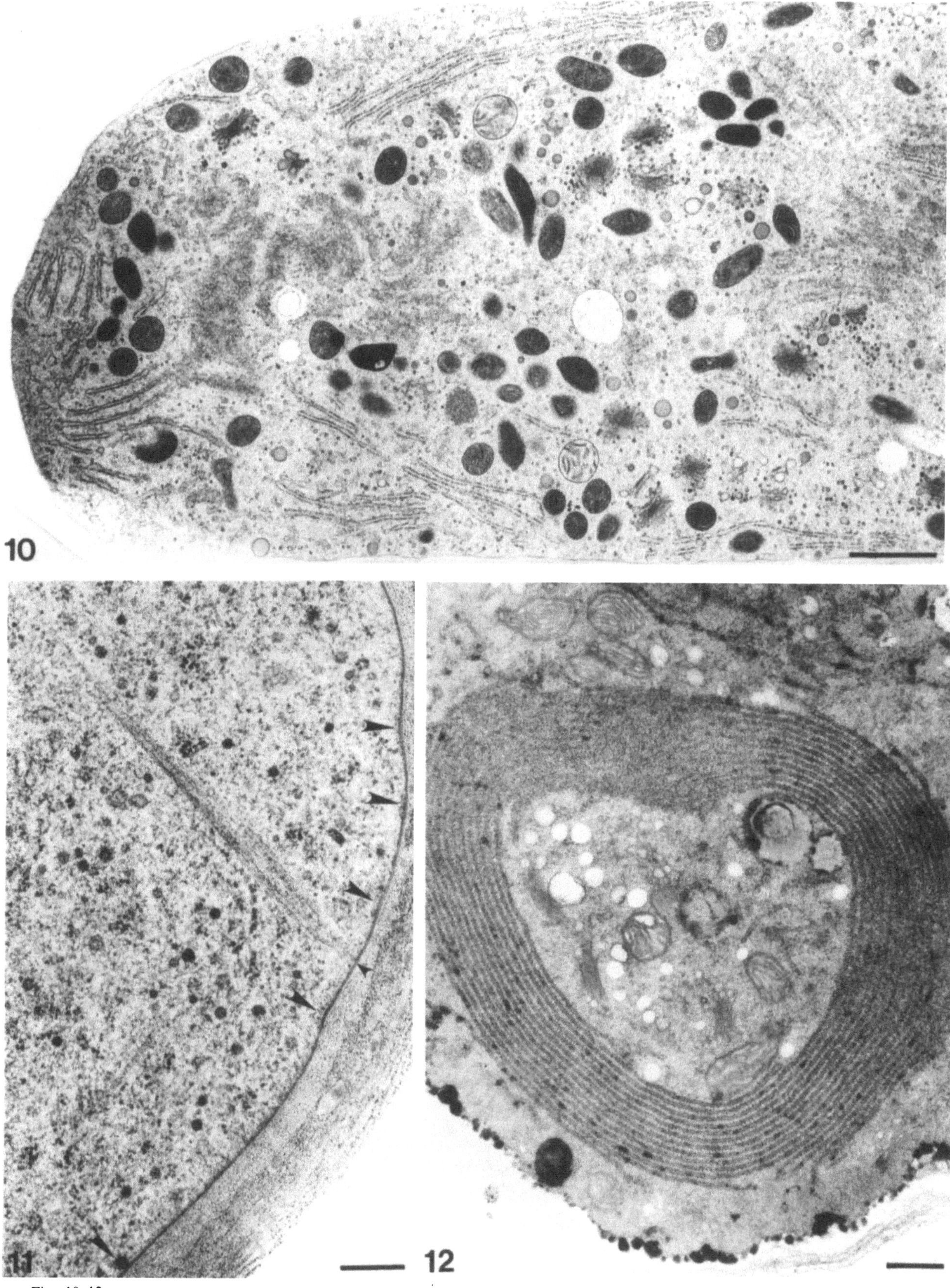

Figs. 10–12

dles in maize coleoptile cells treated with cytochalasin.

*In vitro* studies show that cytochalasin binds to F-actin and slows or inhibits polymerization (*e.g.*, MACLEAN-FLETCHER and POLLARD 1980, BROWN and SPUDICH 1981, BONDER and MOOSEKER 1986; see also COOPER 1987). It is unclear, however, exactly how cytochalasin interacts with existing MFs *in vivo*. WEBER *et al.* (1976) suggest that in cultured animal cells, one of the effects of cytochalasin may be to interfere with the anchoring of actin to the PM. In pollen tubes, though, we observe what appears to be direct attachment of the large MF bundles to the PM only after cytochalasin treatment and not under control conditions.

Staining of formaldehyde-fixed pollen tubes with RP under control conditions reveals an extensive system of actin filaments, a finding previously reported for other species (PERDUE and PARTHASARATHY 1985, PIERSON *et al.* 1986). Staining *N. alata* pollen tubes with RP after cytochalasin treatment results in little or no fluorescence above a diffuse background glow, as was found in pollen tubes of several other species (PERDUE and PARTHASARATHY 1985), but staining with fluorescently labelled anti-actin after cytochalasin treatment reveals that actin bundles are still present in abundance (X. TANG and P. K. HEPLER, in preparation), and indicates that the large bundles observed at the EM level are indeed composed of actin. This raises the possibility that actin bound to cytochalasin may not bind phalloidin as efficiently as does untreated actin, so that RP may not be the best actin indicator under all circumstances.

Capture of at most one vesicle in the process of fusing with the PM indicates a rapid rate of fusion. HOCH and HOWARD (1980) made similar observations with fungal hyphae, another tip growing system, where many vesicles could be seen in the process of fusing with the PM after CF, but few or none after RF-FS. In pollen tubes, the rapid loss of large vesicles from the tip after cytochalasin treatment indicates that vesicle fusion is unaffected. Growth of the pollen tube

ceases in the presence of cytochalasin because the lack of cytoplasmic streaming cannot replenish vesicles to the tip rather than because of a direct effect on vesicle fusion, a finding also reported by PICTON and STEER (1981).

PICTON and STEER (1983), in a study of vesicle production and membrane recycling in growing pollen tubes, failed to observe coated areas or coated vesicles associated with the PM. In the present study, RF-FS preserved coated vesicles and coated areas of the PM in abundance, and they were present both before and after cytochalasin treatment. This indicates that membrane uptake continues to occur, at least up to 45 min maximum treatment time, even though vesicle fusion has ceased.

MT-associated MFs are seen in a variety of plant cells especially after RF-FS (TIWARI *et al.* 1984, LANCELLE *et al.* 1986, 1987) but also occasionally after CF (FRANKE *et al.* 1972, SEAGULL and HEATH 1979, 1980; HARDHAM *et al.* 1980, TRAAS *et al.* 1985). Their composition and function are uncertain, but there is some evidence that they may be actin (TRAAS *et al.* 1987). Structurally they appear unaffected by cytochalasin treatment. SEAGULL and HEATH (1980), however, report that their numbers decrease after cytochalasin treatment in radish root hairs, especially near the tip of the hair. In pollen tubes, there are very few MTs near the growing tip (FRANKE *et al.* 1972, DERKSEN *et al.* 1985, LANCELLE *et al.* 1987), so it is difficult to say whether or not the numbers of MT-associated MFs are affected in this area of the pollen tube.

The striking effect of cytochalasin treatment on the tonoplasts of some, but not all, of the vacuoles in the pollen tube is unexplained. Cytochalasins are known to bind to various membrane proteins, and in particular CB has been shown to disrupt hexose transport in a variety of cells (*e.g.*, ATLAS and LIN 1978, LIN 1978). The effect we have reported here is apparently unrelated to hexose transport, however, since both CB and CD have the same structural effect, and CD is reported not to affect hexose transport (ATLAS and LIN 1978).

---

Figs. 13–18. Micrographs from pollen tubes treated with cytochalasin and prepared by RF-FS

Fig. 13. Low magnification micrograph of the pollen tube cytoplasm shows the stacked RER that is not present under control conditions (cf. Fig. 2), but is present throughout the cytoplasm in the presence of cytochalasin. Bar, 1 µm

Figs. 14–17. The effect of cytochalasin on some, but not all, vacuole membranes. Tangential sections through the tonoplast show a crystalline rearrangement of some unknown component (large arrowheads). In other areas this component appears to extend outward from the membrane and bridge to other membranes or to nearby short bundles of MFs (small arrowheads). Figs. 14–16: Bar, 0.25 µm; Fig. 17: Bar, 0.1 µm

Fig. 18. Microtubule-associated MFs are still present after 45 min in the presence of cytochalasin, and appear unaffected. Bar, 0.1 µm

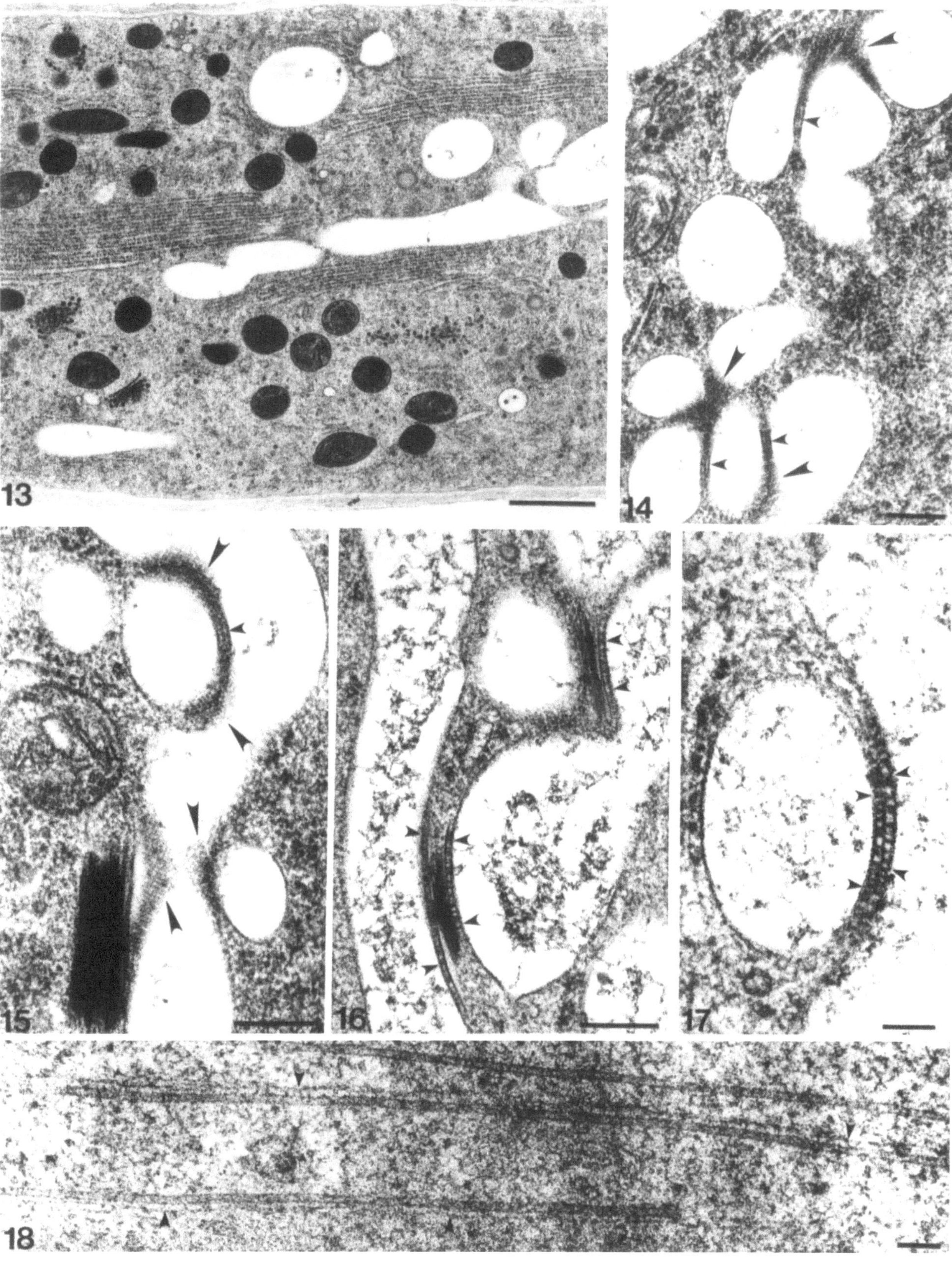

Figs. 13–18

The cytochalasin is apparently binding to some membrane constituent and causing the notable crystalline rearrangement that is common, but not seen in all of the vacuoles present. Short, thick MF bundles may be closely associated, even bridged to, the affected tonoplast. These effects completely disappear upon removal of cytochalasin from the growth medium.

## 5. Conclusion

Pollen tubes are important systems for studies aimed at examining cytoskeletal structure and function. The relative ease with which they can be prepared for ultrastructural analysis by RF-FS makes them ideal for a model system to which a variety of experimental agents can be applied and effects examined at both the LM and EM levels. The superior preservation obtainable by RF-FS results in more reliable information than that obtained by CF alone. Further studies in this direction are in progress.

## Acknowledgements

We thank M. CRESTI for sparking our interest in pollen tube structure and function, and G. MULCAHY for providing plant material. This work was supported by NSF grant DCB 87-02057.

## References

ATLAS SJ, LIN S (1978) Dihydrocytochalasin B: biological effects and binding to 3T3 cells. J Cell Biol 76: 360–370

BONDER EM, MOOSEKER MS (1986) Cytochalasin B slows but does not prevent monomer addition at the barbed end of the actin filament. J Cell Biol 102: 282–288

BROWN SS, SPUDICH JA (1981) Mechanism of action of cytochalasin: evidence that it binds to actin filament ends. J Cell Biol 88: 487–491

CANDE WZ, GOLDSMITH MHM, RAY PM (1973) Polar auxin transport and auxin-induced elongation in the absence of cytoplasmic streaming. Planta 111: 279–296

CONDEELIS JS (1974) The identification of F-actin in the pollen tube and protoplast of *Amaryllis belladonna*. Exp Cell Res 88: 435–439

COOPER JA (1987) Effect of cytochalasin and phalloidin on actin. J Cell Biol 105: 1473–1478

CRESTI M, CIAMPOLINI F, MULCAHY DLM, MULCAHY G (1985) Ultrastructure of *Nicotiana alata* pollen, its germination and early tube formation. Am J Bot 72: 719–727

— HEPLER PK, TIEZZI A, CIAMPOLINI F (1986) Fibrillar structures in *Nicotiana* pollen: changes in ultrastructure during pollen activation and tube emission. In: MULCAHY DL, MULCAHY G, OTTAVIANO E (eds) Biotechnology and ecology of pollen. Springer, New York, pp 283–288

DERKSEN J, PIERSON ES, TRAAS JA (1985) Microtubules in vegetative and generative cells of pollen tubes. Eur J Cell Biol 38: 142–148

FRANKE WW, HERTH W, VANDER WOUDE WJ, MORRE DJ (1972) Tubular and filamentous structures in pollen tubes: possible involvement as guide elements in protoplasmic streaming and vectorial migration of secretory vesicles. Planta 105: 317–341

HARDHAM AR, GREEN PB, LANG JM (1980) Reorganization of cortical microtubules and cellulose deposition during leaf formation in *Graptopetalum paraguayense*. Planta 149: 181–195

HEPLER PK (1981) The structure of the endoplasmic reticulum revealed by osmium tetroxide-potassium ferricyanide staining. Eur J Cell Biol 26: 102–110

HERTH W, FRANKE WW, VAN DER WOUDE WJ (1972) Cytochalasin stops tip growth in plants. Naturwissenschaften 59: 38 a

HESLOP-HARRISON J, HESLOP-HARRISON Y, CRESTI M, TIEZZI A, CIAMPOLINI F (1986) Actin during pollen germination. J Cell Sci 86: 1–8

HOCH HC, HOWARD RJ (1980) Ultrastructure of freeze-substituted hyphae of the Basidiomycete *Laetisaria arvalis*. Protoplasma 103: 281–297

KAMIYA N (1981) Physical and chemical basis of cytoplasmic streaming. Ann Rev Plant Physiol 32: 205–236

LANCELLE SA, CALLAHAM DA, HEPLER PK (1986) A method for rapid freeze fixation of plant cells. Protoplasma 131: 153–165

— CRESTI M, HEPLER PK (1987) Ultrastructure of the cytoskeleton in freeze-substituted pollen tubes of *Nicotiana alata*. Protoplasma 140: 141–150

LIN S (1978) Interaction of cytochalasins with the red blood cell membrane and its associated proteins. In: TANENBAUM S (ed) Cytochalasins. North Holland, Amsterdam, pp 499–520

MACLEAN-FLETCHER S, POLLARD TD (1980) Mechanism of action of cytochalasin B on actin. Cell 20: 329–341

MASCARENHAS JP, LAFOUNTAIN J (1972) Protoplasmic streaming, cytochalasin B, and growth of the pollen tube. Tissue Cell 4: 11–14

MERSEY B, MCCULLY ME (1978) Monitoring the course of fixation of plant cells. J Microsc 114: 49–76

MIRANDA AF, GODMAN GC, TANENBAUM SW (1974) Action of cytochalasin D on cells of established lines. II. Cortex and microfilaments. J Cell Biol 62: 406–423

PERDUE TD, PARTHASARATHY MV (1985) In situ localization of F-actin in pollen tubes. Eur J Cell Biol 39: 13–20

PICTON JM, STEER MW (1981) Determination of secretory vesicle production rates by dictyosomes in pollen tubes of *Tradescantia* using cytochalasin D. J Cell Sci 49: 261–272

— — (1983) Membrane recycling and the control of secretory activity in pollen tubes. J Cell Sci 63: 303–310

PIERSON ES, DERKSEN J, TRAAS JA (1986) Organization of microfilaments and microtubules in pollen tubes grown *in vitro* or *in vivo* in various angiosperms. Eur J Cell Biol 41: 14–18

RATHKE PC, SEIB E, WEBER K, OSBORN M, FRANKE WW (1977) Rod-like elements from actin-containing microfilament bundles observed in cultured cells after treatment with cytochalasin A (CA). Exp Cell Res 105: 253–262

SEAGULL RW, HEATH IB (1979) The effects of tannic acid on the *in vivo* preservation of microfilaments. Eur J Cell Biol 20: 184–188

— — (1980) The differential effects of cytochalasin B on microfilament populations and cytoplasmic streaming. Protoplasma 103: 231–240

TIWARI SC, WICK SM, WILLIAMSON RE, GUNNING BES (1984) Cytoskeleton and integration of cellular function in cells of higher plants. J Cell Biol 99: 63 s–69 s

Traas JA, Braat P, Emons AMC, Meekes H, Derksen J (1985) Microtubules in root hairs. J Cell Sci 76: 303–320

— Doonan JH, Rawlins DJ, Shaw PJ, Watts J, Lloyd CW (1987) An actin network is present in the cytoplasm throughout the cell cycle of carrot cells and associates with the dividing nucleus. J Cell Biol 105: 387–395

Weber K, Rathke PC, Osborn M, Franke WW (1976) Distribution of actin and tubulin in cells and in glycerinated cell models after treatment with cytochalasin B (CB). Exp Cell Res 102: 285–297

Yahara I, Fumiko J, Sekita S, Yoshihira K, Natori A (1982) Correlation between effects of 24 different cytochalasins on cellular structures and cellular events and those on actin *in vitro*. J Cell Biol 92: 69–78

Zhang H-Q, Croes AF (1982) A new medium for pollen germination *in vitro*. Acta Bot Neerl 31: 113–119

Protoplasma (1988) [Suppl. 2]: 76–87



# Pinocytosis and Locomotion of *Amoebae*
# XVII. Different Morphodynamic Forms of Endocytosis
# and Microfilament Organization in *Amoeba proteus*

H. P. KLEIN, BÄRBEL KÖSTER, and W. STOCKEM*

Institute of Cytology, University of Bonn

Received January 14, 1988
Accepted April 26, 1988

Dedicated to Professor Dr. NOBURO KAMIYA on the occasion of his 75th birthday

## Summary

More than twenty substances (salts, amino acids, proteins, and others) were tested with the aim to induce different morphodynamic forms of endocytosis and to study the mode of membrane-microfilament interaction in *Amoeba proteus*. The chemical nature of all applied substances has a distinct influence on the size and shape of pinocytotic invaginations in that channels are formed ranging from $16 \pm 3\,\mu m$ (divalent cations) to $47 \pm 8\,\mu m$ (monovalent cations) in length and from $1.0 \pm 0.4\,\mu m$ (amino acids, proteins) to $6.2 \pm 1.3\,\mu m$ (tobacco mosaic virus, glycerol, ATP and others) in diameter. The addition of inert particles (indian ink) to inducing substances (proteins) gradually transforms channel-pinocytosis into food cup-phagocytosis and results in the formation of large endosomes with diameters of up to $50\,\mu m$.

The motive force for all dynamic activities during membrane invagination and vesiculation or for endosomal changes in shape and size is generated by a cortical microfilament system in close contact with the plasma membrane. The area of contact between the cortex and the cytoplasmic face of the membrane determines the diameter of pinocytotic channels and vacuoles, whereas the contraction degree correlates with the length of channels and number of vacuoles. The results of the present paper are discussed with respect to the general significance of microfilament-plasma membrane interactions for movement phenomena in amoebae.

*Keywords:* Chemical stimulation; Endocytotic activity; Morphodynamic changes; Membrane-microfilament interaction; *Amoeba proteus.*

*Abbreviations:* ADP adenosine-5'-triphosphate; Con A concanavalin agglutinin; DIC differential-interference-contrast; EGTA ethylene glycol-bis (β-aminoethylether) N,N, N', N'-tetracetate; PIBES 1,4-piperazine-N, N'-bis (2-ethanesulfonic acid); REM scanning electron microscope; TEM transmission electron microscope; TMV tabacco mosaic virus.

## 1. Introduction

Large amoebae have developed endocytotic activities of varying morphological appearance and intensity. During normal locomotion mono- and polypodial cells exhibit a continuous uptake of plasma membrane at the uroid region (WOLPERT and O'NEILL 1962). This process was described as permanent pinocytosis (WOHLFARTH-BOTTERMANN and STOCKEM 1966) and has a rather low internalization rate with a complete turnover of the cell surface every 12 hours (STOCKEM 1972, 1973). The addition of cationic substances to the culture medium transforms permanent pinocytosis in a concentration- and pH-dependent manner into induced pinocytosis (MAST and DOYLE 1934) by which up to 50% of the total plasma membrane area can be ingested within 15 to 30 min (CHAPMAN-ANDRESEN 1962, STOCKEM 1973). Besides permanent and induced pinocytosis many amoebae can also display phagocytotic activity by engulfing large food organisms such as flagellates and ciliates (JEON and BELL 1962, SERAVIN 1968). The question whether pinocytosis and phagocytosis in amoebae are basically similar or represent different mechanisms of cell surface ingestion arises controversial discussions in the literature (PRUSCH and

* Correspondence and Reprints: Institut für Cytologie und Mikromorphologie der Universität Bonn, Ulrich-Haberland-Strasse 61 a, D-5300 Bonn, 1, Federal Republic of Germany.

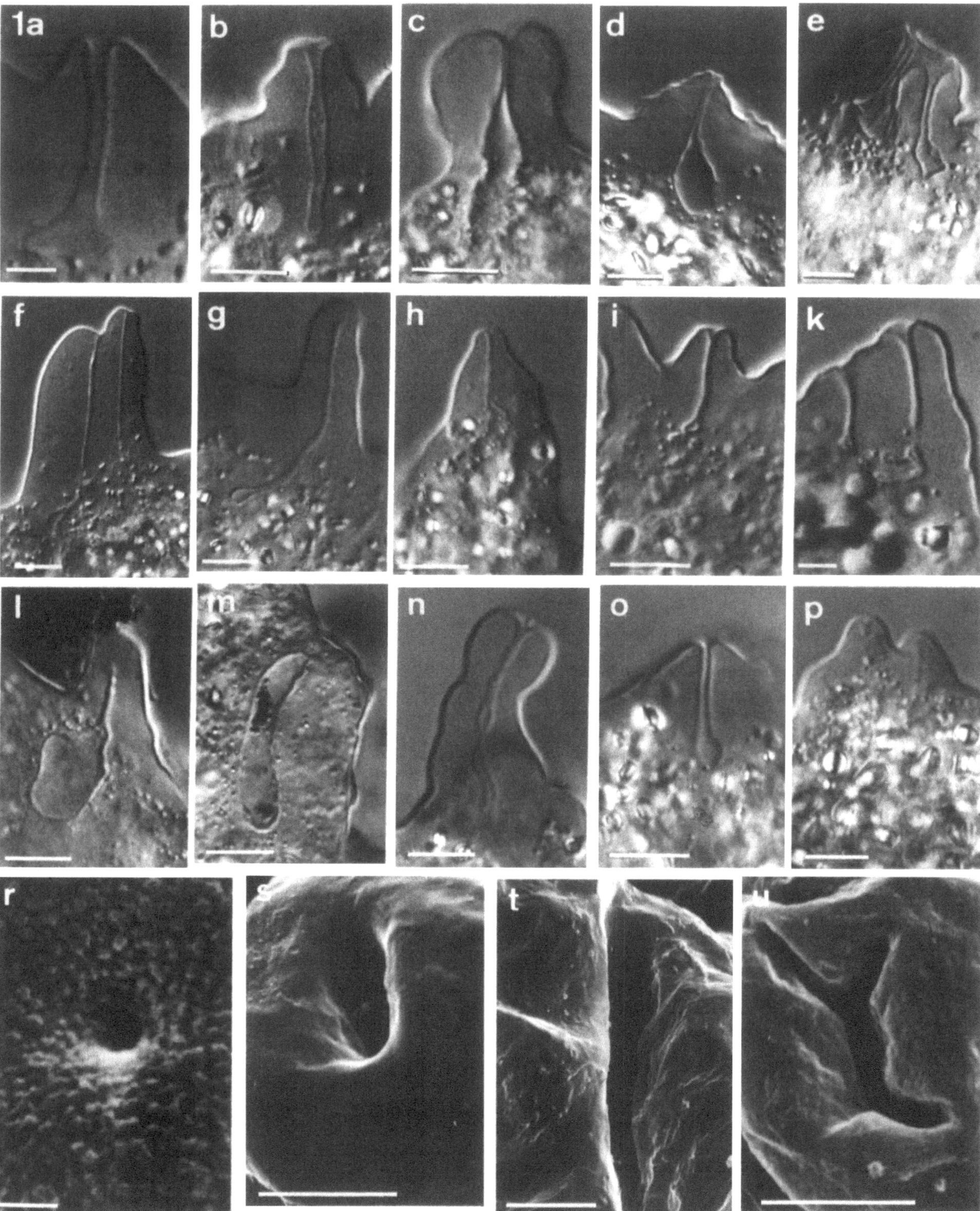

Fig. 1. DIC-light micrographs (*a–p*) and REM-micrographs (*r–u*) showing different morphodynamic forms of endocytotic channels in *A. proteus* after induction with glycerol (*a*), TMV (*b*), ATP (*c*), methionine (*d*), alcian blue (*e*), KCl (*f*), glutamic acid (*g*), NaCl (*h*), ribonuclease (*i*), egg albumin (*k*), ribonuclease/indian ink 50:1 (*l*), egg albumin/indian ink 50:1 (*m*), EGTA (*n*), Con A (*o*), $CaCl_2$ (*p*), egg albumin (*r*), ATP (*s–u*). Scales: *a–p* = 10 µm; *r* = 1 µm; *s–u* = 10 µm

MINCK 1985). Hence, in the present study different forms of pinocytotic and phagocytotic activity were experimentally induced and analyzed with the light and electron microscope to answer this question and to elucidate the organization and functional significance of the microfilament system for these processes.

## 2. Materials and Methods

### 2.1. Material

*Amoeba proteus* was cultured as described by HABEREY and STOCKEM (1971) and starved prior to experiments for 2–3 days in order to obtain cells of a comparable physiological state.

### 2.2. Inducing Substances

The following substances were used for the induction of pinocytosis: 0.0001% alcian blue in aqua bidest. (pH 8.0); 2 mM ATP with 1 mM $CaCl_2$, 3 mM $MgCl_2$, 30 mM KCl in 50 mM Tris-maleate-buffer (pH 7.0); 100 mM $CaCl_2$ in aqua bidest. (pH 6.5); 0.05% Con A in aqua bidest. (pH 6.8); 0.5% egg albumin in 10 mM acetate buffer (pH 4.2); 10 mM EGTA in aqua bidest. (pH 7.2); 125 mM glutamic acid in aqua bidest. (pH 8.0); 0.5% glycerol in aqua bidest. (pH 7.0); 100 mM $InCl_3$ in aqua bidest. (pH 6.5); 100 mM KCl in aqua bidest. (pH 6.5); 100 mM $LaCl_3$ in aqua bidest. (pH 6.5); 100 mM LiCl in aqua bidest. (pH 6.5); 100 mM $MnCl_2$ in aqua bidest. (pH 6.5); 125 mM methionine in aqua bidest. (pH 8.0); 100 mM NaCl in aqua bidest. (pH 6.5); 0.05% poly-L-lysine in aqua bidest. (pH 6.8); 0.02% ribonuclease in aqua bidest. (pH 6.5); 100 mM $SrCl_2$ in aqua bidest. (pH 6.5); 100 mM $TaCl_3$ in aqua bidest. (pH 6.5); tobacco-mosaic-virus (TMV) in aqua bidest. (pH 6.5); 20 mM theophyllin in aqua bidest. (pH 6.0).

Phagocytotic activity was induced by mixtures of 100 mM ribonuclease and indian ink (100:1, 75:1, 50:1, 25:1, 10:1 in V:V) or 0.5% egg albumin and indian ink at corresponding ratios of components, both dissolved in aqua bidest. at pH 8.0 and 4.5, respectively. All light micrographs were made at a Zeiss-photomicroscope I using a differential-interference-contrast equipment (DIC).

### 2.3. Electron Microscopy

For the preservation of microfilaments the following improved fixation solutions were employed:
(1) 1–2% glutaraldehyde in 50 mM s-collidine buffer (pH 7.0) containing 50 mM spermidine phosphate (HAUSER 1978) for 60 min, and
(2) 1–2% glutaraldehyde in 150 mM PIPES-buffer (pH 7.0) containing 0.17% $OsO_4$ and 0.2% phalloidin for 60 min; in some cases the amoebae were briefly rinsed in 0.05% phalloidin before fixation. Postfixation occurred with 0.1% $OsO_4$ in corresponding buffers for 10 min, and dehydration was carried out in ethanol (30%, 50%, 70%, 90%, and 100%); thereafter the specimens were embedded in low-viscosity epoxy resin (SPURR 1969). Ultrathin sections cut with an LKB-ultrotome III were stained with 1% lead citrate and 0.5% uranyl acetate and photographed with a Philips EM 301. For scanning electron microscopy some amoebae were rapidly fixed during intense pinocytosis with 2% $OsO_4$ and 0.5% $HgCl_2$ in aqua bidest. (PARDUCZ 1961) and critical-point-dried before shadowing with gold and carbon. Photographs were taken with a Cambridge Mark II REM.

## 3. Results

### 3.1. Morphodynamic Changes During Channel and Endosome Formation

The tested substances have a variable capacity to induce pinocytosis and cause different patterns of channel distribution at the cell surface (Fig. 6).

Weak-inducing substances such as divalent cations, glycerol, alcian blue or methionine achieve a distinct intensification of permanent pinocytosis at the uroid region in conjunction with a stop of active locomotion (Fig. 6 b). Substances with a medium inducing capacity (glutamic acid, Con A or ATP) can obviously give rise to the formation of endocytotic channels in limited cell body regions besides the uroid, but in contrast to strong-inducing substances such as monovalent and trivalent cations or proteins (Fig. 6 c) these probes never produce invaginations at the total cell surface area.

Distinct variations due to the chemical nature of the tested substances can also be observed with respect to the form and size of pinocytotic channels (Fig. 1). Long (41.0 ± 10 µm) and extremely broad channels (6.2 ± 1.3 µm) were found to be raised by a rather heterogeneous group of substances comprehending glycerol, TMV, ATP and methionine (Fig. 1 a–d). According to REM-studies the profile of the broad channels is not always circular (Fig. 1 s) but can be slit-like (Fig. 1 t) or even heteromorph (Fig. 1 u).

Another channel-type exhibiting a similar length (47.0 ± 8 µm) but a very narrow diameter (1.0 ± 0.4 µm) is displayed after the application of monovalent (KCl, NaCl, LiCl) and trivalent cations ($InCl_3$, $LaCl_3$), glutamic acid or proteins such as ribonuclease and egg albumin (Fig. 1 f–k, r).

Pinocytotic channels of a medium dimension measuring 33.0 ± 5 µm in length and 2.8 ± 0.1 µm in diameter are induced by substances of again different chemical nature, namely alcian blue, EGTA or Con A (Fig. 1 e, n, o), whereas extremely short (16.0 ± 3 µm) and thin

Table 1. *Changes in channel diameter during the optimum phase of egg albumin-induced pinocytosis as a function of indian ink concentration*

| Egg albumin: indian ink in V:V | Channel diameter (µm) |
| --- | --- |
| Pure egg albumin | 1.5 ± 0.3 |
| 100:1 | 3.7 ± 0.5 |
| 75:1 | 5.0 ± 0.4 |
| 50:1 | 7.4 ± 0.6 |
| 25:1 | 11.1 ± 1.1 |
| 10:1 | 22.3 ± 1.7 |

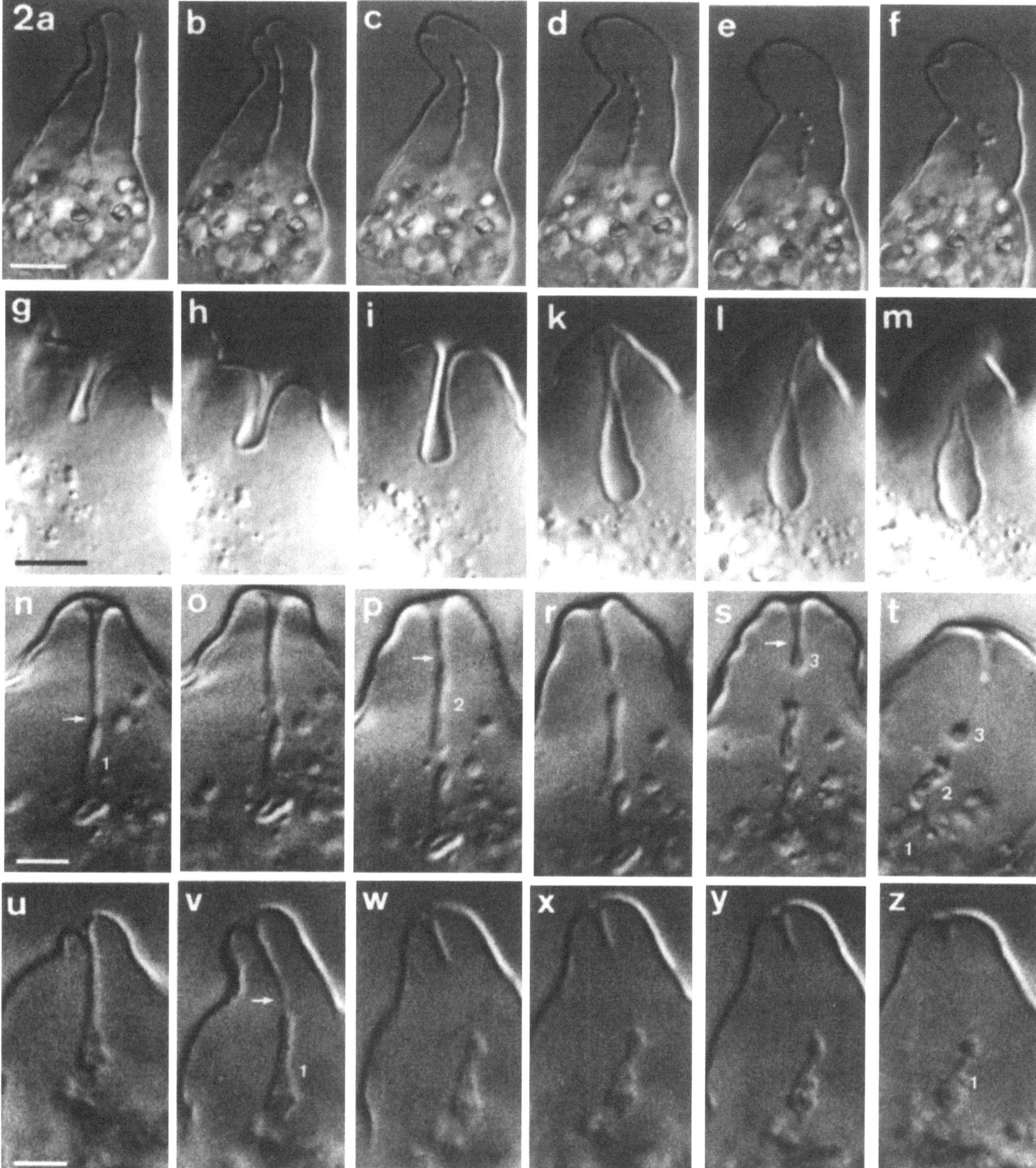

Fig. 2. DIC-light micrographs showing induced pinocytosis in *A. proteus*. *a–f* Induction by ribonuclease results in the formation of long, thin channels and numerous small endosomes. *g–m* Addition of indian ink particles to ribonuclease (1 : 25 V/V) causes a distinct increase in the diameter of the channel and formation of only one large endosome. *n–t* Disintegration (arrows) of a pinocytosis channel into single endosomes (*1, 2, 3*) occurs from the base to the tip. *u–z* Isolated (arrow) endosomes (*1*) show distinct changes in size and shape. Time intervals: *a–f* 4 min; *g–m* 3.5 min; *n–t* 2.5 min; *u–z* 2.5 min; scales: 10 µm

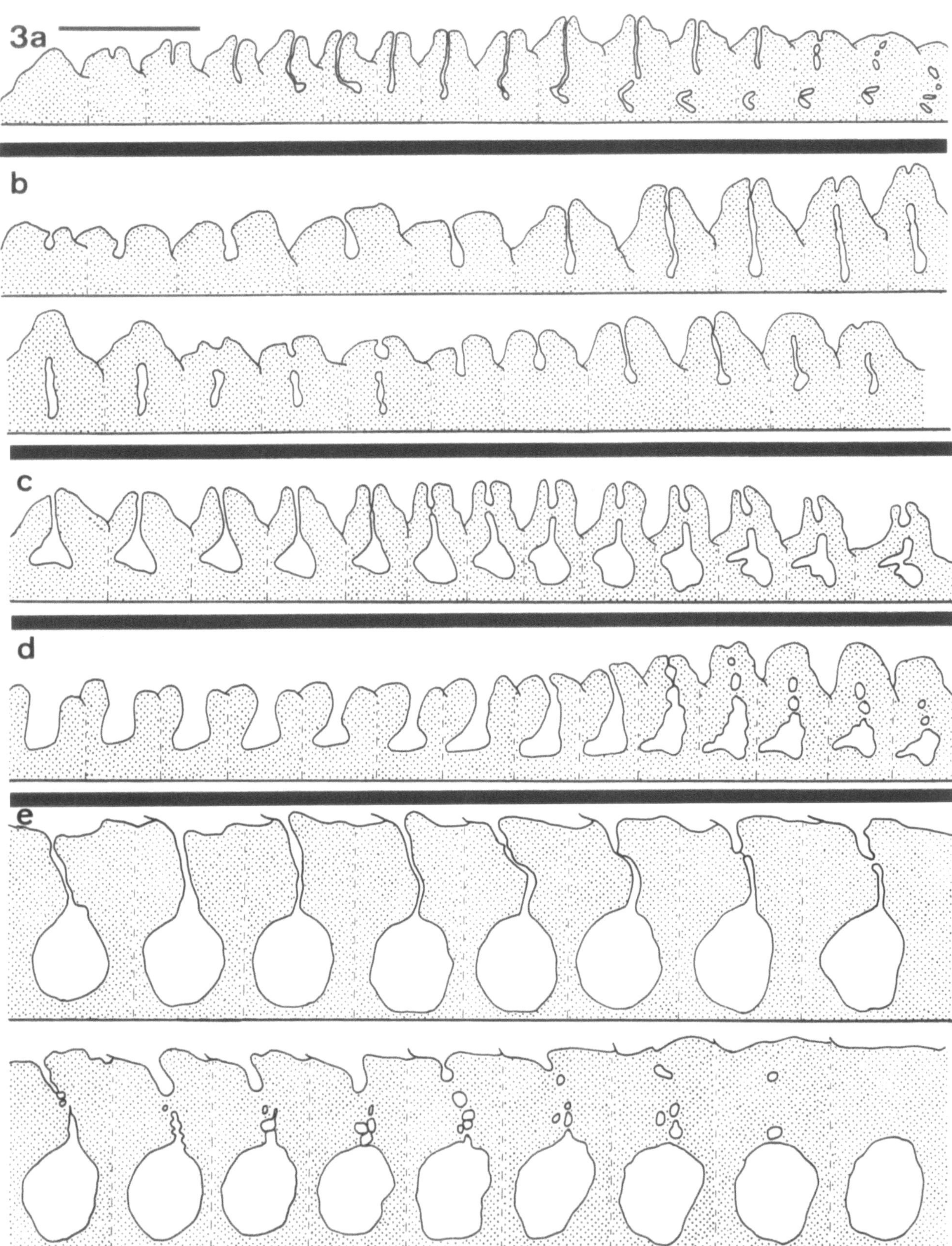

Fig. 3. Single frame sequences to demonstrate morphodynamic changes in the formation of pinocytosis channels and endosomes in reaction to the addition of indian ink particles to different volumes of egg albumin. *a* Egg albumin, *b* egg albumin/indian ink 100:1, *c* egg albumin/ indian ink 50:1, *d* egg albumin/indian ink 25:1, *e* egg albumin/indian ink 10:1. Time intervals: *a* 6.5 min; *b* 4.5 min; *c* 2.5 min; *d* 3 min; *e* 7 min; scale: 50 μm

invaginations ($1.0 \pm 0.4\,\mu m$) were found to arise upon the influence of very weak-inducing substances as represented by divalant cations ($CaCl_2$, $MnCl_2$, $SrCl_2$) (see also Fig. 1 *p*).

The diameter of pinocytotic channels (Table 1) and, thus of the resulting endosomes can be experimentally changed by the addition of inert particles to inducing solutions, *e.g.* by mixtures of indian ink and egg albumin (Fig. 3) or ribonuclease (Fig. 2 *a–m*) at different ratios of components.

Both, photographs and measured values demonstrate that the size of the channels clearly increases with the external concentration of indian ink particles in a rather linear manner until stages are induced which resemble the phagocytotic food-cup formation as caused by prey-organisms (Figs. 1 *l*, *m*, 2 *m*, and 3 *e*). Whereas the size of endosomes arising during experimentally-changed endocytotic activity increases in the same way as the diameter of channels, the number of vacuoles is distinctly reduced at higher particle concentrations until only one large endosome is formed (Fig. 6 *e*).

The fate of endosomes pinched off from pinocytotic channels as followed by cinematografic single frame analysis is characterized by some further morphodynamic events: Independent from the nature of the vacuolar content most primary endosomes start to disintegrate into several secondary ones of a more or less uniform dimension (Fig. 4). Subsequently, the secondary endosomes undergo characteristic changes in size and shape (Figs. 2 *u–z* and 5) in that the diameter continuously decreases within the first few minutes (Table 2).

Whereas the volume loss of secondary endosomes may be mainly due to an osmotic flow of water from the exoplasmic vacuolar compartment into the surrounding cytoplasm, the distinct changes in shape are rather caused by the same contractile system which is also responsible for channel and endosome formation.

### 3.2. Motive Force Generation and Microfilament Organization During Channel and Endosome Formation

When pinocytosis is induced a distinct electron-dense cortical layer appears along the entire cell surface (Fig. 8 *e*). This layer is in close contact with the internal face of the plasma membrane and consists of a three-dimensional, unordered meshwork of 6 nm-microfilaments (Fig. 8 *g*). During pinocytosis the filament layer detaches in places from the plasma membrane (Fig. 8 *f*) with the exception of a central region of fast contact (Figs. 7 *a–c* and 8 *d*). As a consequence of this detach-

Table 2. *Time-dependent changes in endosome volume during egg albumin-induced pinocytosis*. The volume was measured morphometrically by determining the volumetric density

| Time after endosome formation in sec | Volume decrease of endosomes in % |
|---|---|
| 0 | 100 |
| 25 | $98.4 \pm 0.7$ |
| 50 | $96.8 \pm 0.8$ |
| 75 | $95.2 \pm 1.0$ |
| 100 | $93.4 \pm 1.0$ |
| 125 | $92.4 \pm 0.9$ |
| 150 | $90.8 \pm 1.2$ |
| 200 | $86.1 \pm 0.8$ |

ment and further contraction groundplasm is extruded into the cell periphery while the cell organelles are retained by the filament layer (Figs. 7 *c* and 8 *f, h*). Thereby, the characteristic hyaline pseudopodia originate with a central pinocytotic channel (Fig. 8 *a–c*). The diameter of the channels and, consequently, of the formed endosomes depends on the extend of the central contact region, *i.e.* small areas of contact between the microfilament layer and the internal face of the plasma membrane result in small channels and endosomes, and vice versa.

According to combined light and electron microscopical investigations channel formation proceeds discontinuously in that the alternate process of detachment and reformation of filament layers in conjunction with the elongation of the hyaline pseudopodium and the central plasma membrane invagination is repeated several times (Fig. 7 *d*). The result of this "step by step" formation is a multitude of stratified filament layers gradually differing in age and clearly visible under optimum conditions in both, living (Fig. 8 *a–c*) and fixed specimens (Fig. 8 *d, h*). The disintegration of pinocytotic channels into single endosomes always occurs at places where the stratified microfilament system comes into contact with the membrane of the channel, *i.e.* the number of detachment and reformation cycles which the filament layer undergoes determines not only the length of the channels but also the number of endosomes. As a rule, endosomes are always formed from the base to the tip of the channels (Fig. 2 *n–t*).

The ultrastructural organization of the microfilament system along the membrane of the channels and isolated endosomes clearly differs (Fig. 9). Microfilaments surrounding endocytotic channels during the elongation phase as an electron-dense sheath (Fig. 9 *c*) exhibit a strictly parallel orientation to the longitudinal axis

(Fig. 9 *a*); this points to isometric stress forces existing between the base and the tip of the channels. On the other hand, microfilaments surrounding the membrane of endosomes are always more or less disordered as characteristic for isotonic conditions. The meshwork shows, however, various degrees of density which obviously correlate with different stages of contraction (Fig. 9 *b, d*).

## 4. Discussion

The endocytotic uptake of substances from the external environment into the cytoplasm is a widespread physiological process in the animal kingdom (SILVERSTEIN *et al.* 1977). Endocytosis includes pinocytosis and phagocytosis, *i.e.* the incorporation of fluids or solutes and the ingestion of large particles, respectively (ALBERTS *et al.* 1983). Whereas most metazoan cells are capable of exhibiting fluid-phase or receptor-mediated pinocytosis, in higher eucaryotic organisms only a few specialized white blood cells such as macrophages and neutrophil leucocytes are able to phagocytose. Amongst protozoan cells large amoebae probably have developed the most extensive diversity of endocytotic mechanisms (BENNET 1969, STOCKEM 1977). Besides phagocytotic capture of living prey-organisms as the predominant form of feeding (MAST and ROOT 1916, JEON and BELL 1962, JEON and JEON 1983), active locomoting amoebae show also a permanent pinocytotic uptake of membrane at the uroid region; apart of this, under experimental conditions, *i.e.* upon the external application of induction substances this pinocytotic activity can be intensified in limited regions or extended over the total cell surface area (HOLTER 1965).

According to recent investigations, permanent pinocytosis obviously represents the more physiological process for the uptake of molecules and small living cells such as bacteria, whereas induced pinocytosis rather serves as a survival mechanism to clean the contaminated cell surface by the uptake of the membrane-bound material and its eventual successive defecation or cellular budding (KUKULIES *et al.* 1986, CHRISTOFIDOU 1987). Nevertheless, chemically-induced pinocytosis in *A. proteus* proved to be a very suitable model system to study the participation of different physiological parameters in endocytosis under controlled experimental conditions. According to conformable results external calcium ions play a predominant role for the induction of pinocytosis in that surface-bound calcium is displaced by inducing substances (JOSEFSSON 1975, PRUSCH and HANNAFIN 1979, KUKULIES *et al.* 1986). Subsequent changes in the permeability of the plasma membrane and distinct alterations in the ionic equilibrium activate the cortical microfilament system to generate motive force for pinocytotic membrane flow (KLEIN and STOCKEM 1979, TAYLOR *et al.* 1980).

In a recent study PRUSCH and MINCK (1985) demonstrated that calcium serves also as an essential parameter for the induction of optimal phagocytosis in *A. proteus*, and that according to observations of other authors (JEON and JEON 1983, STOCKEM *et al.* 1983) the cortical microfilament system is again responsible for the production of dynamic forces necessary to perform phagocytotic food-cup formation. These conformities are against the opinion that pinocytosis and phagocytosis represent different mechanisms (WILLS *et al.* 1972, KLAUS 1973) which are also controlled by different sources of energy (COHN 1966). Moreover, the possibility to transform channel-pinocytosis into food-cup phagocytosis by combined chemical and simple mechanical stimulation supports the suggestion of BRANDT and PAPPAS (1960) according to which "both processes represent a continuum between extremes of the same cellular function" (PRUSCH and MINCK 1985). The results of the present paper clearly demonstrate that pinocytosis and phagocytosis are both based on the same modified process of pseudopodium formation thus representing largely identical forms of endocytotic membrane incorporation. A circular localized detach-

---

Fig. 4. Single frame sequence to demonstrate the disintegration of primary endosomes into secondary ones after induction of pinocytosis by a mixture of egg albumin and indian ink. Time interval: 2 min; scale: 30 μm

Fig. 5. Single frame sequence to demonstrate changes in size and shape of a single endosome. The first frame compares the initial (blank) with the final stage (hatched). Time interval: 33 min; scale: 10 μm

Fig. 6. Schematic drawing to compare different morphodynamic forms of endocytosis in *A. proteus*. *a* Permanent pinocytosis, *b* intensified permanent pinocytosis, *c* induced pinocytosis, *d* induced pinocytosis modified by low concentrations of inert particles, *e* induced pinocytosis intensified to food-cup phagocytosis by high concentrations of inert particles

Fig. 7. *a–d* Schematic drawing to explain the mode of pinocytotic pseudopodium and channel formation by the successive detachment and reformation of different microfilament layers. *1, 2, 3* Stratified microfilament layers of decreasing age. *Gp* granuloplasm with cell organelles; *Hp* hyaloplasm (ground plasm)

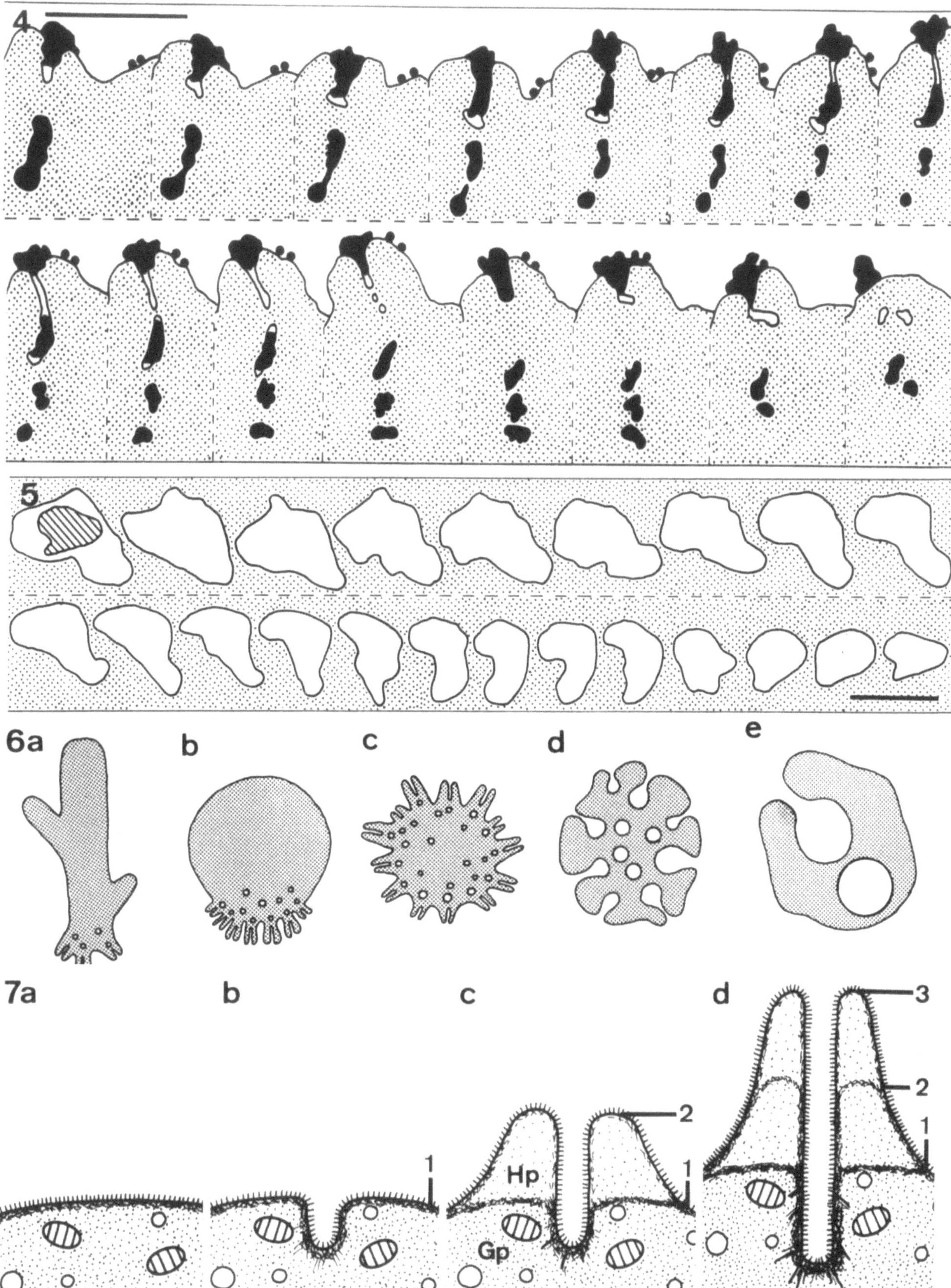


Figs. 4–7

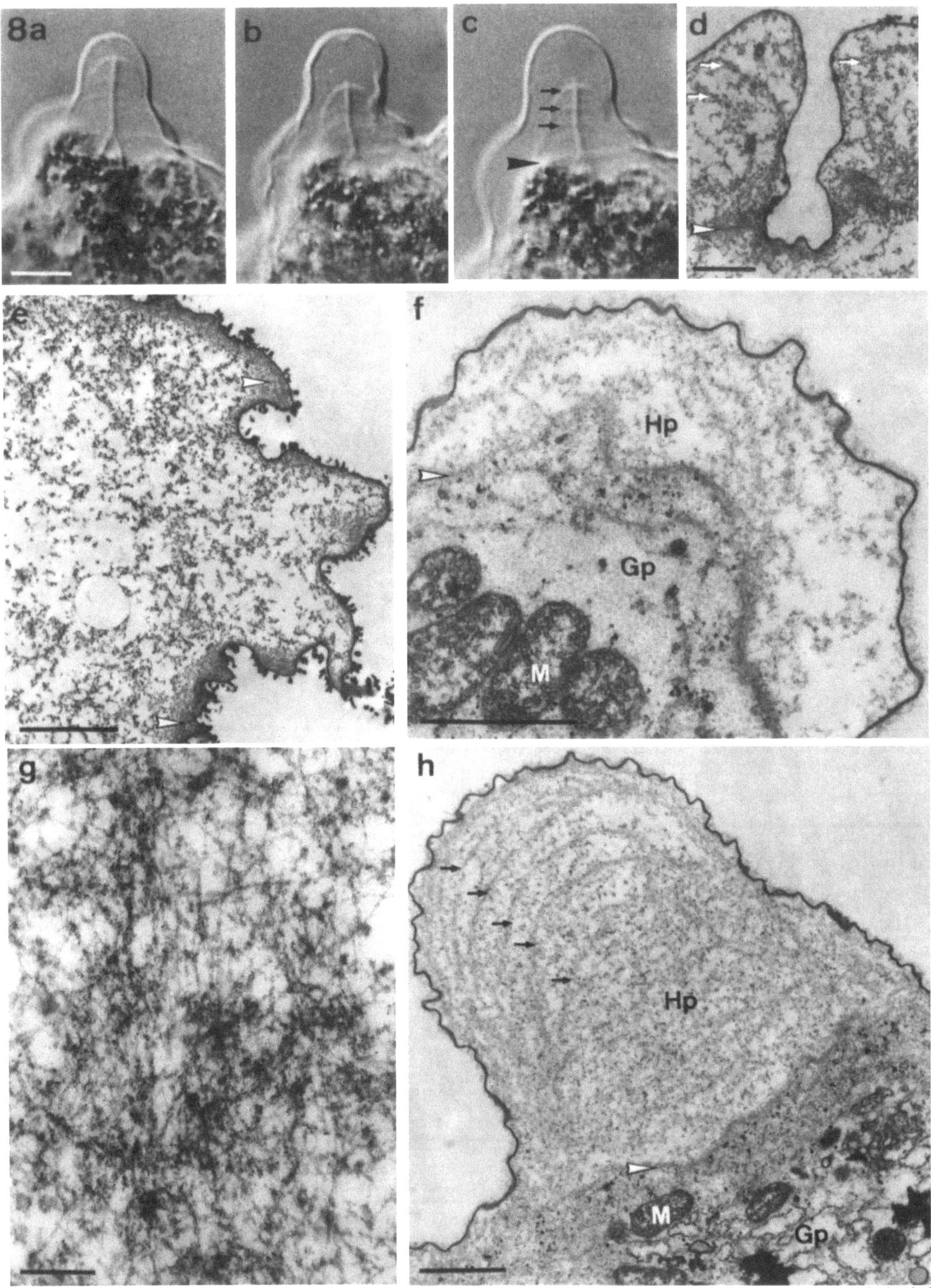

Fig. 8. DIC-light micrographs (*a–c*) and TEM-micrographs (*d–h*) to demonstrate the organization of the microfilament system during induced pinocytosis. The arrows in *c, d,* and *h* point to the stratified microfilament layers after successive detachment from the plasma membrane; the arrowheads in *e* mark the original microfilament layer beneath the plasma membrane immediately after induction of pinocytosis; the arrowheads in *c, d, f,* and *h* indicate the position of the detached original microfilament layer at the border between hyaloplasm (*Hp*) and granuloplasm (*Gp*); *g* higher magnification to show the low degree of order of microfilaments in the membrane attached cortical layer; *f* and *h* pinocytotic pseudopodia cut at a more tangential position so that the channel is not visible; *M* mitochondria. Scales: *a–c* 25 µm; *d* 5 µm; *e* 1 µm; *g* 0.2 µm; *f, h* 3 µm

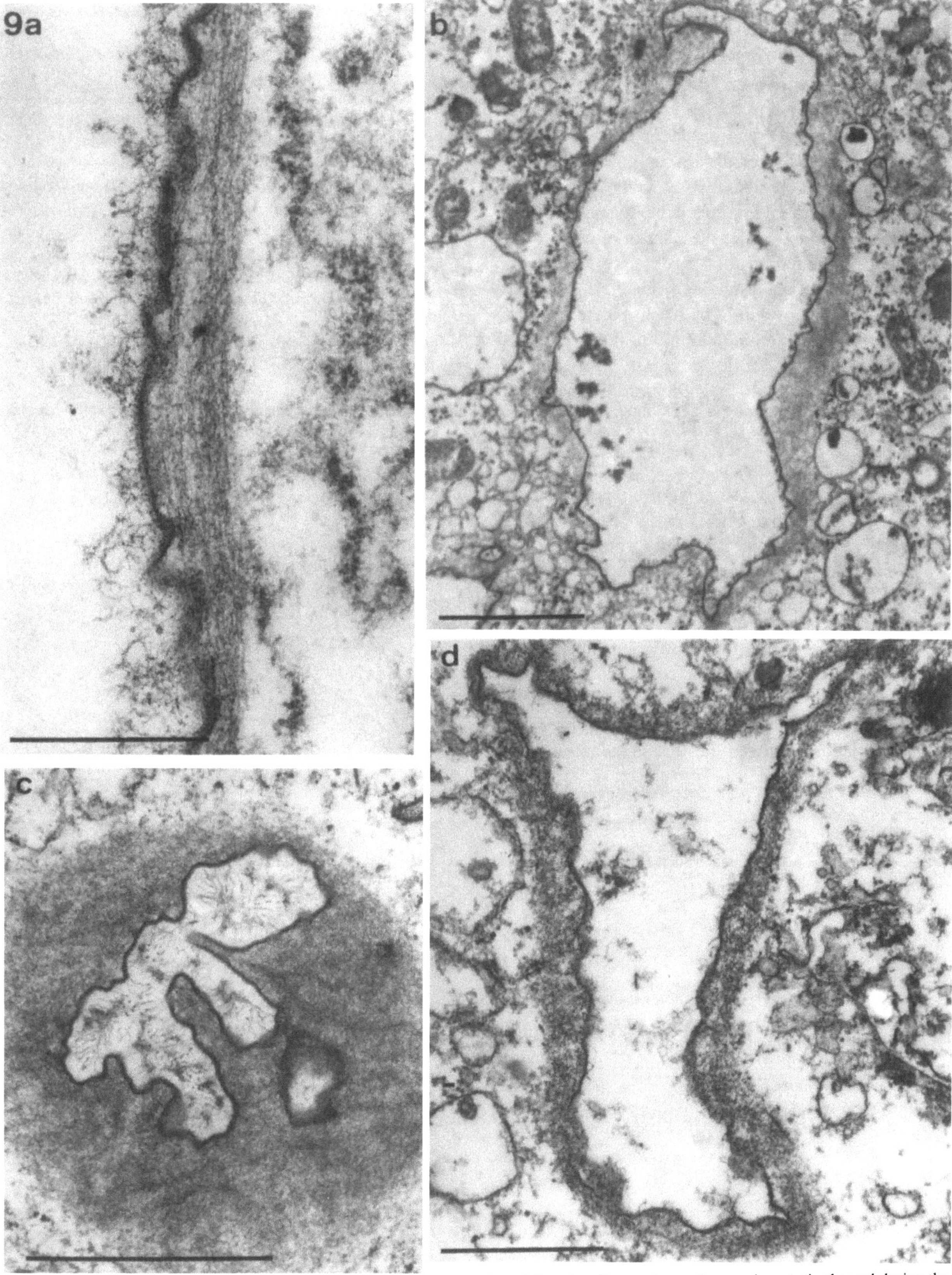

Fig. 9. TEM-micrographs to demonstrate the ultrastructural organization of the microfilament layer along a pinocytotic channel during the elongation phase in a longitudinal (*a*) and cross section (*c*); single endosomes (*b, d*) are surrounded by a microfilament layer showing different degrees of condensation. Scales: *a* 0.5 μm; *b* 5 μm; *c* 1 μm; *d* 2 μm

ment of the microfilament system from the plasma membrane and a more or less extensive area of fast contact within the detached region determine the size of the invaginations and resulting endosomes and, hence, crucially control the diversity of endocytotic manifestations in *A. proteus*.

A still unknown question is related to the physiological significance and molecular mechanism of chemical or mechanical stimulation for the different endocytotic variations. There are strong experimental indications that not only pinocytosis but also phagocytosis can be induced by chemical stimulation. However, chemical inducers of phagocytosis such as prostaglandins (STOCKEM *et al.* 1983) or gluthation (PRUSCH and MINCK 1985) fail to initiate pinocytosis in the same manner as the cationic inducers of pinocytosis are not able to initiate phagocytotic activity. On the other hand, simple mechanical stimulation alone is never sufficient to produce pinocytosis or phagocytosis in *A. proteus* except when combined with chemical induction as demonstrated in this paper. Future investigations are needed to elucidate this apparent contradictions concerning the initial events in pinocytosis and phagocytosis. In this connection it will be important to identify the responsible external receptor sites for both processes and to clarify the trigger mechanism which controls the mode and degree of contact between the microfilament system and the plasma membrane.

## Acknowledgements

This work was supported by a grant from the Deutsche Forschungsgemeinschaft (Sto 126/4-2).

## References

ALBERTS B, BRAY D, LEWIS J, RAFF M, ROBERTS K (1983) Molecular biology of the cell. Garland, New York London

BENNETT H ST (1969) The cell surface: movements and recombinations. In: LIMA-DE-FARIA A (ed) Handbook of molecular cytology. North-Holland, Amsterdam London, pp 1294–1319

BRANDT PW, PAPPAS GD (1960) An electron microscopic study of pinocytosis in amoeba. I. The surface attachment phase. J Biophys Biochem Cytol 8: 675–687

CHAPMAN-ANDRESEN C (1962) Studies on pinocytosis in amoeba. Compt Rend Trav Lab Carlsberg 33: 73–264

CHRISTOFIDOU M (1987) Cytologische Untersuchungen über den Ablauf des Endomembrantransportes bei *Amoeba proteus*. Dipl Arbeit, Math Nat Fak, University Bonn, Federal Republic of Germany

COHN ZA (1966) The regulation of pinocytosis in mouse macrophages: I. Metabolic requirements as defined by the use of inhibitors. J Exp Med 124: 557–571

HABEREY M, STOCKEM W (1971) *Amoeba proteus*: Morphologie, Zucht und Verhalten. Mikrokosmos 60: 33–42

HAUSER M (1978) Demonstration of membrane-associated and oriented microfilaments in *Amoeba proteus* by means of a Schiff base/glutaraldehyde fixative. Cytobiologie 18: 95–106

HOLTER H (1965) Physiologie der Pinocytose bei Amöben. In: WOHLFARTH-BOTTERMANN KE (ed) Sekretion und Exkretion. Springer, Berlin Heidelberg New York, pp 119–146

JEON KW, BELL LGE (1962) Pseudopod and foodcup formation in *Amoeba proteus*. Exp Cell Res 27: 350–352

— JEON MS (1983) Generation of mechanical forces in phagocytosing amoeba: light and electron microscopic study. J Protozool 30: 536–538

JOSEFSSON JO (1975) Studies on the mechanism of induction of pinocytosis in *Amoeba proteus*. Acta Physiol Scand [Suppl] 423: 1–65

KLAUS GG (1973) Cytochalasin B. Dissociation of pinocytosis and phagocytosis by peritoneal macrophages. Exp Cell Res 79: 73–78

KLEIN HP, STOCKEM W (1979) Pinocytosis and locomotion of Amoeba. XII. Dynamics and motive force generation during induced pinocytosis in *A. proteus*. Cell Tissue Res 197: 263–279

KUKULIES J, ACKERMANN G, STOCKEM W (1986) Pinocytosis and locomotion in Amoebae. XIV. Demonstration of two different receptor sites on the cell surface of *Amoeba proteus*. Protoplasma 131: 233–243

MAST SO, DOYLE WL (1934) Ingestion of fluids by *Amoeba*. Protoplasma 20: 553–560

— ROOT FM (1916) Observations on amoeba feeding on rotifers, nematodes and ciliates, and their leaning on the surface-tension theory. J Exp Zool 21: 33–49

PARDUCZ B (1961) Eine neue Schnellfixierung im Dienste der Protistenforschung und des Unterrichts. Annu Mus Nat Hung 2: 5–15

PRUSCH RD, HANNAFIN J (1979) Sucrose uptake by pinocytosis in *Amoeba proteus* and the influence of external calcium. J Gen Physiol 74: 523–535

— MINCK DR (1985) Chemical stimulation of phagocytosis in *A. proteus* and the influence of external calcium. Cell Tissue Res 242: 557–564

SERAVIN LN (1968) The role of mechanical and chemical stimulators on the induction of phagocytotic reactions in *Amoeba proteus* and *Amoeba dubia*. Acta Protozool 6: 97–107

SILVERSTEIN SC, STEINMAN RM, COHN ZA (1977) Endocytosis. Ann Rev Biochem 46: 669–722

SPURR AR (1969) A low-viscosity epoxy resin embedding medium for electron microscopy. J Ultrastruct Res 26: 31–43

STOCKEM W (1972) Membrane-turnover during locomotion of *Amoeba proteus*. Acta Protozool 11: 83–93

— (1973) Pinocytose und Bewegung von Amöben: IV. Mitteilung: Quantitative Untersuchungen zur permanenten und induzierten Endocytose von *A. proteus*. Z Zellforsch 136: 433–446

— (1977) Endoytosis. In: JAMIESON GA, ROBINSON DM (eds) Mammalian cell membranes. Responses of plasma membranes, vol 5. Butterworths, London, pp 151–195

— HOFFMANN HU, GRUBER B (1983) Dynamics of the cytoskeleton in *Amoeba proteus*. Redistribution of microinjected fluorescein-labeled actin during locomotion, immobilization and phagocytosis. Cell Tissue Res 232: 79–96

Taylor DL, Blinks JR, Reynolds S (1980) Contractile basis of ameboid movement. VII. Aequorin luminescence during ameboid movement, endocytosis and capping. J Cell Biol 86: 599–607

Wills EJ, Davies P, Allison AC, Haswell AD (1972) Cytochalasin B fails to inhibit pinocytosis by macrophages. Nature (New Biol) 240: 58–60

Wohlfarth-Bottermann KE, Stockem W (1966) Pinocytose und Bewegung von Amöben. II. Permanente und induzierte Pinocytose bei *Amoeba proteus*. Z Zellforsch 73: 444–474

Wolpert L, O'Neill (1962) Dynamics of the membrane of *A. proteus* studied with labelled specific antibody. Nature 196: 1261–1266

Protoplasma (1988) [Suppl. 2]: 88–94



# Diffusion of Substances in the Cytoplasm and Across the Nuclear Envelope in Egg Cells

YUKIO HIRAMOTO* and ISAMU KANEDA

Biological Laboratory, Tokyo Institute of Technology, Tokyo

Received February 8, 1988
Accepted March 19, 1988

Dedicated to Professor Dr. NOBURO KAMIYA on the occasion of his 75th birthday

## Summary

The diffusion of substances in the cytoplasm and across the nuclear envelope was measured by recording changes in the fluorescence distribution in the protoplasm in sand dollar eggs and starfish and mouse oocytes microinjected with fluorescent substances. The diffusion rates of fluorescein (M.W.: 408.5) and FITC-dextran (M.W.: 39,000) in sand dollar egg cytoplasm were 20–30%, respectively of the diffusion rates of these substances in water. In starfish oocytes, FITC-dextran with a molecular weight of 9,000 passed through the nuclear envelope, while FITC-dextran with a molecular weight of 39,000 did not. In mouse oocytes, FITC-dextran with a molecular weight of 64,400 passed through the nuclear envelope, while FITC-dextran with a molecular weight of 156,900 did not.

*Keywords:* Diffusion rate in cytoplasm; Permeability of nuclear envelope; Fluorescence microscopy; Microinjection; Egg cells.

## 1. Introduction

Quantitative investigations on the transport of substances in the protoplasm are important in cell biology, because intracellular substances play their roles at the sites specific to them in the protoplasm. The protoplasm is considered to be a complex system consisting of a liquid phase (cytosol) in which various organic and inorganic substances are dissolved, cytoskeletal structures responsible for visco-elastic properties (*cf.* CRICK and HUGHES 1951, HIRAMOTO 1969, 1987, SATO *et al.* 1983) and cell organelles. Some substances are transported in the protoplasm by diffusion and others are by active processes involving energy supply.

In the present study, we determined the diffusion rates of some substances in the protoplasm by fluorescence microscopy using fluorescent substances microinjected into the cell as probes. We also determined the permeability of the nuclear envelope by observing the behaviour of fluorescently labeled molecules of various sizes microinjected into the cytoplasm or the nucleus.

## 2. Materials and Methods

### 2.1. Materials

Unfertilized eggs were obtained from the sand dollar, *Clypeaster japonicus* by injecting seawater containing 1 mM acetyl choline into the body cavity. Eggs were washed three times in artificial seawater (ASW). In fertilized eggs, the fertilization membrane and the hyaline layer were removed by treating the eggs with 1 M urea solution for 1–2 minutes shortly after insemination and then kept in ASW. Oocytes of the starfish, *Asterina pectinifera* were released into ASW by cutting the ovaries with scissors, and washed twice in ASW. Oocytes of the randomly bred ICR mouse, *Mus musculus* were released by cutting the ovaries with scissors into phosphate buffer solution (PBS) containing 0.8 g NaCl, 0.02 g KCl, 1.15 g $Na_2HPO_4$, 0.1 g $CaCl_2$, and 0.1 g $MgCl_2 \cdot 6H_2O$ per l (pH 7.4–7.45) supplemented with 25 µM dibutylyl cyclic AMP to prevent the breakdown of germinal vesicles.

### 2.2. Principles of Measurements

We tried to determine the diffusion coefficient of the substance in the cytoplasm by comparing quantitatively the diffusion rate of the substance microinjected in the cytoplasm with the diffusion rate of the same substance injected in PBS. By using fluorescent substances as probes, the change in the spatial distribution of the substance injected in the cytoplasm or in PBS could be recorded with a video system connected with a fluorescence microscope. If the distribution pattern of the substance in the cytoplasm at $t_1$ after the microinjection is the same as that in PBS at $t_n$ after the injection, it may be concluded

---

* Correspondence and Reprints: Biological Laboratory, University of the Air, Wakaba, Chiba 260 Japan

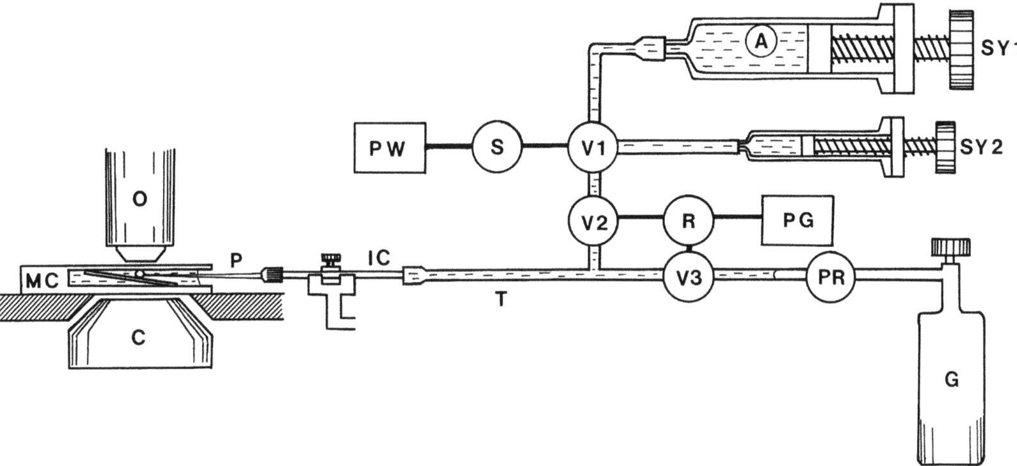

Fig. 1. Apparatus for pressure microinjection. *A* Air bubble, *C* condenser lens, *G* gas bomb, *IC* instrument collar, *MC* manipulation chamber, *O* objective lens, *P* micropipette, *PG* pulse generator, *PR* pressure regulator, *PW* electric power, *R* relay, *S* switch, *SY 1* and *SY 2* syringes, *T* polystylene tubing, *V 1, V 2,* and *V 3* electromagnetic valves, For explanation, see text

that the diffusion coefficient of this substance in the cytoplasm is $t_n/t_l$ times the diffusion coefficient in PBS, provided that the distribution pattern of the substance in the cytoplasm is measured shortly after the microinjection when the concentration of the injected substance near the cell surface is negligibly small.

In order to determine the size of the molecule which passes through the nuclear envelope, we microinjected fluorescein-isothiocyanate-labeled dextran (FITC-dextran) with various molecular weights into the cytoplasm or the nucleus, and checked whether or not the injected molecules passed through the nuclear envelope into or out of the nucleus after a definite time considered to be sufficiently large for thorough diffusion of the injected molecules within the cytoplasm or the nucleus.

### 2.3. Microinjection

Aqueous solutions of fluorescein (K-salt, M.W.: 408.5) and FITC-dextran (FD-4, FD-10 s, FD-40, FD-70 s, and FD-150, Sigma, St. Louis) were used for microinjection. Molecular weights of the above FITC-dextran are 4,100, 9,000, 39,000, 66,400, and 156,900, respectively. FITC-dextran solutions to be injected were prepared so that their concentrations of FITC may be 1 or 0.5 mM.

We used a pressure microinjection apparatus in which the solution can be microinjected within a fraction of second. As shown in Fig. 1, pressure can be applied to the micropipette (P) from three different pressure sources, a gas bomb (G) connected with pressure regulator (PR), large syringe (SY 1) with an air bubble (A) in it and small syringe (SY 2). The plunger of each syringe can be moved by a screw device for precise control of the pressure. These pressure sources are connected with polystylene tubing (T) to the instrument collar (IC) holding the micropipette via electromagnetic valves (V 1, V 2, and V 3). The inside of the tubing (T) and instrument collar (IC) is filled with water except for a part close to the pressure regulator (PR). The opening and closing of V 1 are controlled by a manual switch (S) and V 2 and V 3 are operated by switches in a relay (R) controlled by pulse generator (PG).

The solution to be microinjected was loaded in the micropipette (P) from the tip by a negative pressure generated by SY 2. When an

appropriate amount of solution was loaded in the micropipette, the pressure inside the injection system was adjusted by SY 1 so that the solution may neither flow out of nor flow into the micropipette through the tip. In the case of microinjection, a high pressure (2–5 kg/cm$^2$) from the pressure regulator (PR) was applied to the injection system for a definite duration (10 to 600 msec) pre-determined by the pulse generator (PG). In experiments to determine the diffusion coefficient, the duration of microinjection was 10–30 msec because the duration should be sufficiently short as compared with the time after the microinjection when the spacial distribution patterns of the injected substance was recorded. Immediately after the high pressure was applied, the pressure system was connected to SY 1 which stopped the flow of the solution from the micropipette. The duration of the application of high pressure was adjusted so that the volume of the injected solution may be 0.5 to 20 pl.

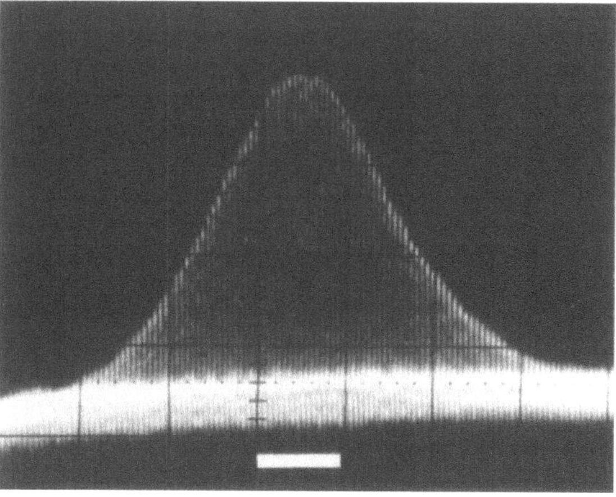

Fig. 2. Intracellular distribution of fluorescence intensity of FITC-dextran microinjected into a sand dollar egg. The distribution of light intensity is shown by the oscillographic record of the output voltage of the video monitor. Bar corresponds to 10 μm on the specimen plane. The diameter of the egg is about 120 μm

Cells were fixed in position in a wedge-shaped space in manipulation chamber (MC) devised by Hiramoto (*cf.* Hiramoto 1971, Kishimoto 1986). Manipulation of the micropipette was made with an oil-pressure controlled micromanipulator (MO-102, Narishige, Tokyo).

### 2.4. Recording and Measurement

In experiments to determine the diffusion coefficients of substances in the cytoplasm, fluorescence patterns of the cell at various times after the microinjection of fluorescent substances were recorded with video system consisting of a high-sensitivity video camera (Panasonic WV-1900), a video-type recorder (Panasonic AG-6010) and a video-monitor (Panasonic WV-5360) connected with a fluorescence microscope (Nikon S equipped with EFD 2).

The spatial distribution of fluorescence intensity across a line passing through the point of microinjection was displayed on the CRT of an oscilloscope (LBO-524, Leader, Tokyo) connected to the output of the video-tape recorder during playing back of the record (*cf.*, Fig. 2). The fluorescence distribution pattern was recorded in the cell at an early time (2–5 sec in the case of fluorescein and 5–10 sec in the case of FITC-dextran) after the microinjection, in which the fluorescence intensity at the peripheral part of the cell was still small. This pattern was compared with patterns of fluorescence distribution in PBS recorded at various times after the injection of the same solution in the PBS, in order to find a pattern that closely resembled the pattern obtained in the cell. In this procedure, the amplitudes of the oscillographic records were normalized to the same height, because the light was more or less attenuated by the absorption and scattering in the cytoplasm and because the amount of the injected substance was not exactly the same in all the experiments. From the ratio of the times required to attain the same fluorescence distribution patterns in the cell and in PBS, the relative diffusion rate was obtained. Because the fluorescence distribution was recorded with the video-tape recorder at 30 frames per second and because the duration required for the microinjection was 10–30 msec, the time resolution (about 1/30 sec) was fairly high to determine $t_1$ and $t_n$ mentioned above. The diffusion coefficient of the substance in the cytoplasm was expressed as a percentage of the diffusion coefficient in PBS, which is practically the same as that in pure water.

In order to determine the critical size of molecule which can pass through the nuclear envelope, fluorescence micrographs of the cell were taken with a microscope camera (Nikon UFX-II) using 35 mm Kodak Tri X film, after a sufficient time (10–50 min depending on the molecular weight of the injected FITC-dextran) elapsed for the thorough diffusion of the injected FITC-dextran within the cytoplasm or the nucleus (germinal vesicle) of the oocyte. Whether or not the microinjected FITC-dextran was permeable to the nuclear envelope was checked by the micrographs.

Experimental temperatures were 25–27 °C.

## 3. Results

### 3.1. Diffusion Coefficients of Fluorescein and FITC-dextran in the Protoplasm

The diffusion coefficient of fluorescein (K-salt, M.W.: 408.5) in the cytoplasm of the unfertilized egg of the sand dollar, *Clypeaster japonicus* was determined to be 24 per cent of the diffusion coefficient of the same substance in PBS on the average. Similar values (29 per cent on the average) were obtained for the diffusion coefficient in the cytoplasm of the fertilized egg (5–10 minutes after fertilization). No significant difference in the difusion rate in the cytoplasm was found between the fertilized egg and the unfertilized eggs. Although we did not determine the absolute value of the diffusion coefficient of fluorescein in water, it may be similar to that of eosin, $4 \times 10^{-6} \, \text{cm}^2/\text{sec}$ (*cf.* Hodgkin and Keynes 1956). If this value is taken as the diffusion coefficient of fluorescein in water, the values of the diffusion coefficients in the cytoplasm are $1.0 \times 10^{-6} \, \text{cm}^2/\text{sec}$ for the unfertilized egg and $1.2 \times 10^{-6}/\text{sec}$ for the fertilized egg.

The diffusion coefficient of FITC-dextran (FD-40) in the cytoplasm of the fertilized egg was determined to be 21 per cent of the diffusion coefficient of the same substance. Because the diffusion coefficient of FD-40 (MW: 39,000) in water is $4.9 \times 10^{-7} \, \text{cm}^2/\text{sec}$ judging from Granath and Kvist's data (1967), the diffusion coefficient of this substance in the cytoplasm is calculated to be $1.0 \times 10^{-7} \, \text{cm}^2/\text{sec}$. The percentage value for FD-40 mentioned above is only slightly smaller than the above value for fluorescein, while the absolute value of the diffusion rate of fluorescein was several times as large as that of the rate of FITC-dextran. Most of the percentage values for fluorescein and FITC-dextran in unfertilized and fertilized eggs were in the range of 20 to 30 per cent.

### 3.2. The Permeability of the Nuclear Envelope

Figure 3 shows the results of an experiment in which FITC-dextran (FD-10 s) was microinjected into the nucleus (germinal vesicle) of a starfish oocyte (*a, b,* and *c*) and the results of an experiment in which the same substance was microinjected into the cytoplasm of another starfish oocyte (*d, e,* and *f*). It is noted that a considerable amount of FITC-dextran diffused out of the nucleus (Fig. 3 *c*) and diffused into the nucleus (Fig. 3 *f*) within 10.5 min after the microinjection. Passage of FD-4 molecules across the nuclear envelope was also observed in starfish oocytes after they were microinjected into the cytoplasm or the nucleus.

Figure 4 shows similar experiments in starfish oocytes in which FITC-dextran (FD-40) was microinjected into the germinal vesicle (*a, b,* and *c*) and those in which the same substance was microinjected into the cytoplasm (*d, e,* and *f*). After microinjection of FD-40 into the germinal vesicle, no fluorescence is detected in the cytoplasm even at 34 minutes after the microinjection (Fig. 4 *c*). The apparent fluorescence at the region of

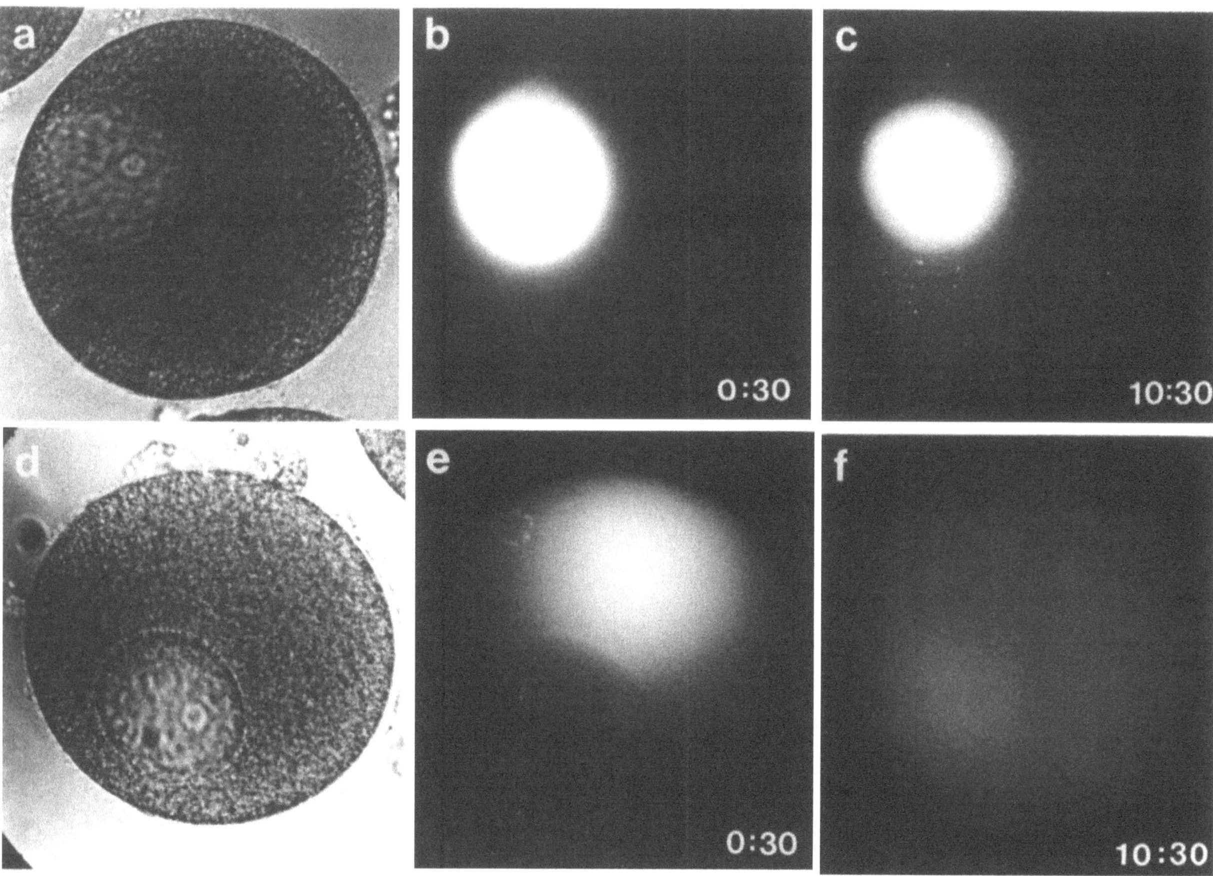


Fig. 3. Diffusion of FITC-dextran (FD-10) across the nuclear envelope in starfish oocyte. FITC-dextran solution was microinjected into the germinal vesicle (a, b, and c) or the cytoplasm (d, e, and f) in a starfish oocyte. a and d Transmitted-light micrographs before the microinjection, b, c, e, and f Fluorescence micrographs after the microinjection. Time (min : sec) after the microinjection is shown in each fluorescence micrograph. Bar indicates 100 μm

the germinal vesicle in Fig. 4 f may be due to the fluorescence in the cytoplasm on upper and lower sides of the nucleus. Similar results were obtained after microinjection of FD-70 s into the cytoplasm or into the germinal vesicle. From the above results, it may be concluded that FD molecules with molecular weight 9,000 or smaller pass through the nuclear envelope of the starfish oocyte while FD-40 molecules with molecular weight 39,000 or larger do not.

Figure 5 shows an experiment in which FD-70 s was microinjected into the germinal vesicle of a mouse oocyte. In this case, FD-70 molecules diffuse out of the germinal vesicle into the cytoplasm within 10 minutes after the microinjection. FD-70 s microinjected into the cytoplasm diffused into the germinal vesicle in mouse oocytes (data not shown).

Figure 6 shows an experiment in a mouse oocyte in which FD-150 was microinjected into the germinal vesicle (a, b, and c) or into the cytoplasm (d, e, and f).

It is noted that the microinjected molecules do not pass through the nuclear envelope of the germinal vesicle. The above results may indicate that FD-70 s molecules can pass through the nuclear envelope of the mouse oocyte while FD-150 cannot.

## 4. Discussion

Various methods have been reported to determine the diffusion coefficients of substances in the protoplasm, e.g. by measuring the intracellular distribution of dyes, radioactive substances and fluorescent substances introduced into the cell. Recently, the fluorescence recovery after photobleaching (FRAP) method has been used by many investigators in determining the mobility and the diffusion rate of substances in living cells (cf., Wojcieszyn et al. 1981, Kreis et al. 1982, Wang et al. 1982, Salmon et al. 1984, Mastro et al. 1984, Luby-Phelps et al. 1985, 1986). In this method, the rate of

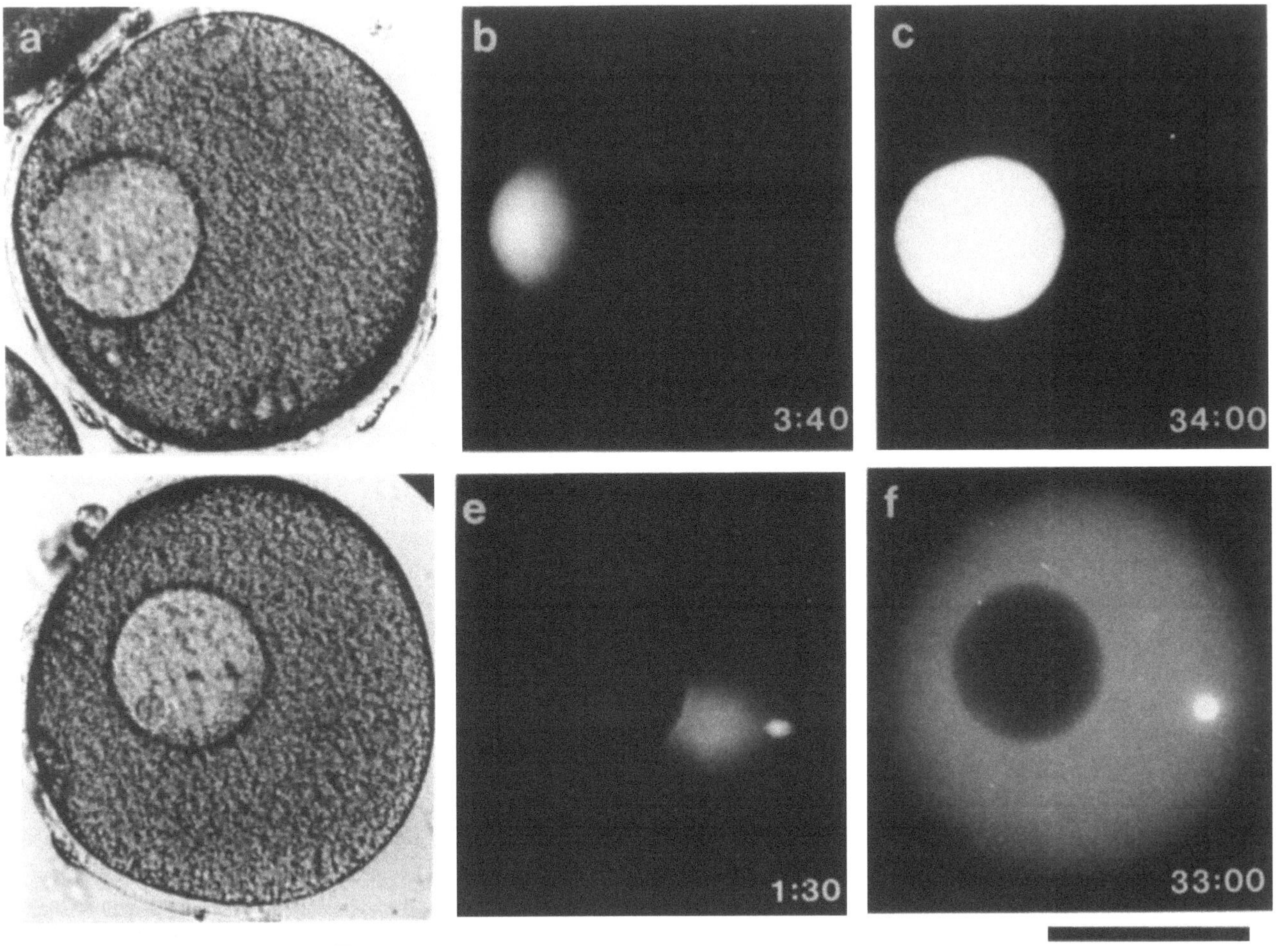

Fig. 4. Impermeability of FITC-dextran (FD-40) through the nuclear envelope in starfish oocytes. FITC-dextran solution was microinjected into the germinal vesicle (*a*, *b*, and *c*) or the cytoplasm (*d*, *e*, and *f*). *a* and *d* Transmitted-light micrographs before the micro-injection, *b*, *c*, *e*, and *f* Fluorescence micrographs after the microinjection. Time (min : sec) after the microinjection is shown in each fluorescence micrograph. Bar indicates 100 μm

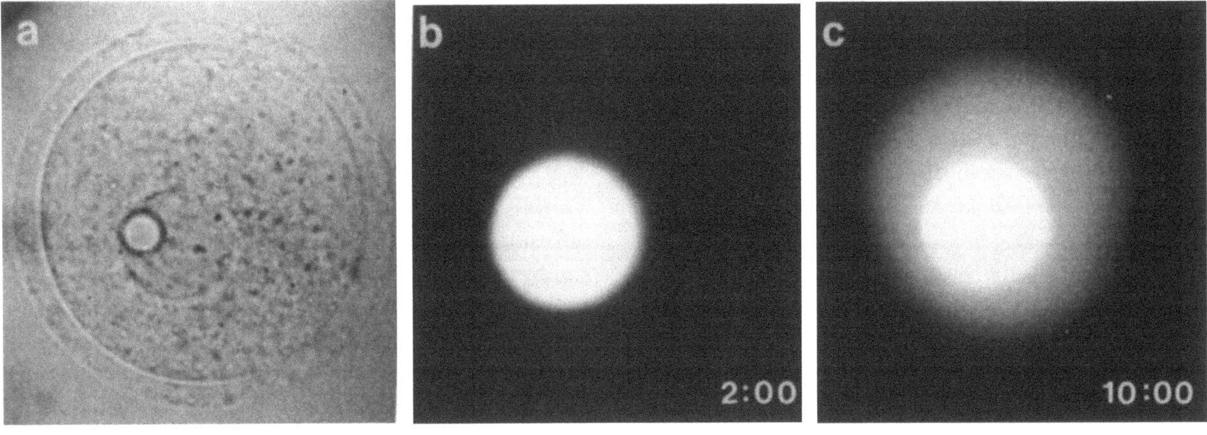

Fig. 5. Diffusion of FITC-dextran (FD-70) across the nuclear envelope in a mouse oocyte. FITC-dextran was microinjected into the germinal vesicle. *a* Transmitted-light micrograph of the oocyte before the microinjection, *b* and *c* Fluorescence micrographs taken 2 min and 20 min after microinjection, respectively. Bar indicates 100 μm

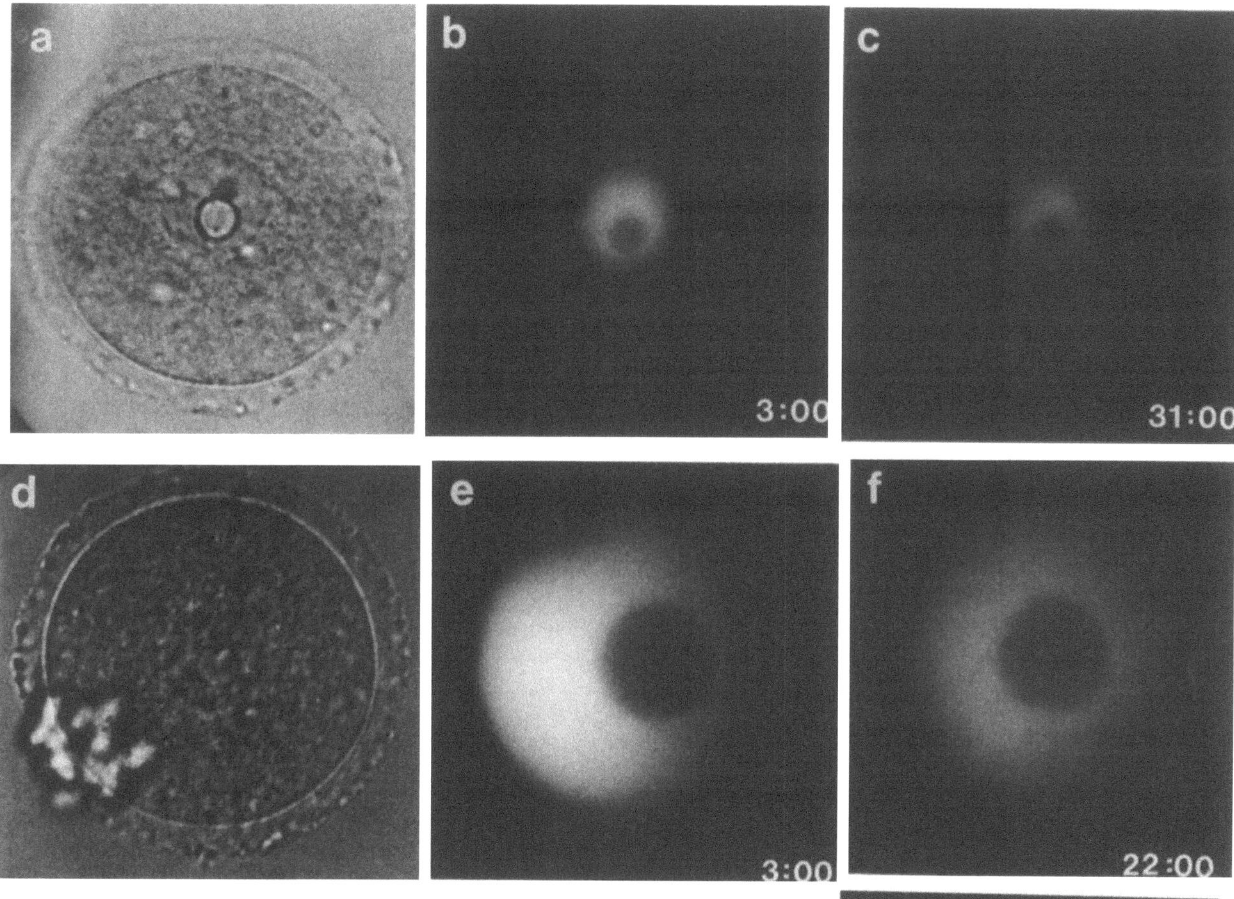

Fig. 6. Impermeability of FITC-dextran (FD-150) across the nuclear envelope in mouse oocytes. FITC-dextran was microinjected into the germinal vesicle (*a*, *b*, and *c*) or the cytoplasm (*d*, *e*, and *f*). *a* and *d* Transmitted-light micrographs before the microinjection, *b*, *c*, *e*, and *f* Fluorescence micrographs after the microinjection. Time after microinjection (min : sec) is shown in each fluorescence micrograph. Bar indicates 100 μm

fluorescence recovery after laser photobleaching of a localized region of the cell, previously microinjected with a fluorescent substance, is used to determine the diffusion coefficient.

In the present study, we present a simple method to determine the rate of transport of substance in the protoplasm within a short period by injecting fluorescently-labeled substances into the cell. The principle of the measurement is based on the fact that the spatial distribution of the substance is determined by the diffusion coefficient and the time after the substance is applied if the space for diffusion is infinitely large, and that the diffusion coefficient and the time are inversely related to the diffusion process. Therefore, it is possible to determine the diffusion coefficients of any substance by simply comparing their spatial distributions with those of a substance with a known diffusion coefficient. Possible errors due to the simplified assumptions on the optical system in various other methods to deter-

mine the diffusion coefficient by comparing the spacial distribution of diffusing substance recorded experimentally with that expected theoretically are not necessary to be considered because spatial distributions determined experimentally by the same optical system are merely compared in the present method to determine the diffusion coefficient. Despite the presence of the cell membrane limiting the transport of substances at the cell surface, the method may be applied to determine the diffusion coefficient of substances in the cytoplasm, if the distribution pattern is taken at an early time after the application of the substance, where the pattern is scarcely affected by the presence of the cell membrane.

In the present study, the diffusion coefficients of substances in the cytoplasm was 20–30% of the diffusion coefficient in water for both fluorescein and FITC-dextran, respectively, while the absolute values of the diffusion coefficients are different between these substances. This fact may indicate that the diffusion rates of these substances are determined simply by the viscosity of the cytosol which is 3 to 5 times those in water. Similar values of the percentage were reported by previous investigators using various methods (*e.g.*, Hodgkin and Keynes 1956 in squid giant axons, Horowitz and Moore 1974 and Paine *et al.* 1975 in amphibian eggs, Okuno 1977 and Salmon *et al.* 1984 in echinoderm embryonic cells, and Mastro *et al.* 1984 and Luby-Phelps *et al.* 1985, 1986 in mammalian cells). Luby-Phelps *et al.* (1986), however, reported that the percentage depends on the molecular size of the substance. Small values of the ratio of the diffusion coefficient in the cytoplasm to that in water were reported by Wojcieszyn *et al.* (1981) in mammalian fibroblasts and by Kreis *et al.* (1982) in chicken gizzard cells. Possibly in some substances, diffusion in the protoplasm depends not only on the viscosity of the cytosol but also on various other factors, such as meshwork size of the cytoskeletal structures, molecular shape, electric charge, and trapping of molecules in the protoplasm. The permeability of the nuclear envelope was determined by the diffusion of FITC-dextrans with various molecular weights microinjected into the cytoplasm or the nucleus in the present study. Dextrans are considered to be suitable substances for the present experiment to determine the size of the nuclear pore, because they do not interact significantly with intracellular components and because they have only a low electric charge (*cf.*, Luby-Phelps *et al.* 1986). The Strokes radii of FITC-dextran molecules were estimated from Granath and Kvist's data (1967) to be 2.2 nm for 9 kD (FD-10 s), 4.4 nm for 39 kD (FD-d 40), 5.7 nm for 66.4 kD (FD-70 s), and 8.4 nm for 156.9 kD (FD-150). From the above values and the results of the present study it seems likely that the pore diameters for free diffusion of molecules are larger than 4.4 nm but smaller than 8.8 nm in starfish oocytes and larger than 11.4 nm but smaller than 16.6 nm in mouse oocytes. These results are compatible with the ultrastructure of the pore complexes of the nuclear envelope (*cf.*, Franke *et al.* 1981) and previous reports on the permeability of the nuclear envelope (Franke 1974, Horowitz and Moore 1974, Paine *et al.* 1975, Franke *et al.* 1981). Minor difference in the critical size in these reports may be due to the differences in cell types and species. The critical molecular size for transport across the nuclear envelope in the present study may indicate the pore size for passive diffusion. It seems likely that the transport of larger molecules such as mRNA across the nuclear envelope involves active processes including structural changes in the pore complexes.

## References


Franke WW (1974) Structure, biochemistry, and functions of the nuclear envelope. Int Rev Cytol [Suppl] 4: 71–236

— Scheer V, Krohne G, Jarasch E-D (1981) The nuclear envelope and the architecture of the nuclear periphery. J Cell Biol 91: 39 s–50 s

Granath KA, Kvist BE (1967) Molecular weight distribution analysis by gel chromatography on sephadex. J Chromatogr 28: 69–81

Hiramoto Y (1969) Mechanical properties of the protoplasm of the sea urchin egg. I. Unfertilized egg. Exp Cell Res 56: 201–209

— (1971) Micromanipulation I. Saiho (Cell) 3: 22–27

— (1987) Evaluation of cytochemical properties. In: Bereiter-Hahn H, Anderson R, Reif W-E (eds) Cytomechanics. Springer, Berlin Heidelberg New York Tokyo, pp 31–46

Hodgkin AL, Keynes RD (1956) Experiments on the injection of substances into squid giant axon by means of a microsyringe. J Physiol 131: 592–616

Horowitz SB, Moore LC (1974) The nuclear permeability, intracellular distribution, and diffusion of inulin in the amphibian oocyte. J Cell Biol 60: 405–415

Kishimoto T (1986) Microinjection and cytoplasmic transfer in starfish oocytes. Methods Cell Biol 27: 379–394

Kreis TE, Geiger B, Schlessinger J (1982) Mobility of microinjected rhodamine actin within living chicken gizzard cells determined by fluorescence photobleaching recovery. Cell 29: 835–845

Luby-Phelps KL, Lanni KF, Tavlor DL (1985) Behaviour of a fluorescent analog of calmodulin in living 3T3 cells. J Cell Biol 101: 1245–1256

— Taylor DL, Lanni F (1986) Probing the structure of cytoplasm. J Cell Biol 102: 2015–2022

Mastro AM, Babich MA, Tavlor WD, Keith AD (1984) Diffusion of a small molecule in the cytoplasm of mammalian cells. Proc Natl Acad Sci USA 81: 3414–3418

Okuno M (1977) Doctoral thesis (University of Tokyo)

Paine PL, Moore LC, Horowitz SB (1975) Nuclear envelope permeability. Nature 254: 109–114

Salmon ED, Saxton WM, Leslie RJ, Karow ML, McIntosh JR (1984) Diffusion coefficient of fluorescein-labeled tubulin in the cytoplasm of embryonic cells of a sea urchin: video image analysis of fluorescence redistribution after photobleaching. J Cell Biol 99: 2157–2164

Sato M, Wong TZ, Allen RD (1983) Rheological properties of living cytoplasm: a preliminary investigation of squid axoplasm (*Loigo pealei*). Cell Motil 4: 7–23

Wang YL, McNeil PL, Ware BR, Taylor DL (1982) Mobility of cytoplasmic and membrane-associated actin in living cells. Proc Natl Acad Sci USA 79: 4660–4664

Wojcieszyn JW, Schlegel ES, Jacobson KA (1981) Diffusion of injected macromolecules within the cytoplasm of living cells. Proc Natl Acad Sci USA 78: 4407–4410

Protoplasma (1988) [Suppl. 2]: 95–103


# Cytoskeletal Ultrastructure of Phragmoplast-Nuclei Complexes Isolated from Cultured Tobacco Cells

T. Kakimoto and H. Shibaoka*

Department of Biology, Faculty of Science, Osaka University, Toyonaka

Received February 12, 1988
Accepted March 19, 1988

Dedicated to Professor Dr. Noburo Kamiya on the occasion of his 75th birthday

## Summary

A preparation of phragmoplasts with a purity of about 40% was isolated from tobacco BY-2 cells whose cell cycle was synchronized by treatment with aphidicolin and propyzamide. Each isolated phragmoplast was associated with the daughter nuclei and the association was maintained after the phragmoplast lost either its actin filaments or its microtubules. Actin filaments identified by heavy meromyosin arrowhead decoration were abundant in the isolated phragmoplasts. Most of the filaments were oriented perpendicularly or nearly perpendicularly to the equatorial plate and in about 80% of the filaments thus oriented heavy meromyosin arrowheads pointed away from the plate. Abundant vesicles were present in the isolated phragmoplast-nuclei complex and they were associated with phragmoplast microtubules.

*Keywords:* Actin filaments; Microtubules, Phragmoplast (isolated); Tobacco BY-2 cell.

*Abbreviations:* 2,4-D 2,4-dichlorophenoxyacetic acid; DAPI 4′,6′-diamidino-2-phenylindole; DMSO dimethylsulfoxide; EGTA ethyleneglycol-bis(β-aminoethyl ether)-N,N,N′,N′-tetraacetic acid; PIPES piperazine-N,N′-bis(2-ethanesulfonic acid); PMSF phenylmethylsulfonyl fluoride.

## 1. Introduction

The phragmoplast is composed primarily of microtubules (Ledbetter and Porter 1963). Recently, the presence in the phragmoplast of cytoskeletal elements other than microtubules has been demonstrated. Observations using fluorescent dye-labeled phallotoxins have revealed the presence in the phragmoplast of actin filaments (Clayton and Lloyd 1985, Gunning and Wick 1985, Kakimoto and Shibaoka 1987 a, Palevitz 1987, Schmit and Lambert 1985, Seagull *et al.* 1987, Traas *et al.* 1987), while observations using an antibody against intermediate filament antigen have indicated the presence of an intermediate-filament-like component (Dawson *et al.* 1985).

Although knowledge of the cytoskeletal components in the phragmoplast is accumulating, the interrelationships between these components and the role of each component in the formation of cell plate remain to be clarified. The mitotic apparatus is also composed of various cytoskeletal components. In studies on the mechanism of mitosis, the isolated mitotic apparatus has often proved to be a useful tool (Sakai 1978). It was used successfully in the characterization of spindle ATPase (Pratt *et al.* 1980) and in the clarification of the structural relationship between spindle microtubules and dynein (Hirokawa *et al.* 1985).

Thus, in the hope that isolated phragmoplasts would be useful for studies of the biochemical and ultrastructural nature of the phragmoplast, we attempted to isolate phragmoplasts. Tobacco BY-2 cells were chosen as a source of phragmoplasts because of their short doubling time (Kato *et al.* 1972). To induce synchronized formation of phragmoplasts, cells were treated first with an inhibitor of DNA synthesis and then with an inhibitor of microtubules. Aphidicolin was used as the inhibitor of DNA synthesis because this drug has proved useful for the induction of synchronization of the cell cycle in tobacco BY-2 cells (Nagata *et al.*

---

* Correspondence and Reprints: Department of Biology, Faculty of Science, Osaka University, Machikaneyama-cho, Toyonaka 560, Japan.

1982). Propyzamide was used as the inhibitor of microtubules. The effect of this drug on mitosis has been reported to be readily reversible (Izumi *et al.* 1983).

## 2. Materials and Methods

### 2.1. Plant Material

Tobacco BY-2 cells (*Nicotiana tabacum* "Bright Yellow 2") were cultured in suspension in 2.4-D medium (modified Linsmaier and Skoog's medium supplemented with 3% sucrose and 0.2 mg/l 2.4-D, pH 5.8) at 26 °C in the dark. They were subcultured every 7 days (Nagata *et al.* 1981).

### 2.2. Synchronization of the Cell Cycle

To induce synchronization of the cell cycle, cells were treated with aphidicolin by the method of Nagata *et al.* (1982) and then with propyzamide. About 5 ml of 7-day-old subcultured cells (approximately $2.5 \times 10^6$ cells) were suspended in 100 ml of 2,4-D medium which contained 5 µg/ml aphidicolin (Sigma Chemical Co., St. Louis MO) and cultured for 24 hours. Aphidicolin was diluted from a 10 mg/ml stock solution made up in 100% DMSO. DMSO at 0.1% or lower showed no effect either on microtubules or on cell division. Cells were collected and washed with 2,4-D medium. The washed cells were transferred to fresh 2,4-D medium and cultured. Cells were sampled at appropriate time intervals and stained with propionic orcein for examination of the mitotic index. Three to four hours after the termination of treatment with aphidicolin, when the mitotic index began to increase, the cultures were divided into two portions and propyzamide was added to the one portion at a final concentration of 6 µM, while nothing was added to the other portion. Propyzamide was diluted from a 10 mM stock solution made up in 100% DMSO. Cells were cultured further in the presence and absence of propyzamide. In the absence of propyzamide, the mitotic index continued to increase for 5–7 hours after the termination of treatment with aphidicolin, but it began to decrease thereafter. Six to seven hours after the start of treatment with propyzamide (*i.e.*, 9–11 hours after the termination of treatment with aphidicolin), when the mitotic index of the cells cultured in the absence of propyzamide had decreased to below 15%, the cells cultured in the presence of propyzamide were harvested and washed with 2,4-D medium. The washed cells were transferred to fresh 2,4-D medium and cultured. Cells were sampled at appropriate time intervals for a determination of the percentage of cells whose chromosomes were not aligned at the equatorial plane (cells arrested at metaphase + cells in prophase), the percentage of cells whose chromosomes were aligned at the equatorial plane (cells in normal metaphase), and the percentage of cells at anaphase and telophase.

### 2.3. Isolation of Phragmoplasts

The cells treated with aphidicolin and then with propyzamide were harvested 70 minutes after the termination of treatment with propyzamide and incubated for 1 hour at room temperature in a solution of wall-digesting enzymes which contained 1% Cellulase Onozuka RS (Yakult Honsha Co., Tokyo), 0.1% Pectolyase Y 23 (Seishin Pharmaceutical Co., Tokyo), and 0.38 M sorbitol (pH 5.6). Protoplasts released from the cells were collected and washed with a solution of 0.35 M mannitol and transferred to lysis buffer A (100 mM PIPES, pH 7.0, 1 mM MgCl₂, 20 mM KCl, 0.3 M mannitol, 0.3 mM PMSF, 1 µM taxol, 50 µg/ml leupeptin, 0.1% Triton X-100, 0.1 µM rhodamine-phalloidin, 0.2 mg/ml tropomyosin, 0.5 mg/ml heavy meromyosin, 0.15 mM spermine, 0.5 mM spermidine, 1 mM dithiothreitol) or lysis buffer B (100 mM PIPES, pH 7.0, 1 mM MgCl₂, 20 mM KCl, 0.3 M mannitol, 0.3 mM PMSF, 5 mM EGTA, 50 µg/ml leupeptin, 0.1% Triton X-100, 1 mM dithiothreitol) to dissolve the plasma membranes of the protoplasts. The protoplast lysate was centrifuged at $300 \times g$ for 10 minutes and the resulting pellet was subjected to centrifugation in a discontinuous Percoll (Pharmacia, Uppsala, Sweden) density gradient. After centrifugation at $600 \times g$ for 20 minutes, phragmoplasts were retrieved from the 45–50% interface of the Percoll gradient. Taxol or EGTA was added to the lysis buffer to preserve microtubules, and tropomyosin was added to preserve actin filaments (Kakimoto and Shibaoka 1987 b). Heavy meromyosin was used to identify actin filaments (Ishikawa *et al.* 1969), and spermine and spermidine to stabilize nuclei (Willmitzer and Wagner 1981). Rhodamine-phalloidin was added to lysis buffer A to stabilize actin filaments and allow us to check whether or not actin filaments in isolated phragmoplasts can withstand the fixatives used for preparing specimens for electron microscopy.

### 2.4. Fluorescence Microscopy

Cells or isolated phragmoplasts were fixed with 3.1% formaldehyde in potassium phosphate buffer (50 mM, pH 7.0), which contained 1 mM MgCl₂, 5 mM EGTA and 0.3 mM PMSF, for 40 minutes. They were stained with mouse monoclonal anti-chick brain α-tubulin as described previously (Kakimoto and Shibaoka 1987 b), and then with 0.5 µg/ml DAPI solution. Patterns of staining with anti-tubulin antibody were examined by the procedure described previously (Kakimoto and Shibaoka 1987 a, b) and patterns of staining with DAPI were examined by the procedure described by Katsuta and Shibaoka (1988).

### 2.5. Electron Microscopy

For electron microscopy, phragmoplasts isolated with lysis buffer A were used. The isolated phragmoplasts were washed with PIPES-Mg buffer (100 mM PIPES, pH 7.0, 1 mM MgCl₂, 20 mM KCl, 0.3 M mannitol, 0.3 mM PMSF) that contained 0.1% lysine, and then with PIPES-Mg buffer. Washing with the lysine solution was designed to stabilize actin filaments (Kakimoto and Shibaoka 1987 a). The washed phragmoplasts were fixed, dehydrated, stained with hafnium chloride, and embedded in Spurr's resin by the procedure described previously (Kakimoto and Shibaoka 1987 b).

The ultrastructure of the phragmoplasts was also examined by freeze-substitution electron microscopy (Van Harreveld and Crowell 1964). The isolated phragmoplasts were rapidly frozen by bringing them into contact with a copper block that had been precooled with liquid nitrogen. The frozen phragmoplasts were transferred into a solution of 1% osmium tetroxide in acetone cooled in dry ice-acetone and left for 72 hours at − 80 °C. The temperature of the fixed phragmoplasts was raised stepwise until it reached that of the room. After sitting at room temperature for 30 minutes, the fixed phragmoplasts were stained *en bloc* in 0.1% hafnium chloride for 1 hour (Hatae *et al.* 1984) and embedded in Spurr's resin. Sections were cut and stained with uranyl acetate and lead citrate, and were examined with an electron microscope (JEM-100C, Jeol, Tokyo).

## 3. Results

### 3.1. Effects of Propyzamide on the Cell Cycle

Three-day-old subcultured cells were cultured in 2,4-D medium that contained propyzamide. As has been reported recently (Akashi *et al.* 1988), propyzamide causes a disruption of microtubules and thereby arrests chromosomes in metaphase in tobacco BY-2 cells. In the presence of propyzamide, chromosomes condense to assume a metaphase-like appearance, but they are not aligned in an equatorial plate. In cells treated with 6 µM propyzamide for 6 hours, only fragmented microtubules or none at all were present. The mitotic index of 3-day-old subcultured cells was about 5%. In the presence of propyzamide at 6 µM or higher, the index increased with time and reached about 25% after 9 hours (Fig. 1). After about 6 hours, however, adjacent chromosomes began to fuse to produce multi-nucleate cells. Extending the treatment with propyzamide for more than 6 hours seems to be ineffective in increasing the number of cells that resume the progression through the cell cycle in synchrony.

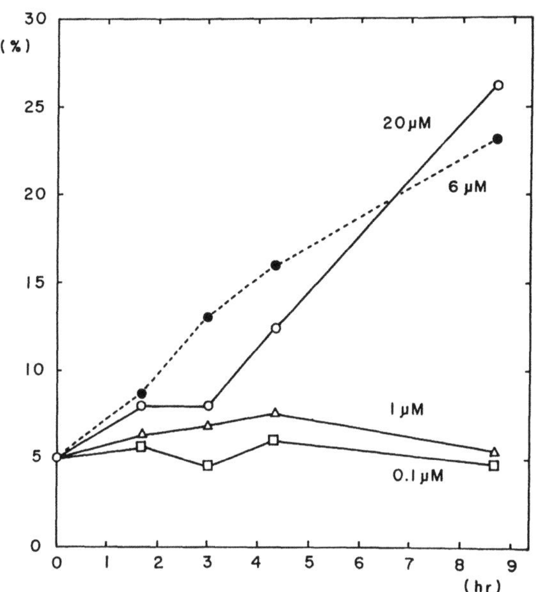

Fig. 1. Changes in the mitotic index of tobacco BY-2 cells cultured in 2,4-D medium that contained various concentrations of propyzamide. The abscissa indicates time after the start of treatment with propyzamide

### 3.2. Effect of Aphidicolin on the Cell Cycle

Using the procedure of Nagata *et al.* (1982), we cultured 7-day-old subcultured cells in 2,4-D medium that contained 5 µg/ml aphidicolin for 24 hours, and then transferred them to fresh 2,4-D medium and continued their incubation. Treatment with aphidicolin decreased

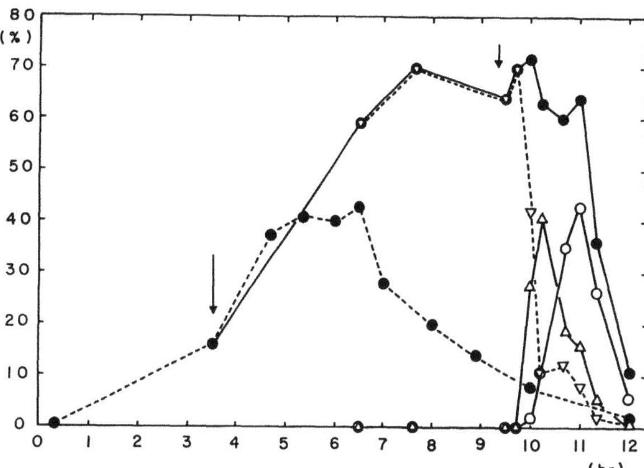

Fig. 2. Change in the mitotic index of tobacco BY-2 cells after treatment with aphidicolin and the effect of propyzamide on it. - - - ●- - - Mitotic index of cells cultured in 2,4-D medium that contained aphidicolin (5 µg/ml) for 24 hours and then in 2,4-D medium. —●— Mitotic index of cells cultured in 2,4-D medium that contained aphidicolin (5 µg/ml) for 24 hours, then in 2,4-D medium, then in 2,4-D medium that contained propyzamide (6 µM), and again in 2,4-D medium. Treatment with propyzamide was started at the time indicated by a long arrow and terminated at the time indicated by a short arrow. - - -▽- - - Percentage of cells whose chromosomes are not arranged at the equatorial plate. —△— Percentage of cells whose chromosomes are arranged at the equatorial plate. —○— Percentage of cells in anaphase + telophase. The abscissa indicates time after the termination of treatment with aphidicolin

the mitotic index of tobacco culture cells to zero, but after the cells were transferred to the medium that contained no aphidicolin the mitotic index increased with time and reached a peak of about 45% (Fig. 2). The percentage of cells having phragmoplasts, however, did not exceed 15% (data not shown).

### 3.3. Synchronization of the Cell Cycle

The results obtained demonstrated that both propyzamide and aphidicolin were useful for inducing synchronization of the cell cycle, but the degree of synchronization induced by propyzamide or aphidicolin alone was far from satisfactory for our purposes. Thus, we attempted to induce a higher degree of synchronization by treating cells with both aphidicolin and propyzamide according to the procedure described in Materials and Methods.

Propyzamide increased the mitotic index of aphidicolin-pretreated cells to about 70% (Fig. 2). The increase in mitotic index was due to the increase in the number of cells whose chromosomes were not aligned in an equatorial plate (Fig. 2). The number of such cells began to decrease about 40 minutes after the termination of treatment with aphidicolin, while the number

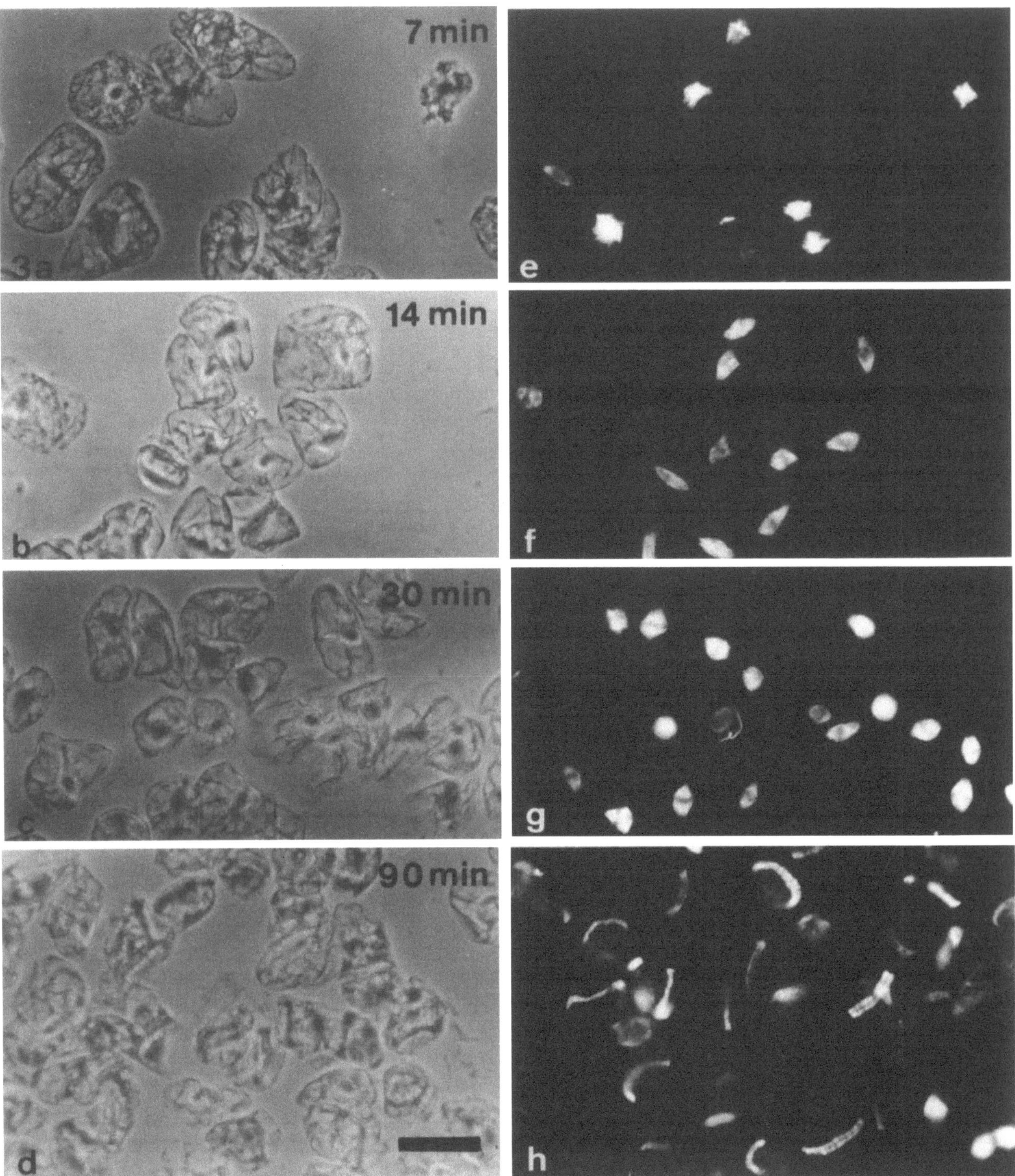



Fig. 3. Synchronized formation of phragmoplasts after the termination of the treatment with propyzamide described in the legend to Fig. 2. After the termination of treatment with propyzamide, cells were sampled at appropriate time intervals and stained for microtubules. Time after the termination of treatment with propyzamide is given in minutes in the photographs. *a—d* Phase contrast micrographs. *e–h* Immunofluorescence micrographs. Bar = 50 μm

of cells whose chromosomes were aligned in an equatorial plate increased (Fig. 2). The percentage of the latter cells reached a peak of 40% after 60 minutes. After 40 minutes, anaphase cells began to appear, and after about 100 minutes the percentage of cells in anaphase + telophase reached a peak of 43% (Fig. 2). At this point the proportion of cells having the phragmoplast was about 30%.

Soon after the termination of treatment with propyzamide, microtubules began to polymerize. As shown in Fig. 3 e, microtubules were observed after as little as 7 minutes, showing that the effects of propyzamide can be reversed rapidly. After 14 minutes, the mitotic spindles became evident (Fig. 3 f). After 90 minutes, phragmoplasts were observed at a high frequency (Fig. 3 h). Similar experiments using subcultured cells of various ages gave almost identical results, but when young subcultured cells, such as 3-day-old cells, were used, treatment with aphidicolin for 20–22 hours gave better results than treatment for 24 hours.

### 3.4. Isolation of Phragmoplasts

It takes about 60 minutes to digest cell walls and release protoplasts, and the cell cycle continues to proceed in cells in the solution of wall-digesting enzymes. However, since the progression through the cell cycle slows down as digestion of the cell walls progresses, treatment with the enzyme solution was started 30 minutes before the percentage of cells with phragmoplasts reached a peak (*i.e.*, about 70 minutes after the termination of treatment with propyzamide). This procedure gave the maximum yield of isolated phragmoplasts.

Released protoplasts were lysed and phragmoplasts were collected from the lysate by centrifugation. The phragmoplast-rich pellet also contained resting nuclei and mitotic spindles, and the purity of the preparation of phragmoplasts ranged from 20 to 30%. The purity was increased to about 40% by subjecting the resuspended phragmoplast-rich pellet to differential gradient centrifugation in Percoll.

Since cell walls are thinner in 3-day-old subcultured cells than in 7-day-old cells, preparation of protoplasts is much easier with the former than with the latter cells.

### 3.5. Association of Daughter Nuclei with Phragmoplasts

The isolated phragmoplasts were associated with daughter nuclei (Fig. 4). The phragmoplasts isolated with lysis buffer B showed no rhodamine-phalloidin staining, but were accompanied by daughter nuclei, indicating that actin filaments are not involved in the

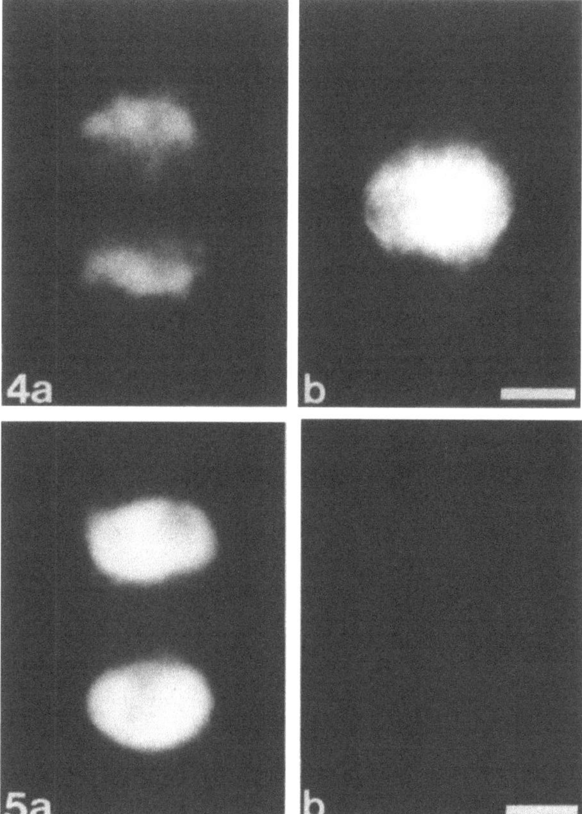

Fig. 4. A phragmoplast-nuclei complex isolated from a tobacco BY-2 cell by treatment with lysis buffer B and double stained with DAPI (*a*) and anti-α-tubulin antibody (*b*). Note that the phragmoplast (*b*) is associated with the daughter nuclei (*a*). Bar = 10 µm

Fig. 5. Phragmoplast-nuclei complexes isolated from tobacco BY-2 cells by treatment with lysis buffer B, incubated in 100 µM $CaCl_2$ for 20 minutes at 0 °C, and double stained with DAPI (*a*) and anti-α-tubulin antibody (*b*). Note that phragmoplast microtubules are destroyed by treatment with $CaCl_2$ (*b*), but a pair of daughter nuclei is not separated by this treatment (*a*). Bar = 10 µm

association between phragmoplast and nuclei. The phragmoplast-nuclei complexes isolated with lysis buffer B were incubated with Ca-PIPES buffer (100 mM PIPES, pH 7.0, 100 µM $CaCl_2$, 1 mM dithiothreitol) at 0 °C for 20 minutes. Microtubules in the phragmoplast disappeared as a result of this treatment (Fig. 5 b), but daughter nuclei were not separated by this treatment (Fig. 5 a). Microtubules do not seem to be involved in maintaining the association between phragmoplast and nuclei.

### 3.6. Ultrastructure of Isolated Phragmoplasts

Actin filaments as well as microtubules were present in the phragmoplast-nuclei complexes isolated with lysis buffer A and fixed by conventional procedures

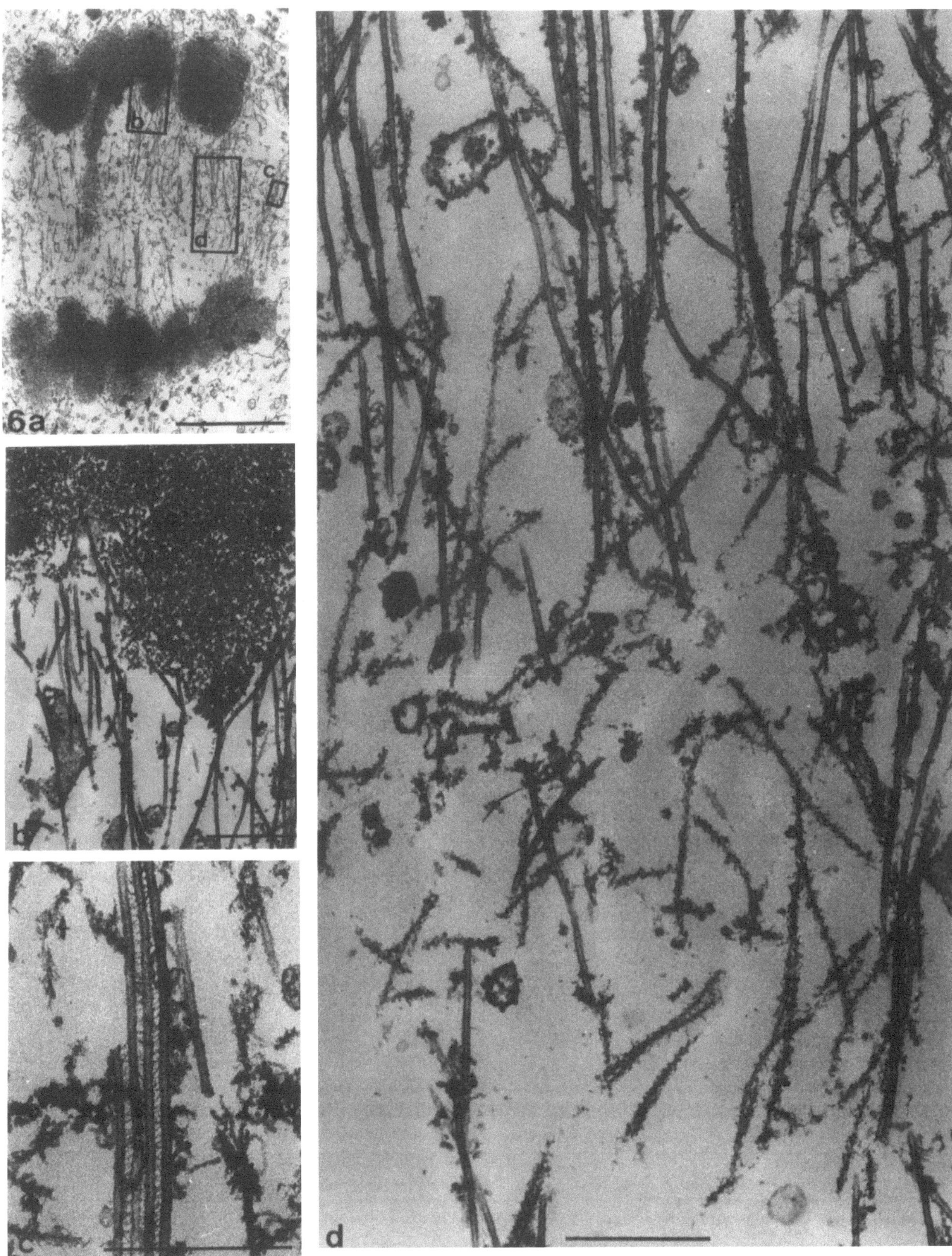

Fig. 6. Electron micrographs of a phragmoplast-nuclei complex isolated from tobacco BY-2 cell by treatment with lysis buffer A. *a* A whole complex. Boxed areas *b, c,* and *d* are shown at higher magnification in Fig. 6 *b, c,* and *d,* respectively. *b* Near the reforming daughter nucleus, abundant microtubules, but no actin filaments, are seen. Some microtubules are associated with thin filaments that are not decorated by heavy meromyosin. *c* Cross-bridges between adjacent microtubules. *d* Near the equatorial plate, actin filaments decorated by heavy meromyosin are seen. They are oriented perpendicularly or nearly perpendicularly to the plate. Note that arrowheads point away from the plate in most of the filaments. *a* Bar = 5 µm, *b–d* Bar = 0.5 µm

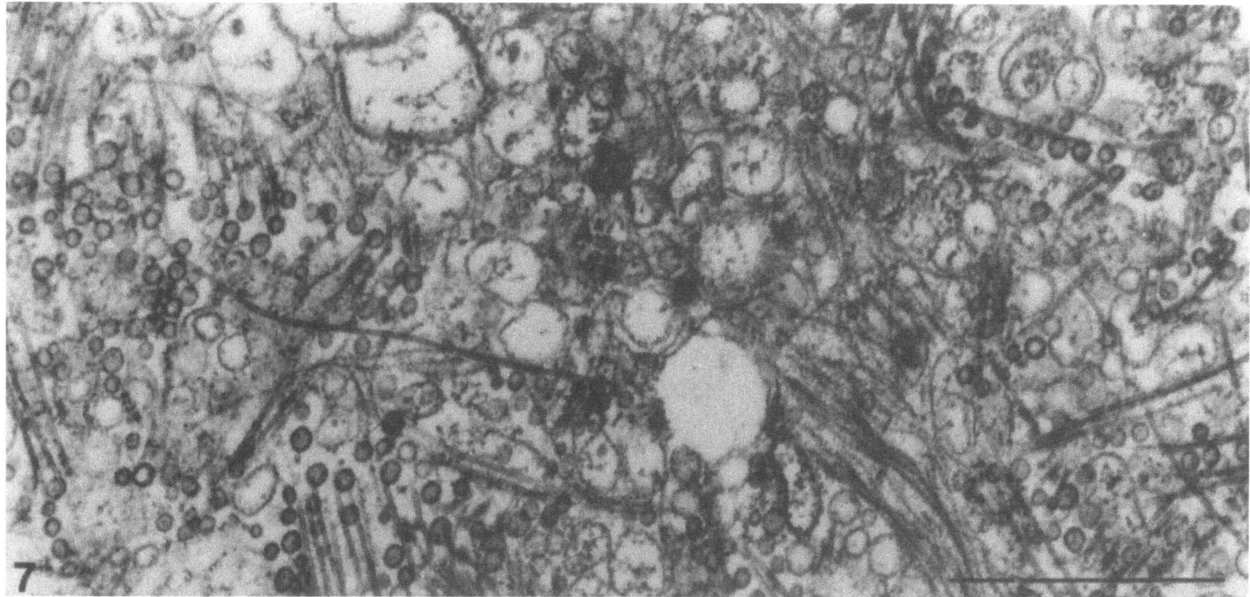

Fig. 7. Electron micrograph of a part of a phragmoplast isolated from a tobacco BY-2 cell by treatment with lysis buffer A and fixed by freeze-substitution method. Abundant, electron-dense small vesicles are seen attached to the phragmoplast microtubule. Bar = 1 μm

(Fig. 6 a, d). The vesicles that are usually observed between phragmoplast microtubules and those at the equatorial plane (Bajer 1968, Hepler and Jackson 1968) were not well preserved (Fig. 6 d). Microtubules extended from the equatorial plate to the reforming daughter nuclei (Fig. 6 a, b), but actin filaments were not present near the daughter nuclei (Fig. 6 b). Actin filaments which bound heavy meromyosin and formed arrowheads complexes were abundantly distributed near the equatorial plate (Fig. 6 d). Most of them were arranged in parallel or nearly in parallel to the spindle axis, although they were not as neatly arranged as microtubules (Fig. 6 d). The direction of myosin arrowheads was examined with 200 actin filaments which were arranged in parallel or nearly in parallel to the spindle axis. As the shapes of arrowheads were not distinct in 50 filaments out of the 200, we determined the direction of the arrowheads with 150 filaments, with the result that 77% of them (115 filaments) had arrowheads pointing away from the cell plate. Crossbridges were often observed between adjacent microtubules (Fig. 6 c). Thin filaments which were not decorated with heavy meromyosin were also present in association with microtubules (Fig. 6 b).

Although vesicles were only poorly preserved by the conventional method, they were well preserved by the freeze-substitution method. Abundant vesicles which appeared to be attached to microtubules were observed (Fig. 7).

## 4. Discussion

Treatment with aphidicolin increased the mitotic index of tobacco culture cells, but it did not increase the percentage of cells which synchronously formed phragmoplasts to a level high enough for our purpose. This failure is due, at least partly, to the distance in the cell cycle between the phase at which aphidicolin arrests the cycle and the phase at which formation of phragmoplasts occurs. Many events intervene between these two phases and as the cells complete these various events the synchronization of the cell cycle, induced by aphidicolin, may be disturbed. The phase at which inhibitors of microtubules arrest the cycle is close to the phase at which formation of phragmoplasts occurs. Thus, in the anticipation that the use of a microtubule inhibitor would improve the method for inducing the synchronized formation of phragmoplasts, we examined the effects of propyzamide. Although propyzamide alone did not give a satisfactory result, the use of this drug in combination with aphidicolin increased the percentage of cells having phragmoplasts to about 30% and allowed us to isolate a preparation of phragmoplasts with a purity of about 40%.

Each isolated phragmoplast was associated with the daughter nuclei. A firm association of the daughter nucleus with the phragmoplast was demonstrated by Dawson et al. (1985) who observed that a phragmoplast isolated from an onion root meristematic cell during cell-squashing was still attached to one of the daughter nuclei. The association between phragmoplast and nuclei was observed even in complexes which showed neither the staining of microtubules nor of actin filaments. This result suggests the involvement of cytoskeletal components other than microtubules and actin filaments in the association of the daughter nuclei with the phragmoplast. In this connection, it is noteworthy that phragmoplasts bind antibody raised against intermediate filament antigen (Dawson *et al.* 1985) and that thin filaments which are not decorated by heavy meromyosin are present in the isolated phragmoplast (Fig. 6 *b*). We should examine whether such thin filaments have some connection with intermediate filaments or not.

Using isolated phragmoplast-nuclei complexes, we confirmed the presence in the phragmoplast of actin filaments by electron microscopy. Most of the actin filaments in the phragmoplast were arranged in parallel or nearly in parallel to the spindle axis and in about 80% of the filaments thus arranged heavy meromyosin arrowheads pointed away from the equatorial plate, indicating that the polarity of the filaments was opposite on two side of the plate. This result and the well-known fact that arrowheads on microfilaments in Characean plant cells point in the direction opposite to that of cytoplasmic streaming (Kersey *et al.* 1976) suggest the possible involvement of phragmoplast actin filaments in the transport of vesicles toward the site of formation of the cell plate.

To analyze the mechanism of transport of vesicles toward the site of formation of the cell plate, it is also necessary to examine the relationship between vesicles and phragmoplast microtubules. Although the vesicles in the isolated phragmoplast were not well preserved by the conventional method of fixation, they were preserved in satisfactory condition by the freeze-substitution method. Thus, it appears that the freeze-substitution method is superior to the conventional method for studies of the relationship between vesicles and phragmoplast microtubules. However, the freeze-substitution method does not seem to give a satisfactory result in the case of the phragmoplast *in situ*, perhaps because this method is useful only for fixing specimens of 10 μm thickness or less (Van Harreveld and Crowell 1964) and the phragmoplast is usually located in the center of vacuolated cells. The isolated phragmoplasts should be of great utility for studies of the relationship between vesicles and phragmoplast microtubules by freeze-substitution electron microscopy.

The preparation of phragmoplasts obtained by the pro-

cedure described in the present paper will be of considerable value in microscopic studies of the phragmoplast, but will not be of any use in biochemical studies, because of its low purity. We need to improve the method for inducing the synchronized formation of phragmoplasts and to devise a method for separating phragmoplasts from the contaminating nuclei and spindles in the preparation. The daughter nuclei associated with the isolated phragmoplast make it difficult to use our preparation of phragmoplasts for biochemical studies. We must develop a method for separating daughter nuclei from phragmoplasts. In this connection, studies on the thin filaments which are not decorated with heavy meromyosin in the phragmoplast will be of crucial importance.

## Acknowledgements

This work was supported in part by Grants-in-Aid for Scientific Research (62480009) and Cooperative Research (62300004) from the Ministry of Education, Science and Culture of Japan. Taxol was kindly supplied by the Natural Products Branch, Division of Cancer Treatment, National Cancer Institute (Bethesda, Maryland) and propyzamide by Sumitomo Chemical Co. (Takarazuka, Hyogo). Tropomyosin (from rabbit skeletal muscle) was kindly supplied by Dr. M. Yasui of the Department of Biology, Osaka University, and heavy meromyosin by Dr. Y. Miyata of the same department, to whom we convey our thanks.

## References

Akashi T, Izumi K, Nagano E, Enomoto M, Mizuno K, Shibaoka H (1988) Effects of propyzamide on tobacco cell microtubules *in vivo* and *in vitro*. Plant Cell Physiol 29 (in press)

Bajer A (1968) Fine structure studies on phragmoplast and cell plate formation. Chromosoma 24: 383–417

Clayton L, Lloyd CW (1985) Actin organization during the cell cycle in meristematic plant cells. Exp Cell Res 156: 231–238

Dawson PJ, Hulme JS, Lloyd CW (1985) Monoclonal antibody to intermediate filament antigen cross-reacts with higher plant cells. J Cell Biol 100: 1793–1798

Gunning BES, Wick SM (1985) Preprophase bands, phragmoplasts, and spatial control of cytokinesis. J Cell Sci [Suppl] 2: 157–179

Hatae T, Okuyama K, Fujita M (1984) Visualization of the cytoskeletal elements in tissue culture cells by bloc-staining with hafnium chloride after rapid freezing and freeze-substitution fixation. J Electron Microsc 33: 186–190

Hepler PK, Jackson WT (1986) Microtubules and early stages of cell-plate formation in the endosperm of *Haemanthus katherinae* Baker. J Cell Biol 38: 437–446

Hirokawa N, Takemura R, Hisanaga S (1985) Cytoskeletal architecture of isolated mitotic spindle with special reference to microtubule-associated proteins and cytoplasmic dynein. J Cell Biol 101: 1858–1870

Ishikawa H, Bischoff R, Holtzer H (1969) Formation of arrowhead complexes with heavy meromyosin in a variety of cell types. J Cell Biol 43: 312–328

Izumi K, Kanazawa H, Saho T (1983) Colchicine-like effect of propyzamide on the root meristematic cells of *Vicia faba* L. J Fac Sci Hokkaido Univ Ser V 13: 17–24

Kakimoto T, Shibaoka H (1987 a) Actin filaments and microtubules in the preprophase band and phragmoplast of tobacco cells. Protoplasma 140: 151–156

— — (1987 b) A new method for preservation of actin filaments in higher plant cells. Plant Cell Physiol 28: 1581–1585

Kato K, Matsumoto T, Koiwai A, Mizusaki S, Nishida K, Noguchi M, Tamaki E (1972) Liquid suspension culture of tobacco cells. In: Proc IV IFS: Ferment Technol Today, pp 689–695

Katsuta J, Shibaoka H (1988) Role of the cytoskeleton and the cell wall in nuclear positioning in tobacco BY-2 cells. Plant Cell Physiol 29: 403–413

Kersey YM, Hepler PK, Palevitz BA, Wessells NK (1976) Polarity of actin filaments in Characean algae. Proc Natl Acad Sci USA 73: 165–167

Ledbetter MC, Porter KR (1963) A "microtubule" in plant cell fine structure. J Cell Biol 19: 239–250

Nagata T, Okada K, Takebe I (1982) Mitotic protoplasts and their infection with tobacco mosaic virus RNA encapsulated in liposomes. Plant Cell Rep 1: 250–252

— — — Matsui C (1981) Delivery of tobacco mosaic virus RNA into plant protoplasts mediated by reverse-phase evaporation vesicles (liposomes). Mol Gen Genet 184: 161–165

Palevitz BA (1987) Accumulation of F-actin during cytokinesis in *Allium*. Correlation with microtubule distribution and the effects of drugs. Protoplasma 141: 24–32

Pratt MM, Otter T, Salmon ED (1980) Dynein-like $Mg^{2+}$-ATPase in mitotic spindles isolated from sea urchin embryos (*Strongylocentrotus droebachiensis*). J Cell Biol 86: 738–745

Sakai H (1978) The isolated mitotic apparatus and chromosome motion. Intern Rev Cytol 55: 23–48

Schmit AC, Lambert AM (1985) F-actin dynamics is associated to microtubule function during cytokinesis in higher plants. A revised concept. In: De Brabander M, De Mey J (eds) Microtubules and microtubule inhibitors 1985. Elsevier, Amsterdam, pp 243–252

Seagull RW, Falconer MM, Weerdenburg CA (1987) Microfilaments: dynamic arrays in higher plant cells. J Cell Biol 104: 995–1004

Traas JA, Doonan JH, Rawlins DJ, Shaw PJ, Watts J, Lloyd CW (1987) An actin network is present in the cytoplasm throughout the cell cycle of carrot cells and associates with the dividing nucleus. J Cell Biol 105: 387–395

Van Harreveld A, Crowell J (1964) Electron microscopy after rapid freezing on a metal surface and substitution fixation. Anat Rec 149: 381–386

Willmitzer L, Wagner KG (1981) The isolation of nuclei from tissue-cultured plant cells. Exp Cell Res 135: 69–77

Protoplasma (1988) [Suppl. 2]: 104–115
© by Springer-Verlag 1988

# Structural Analysis of the Sea Urchin Egg Cortex Isolated on a Substratum

S. Yonemura[1,*], S. Tsukita[2], S. Tsukita[2], and I. Mabuchi[1]

[1] Department of Biology, College of Arts and Sciences, University of Tokyo, Komaba, Meguro-ku, Tokyo,
[2] Department of Ultrastructural Research, The Tokyo Metropolitan Institute of Medical Sciences, Honkomagome, Bunkyo-ku, Tokyo

Received March 1, 1988
Accepted May 7, 1988

Dedicated to Professor Dr. Noburo Kamiya on the occasion of his 75th birthday

## Summary

A method for the isolation of the sea urchin egg cortex appropriate for preservation of the cortical structure was sought. In the cortex of an egg at the streak stage isolated by homogenization, the cortical vacuoles tended to be situated apart from the plasma membrane to some extent, making a room which well-developed actin filament meshworks occupied. However, in the intact egg at the streak stage, which was rapid-frozen and freeze-substituted, the vacuoles were observed close to the plasma membrane. Homogenization may have induced the formation of the meshwork. On the other hand, the isolation on a protamine-coated substratum was found to preserve the cortical structure in situ and used for the electron microscopic analysis of the structural changes of the cortical cytoskeleton. Short actin filaments in microvilli of the unfertilized egg cortex were demonstrated clearly for the first time. They were attached to the plasma membrane at their barbed ends and cross-linked with each other by thin strands. A dramatic actin polymerization which occurs soon after fertilization was observed as a formation of a filamentous layer underneath the plasma membrane. Subsequently, this layer reduced in thickness and microvillar actin filament bundles become prominent structures of the cortical cytoskeleton. These results supported previous studies using fluorescence microscopy.

Keywords: Actin; Sea urchin egg; Isolated cortex; Cytoskeleton; Fertilization.

Abbreviations: CIM cortex isolation medium; EGTA ethyleneglycol-bis-(β-aminoethylether) N,N,N′,N′-tetraacetic acid; MOPS 3-(N-morpholino)propanesulfonic acid; PIPES piperazine-N,N′-bis(2-ethanesulfonic acid); S-1 subfragment 1.

## 1. Introduction

Sea urchin eggs can be fertilized artificially and they develop synchronously. Thus, the sea urchin egg has offered a great advantageous model for the study of the mechanism of the fertilization and cleavage. It has been reported that a dramatic polymerization of actin occurs in the sea urchin egg cortex during and after fertilization (reviewed by Vacquier 1981). In a previous paper, we demonstrated that the cortical actin polymerization during fertilization begins at the point of sperm penetration and that the polymerization propagates in a wavelike manner over the entire cortex (Yonemura and Mabuchi 1987). This actin polymerization corresponds to the appearance of the actin filament bundles in the fertilization cone and the core actin filament bundles in elongating microvilli at the electron microscopic level (reviewed by Vacquier 1981, Chandler and Heuser 1981). At cleavage, another type of structure containing actin filaments called the contractile ring is constructed in the cortex beneath the cleavage furrow (reviewed by Schroeder 1975, and by Mabuchi 1986).

Although morphological studies on the cortical cytoskeleton were carried out by various experimental methods, there are still some inconsistencies with each result. Cortices isolated by homogenizing eggs showed a cortical meshwork layer of actin filaments about 0.5–1 μm thick between the plasma membrane and a layer of electron-translucent vacuoles (Begg and Rebhun 1979, Mabuchi et al. 1980, Begg et al. 1982), while in the cortical layer of intact eggs, the vacuoles were observed just beneath the plasma membrane remaining little room for such meshwork layer (Schroeder 1981, Usui and Yoneda 1982). In these images, however, the structure of the cytoskeleton in the cortex was not

* Correspondence and Reprints: Department of Biology, College of Arts and Sciences, University of Tokyo, Komaba, Meguro-ku, Tokyo, 153 Japan.

demonstrated clearly probably due to the existence of the egg cytoplasm. Therefore, it has not been certain whether the cortical cytoskeleton observed in the isolated cortices represents that *in situ*; it might be altered during the course of isolation.

In this study, we analyzed cortical cytoskeleton in the sea urchin egg by electron microscopy in order to re-evaluate its entity. It is widely accepted that the rapid freezing technique using liquid He is highly potent for preserving the dynamic structure at the electron microscopic level (Heuser *et al.* 1979, Tsukita and Yano 1985). Therefore, using this technique combined with freeze-substitution method, we observed the structural changes in the cortical region of the whole eggs during the first cell cycle. We compared the images of the cortices isolated by several methods with those of the freeze-substituted eggs. As a result, the images from the cortices isolated on a substratum were more consistent with those of the freeze-substituted eggs than the cortices isolated by the conventional homogenizing method. Therefore, by using the substratum method, we reexamined the structure of the cortical cytoskeleton of the sea urchin egg.

## 2. Materials and Methods

### 2.1. Sea Urchin Eggs

The sea urchin, *Hemicentrotus pulcherrimus* was used. Eggs and sperm were collected by conventional KCl- or acetylcholine-induced spawning. Eggs were washed with filtered sea water, dejellied at pH 5.0, rewashed with filtered sea water and inseminated. Fertilization membranes were removed by treating eggs with 1 M urea at 1 min after insemination and passing them through a sheet of 74 μm nylon mesh. Eggs were washed with Ca-free artificial sea water (470 mM NaCl, 10 mM KCl, 27 mM MgCl$_2$, 30 mM MgSO$_4$, 10 mM NaHCO$_3$, pH 8.2) three times and cultured in a watch glass at 18–20 °C.

### 2.2. Cortex Isolation

Egg cortices were isolated by two kinds of methods. One is a homogenizing method. The other is a method using protamine sulfate-coated substratum (substratum method).

Homogenizing method: Eggs were washed with about 10 vol of cortex isolation medium (CIM) consisting of 0.8 M glucose, 0.1 M KCl, 2 mM MgCl$_2$, 5 mM ethyleneglycol-bis-(β-aminoethylether) N,N,N′,N′-tetraacetic acid (EGTA), 0.5 mM dithiothreitol, 5 μg/ml leupeptin, 10 mM 3-(N-morpholino)propanesulfonic acid (MOPS) buffer, pH 7.3 at 4 °C. Sometimes glucose was omitted. Washed eggs were resuspended in 10 vol of the CIM and homogenized with a hand-driven Teflon-glass homogenizer. Disruption of eggs was checked with an Olympus phase-contrast microscope (CK-2-TRC-2, Olympus, Tokyo, Japan). When all eggs were disrupted leaving cortical hulls, the suspension was centrifuged at 500 g for 1 min to pellet the cortices. They were washed repeatedly with the CIM by resuspension and brief centrifugation until the supernatant became clear.

Substratum method: Egg cortices were isolated by a method similar to that mentioned in a previous paper (Yonemura and Mabuchi 1987). Epon 812 was solidified as a 2 mm thick plate and was cut to make square chips (about 1 × 1 cm). The chips were washed and coated with 1% protamine sulfate. A small volume of egg suspension was transferred on to the chip. Cortices were isolated by shearing them with a stream of CIM.

### 2.3. Thin Section Electron Microscopy

Isolated cortices were fixed with 2.5% glutaraldehyde, 0.1% tannic acid, 0.8 M glucose, 2 mM MgCl$_2$, 5 mM EGTA, 0.1 M piperazine-N,N′-bis(2-ethanesulfonic acid) (PIPES) buffer, pH 6.8, for 2 hrs at room temperature, then overnight at 0 °C. When cortices were isolated using glucose-free CIM, glucose-free fixative was used. After being washed with 2 mM MgCl$_2$, 5 mM EGTA, 0.1 M PIPES buffer, pH 6.8, cortices were postfixed in ice-cold 1% OsO$_4$ dissolved in the same buffer. After being rinsed with distilled water, the samples were stained with 0.5% aqueous uranyl acetate for 2 hrs at room temperature, dehydrated stepwise with ethanol and embedded in Epon 812. Thin sections were cut with a diamond knife, doubly stained with uranyl acetate and lead citrate and examined in a Phillips EM 400 or JEOL 1200 EX electron microscope at an accelerating voltage of 100 kV.

### 2.4. Myosin S-1 Decoration

Samples were incubated for 1 min at room temperature with rabbit skeletal muscle myosin subfragment-1 (S-1) at a concentration 3 mg/ml in CIM. Then, the samples were rinsed with CIM and processed as described above.

### 2.5. Freeze-Etch Replica Electron Microscopy

The rapid freezing and deep-etch replica methods used in this study have been described in detail elsewhere (Tsukita *et al.* 1982). The egg cortices were isolated by homogenizing method, pelleted, mounted on the specimen holder, and rapidly frozen by being touched against a pure copper block pre-cooled to 4 °K by liquid He using an Eiko freezing apparatus (RF-10, Eiko Engineering, Mito, Japan). The frozen samples were stored in liquid N$_2$ until use.

The frozen sample was then transferred to a cryokit (FTC/LTS-2) fitted to a Sorvall MT-2 ultramicrotome and cut with glass knives at −120 °C to remove the outermost metal-touched surface. The preferable depth of removal was 5–10 μm. After cutting, the sample with a mirror-smooth face was brought into a freeze-etching apparatus (FD-3, Eiko Engineering, Mito, Japan) using a cooled cover. The exposed surface of the sample was deeply etched at −95 °C for 12 min *in vacuo* at 1 × 10$^{-7}$ mmHg followed by rotary shadowing with platinum at an angle of 25° and carbon at 90°. The sample was then immersed in household bleach. The replicas that floated off the samples were washed three times with distilled water and picked up on formvar-filmed grids. The replicas were examined in a JEOL 1200 EX electron microscoope equipped with a tilting stage at an accelerating voltage of 100 kV. Negative images on the electron microscope films were routinely contact-reversed and then printed.

## 3. Results

### 3.1. Evaluation of Methods for Cortex Isolation Using Cortices from Eggs at the Streak Stage

Eggs at the streak stage were used for the comparison of the isolation methods. Figure 1 shows an image of

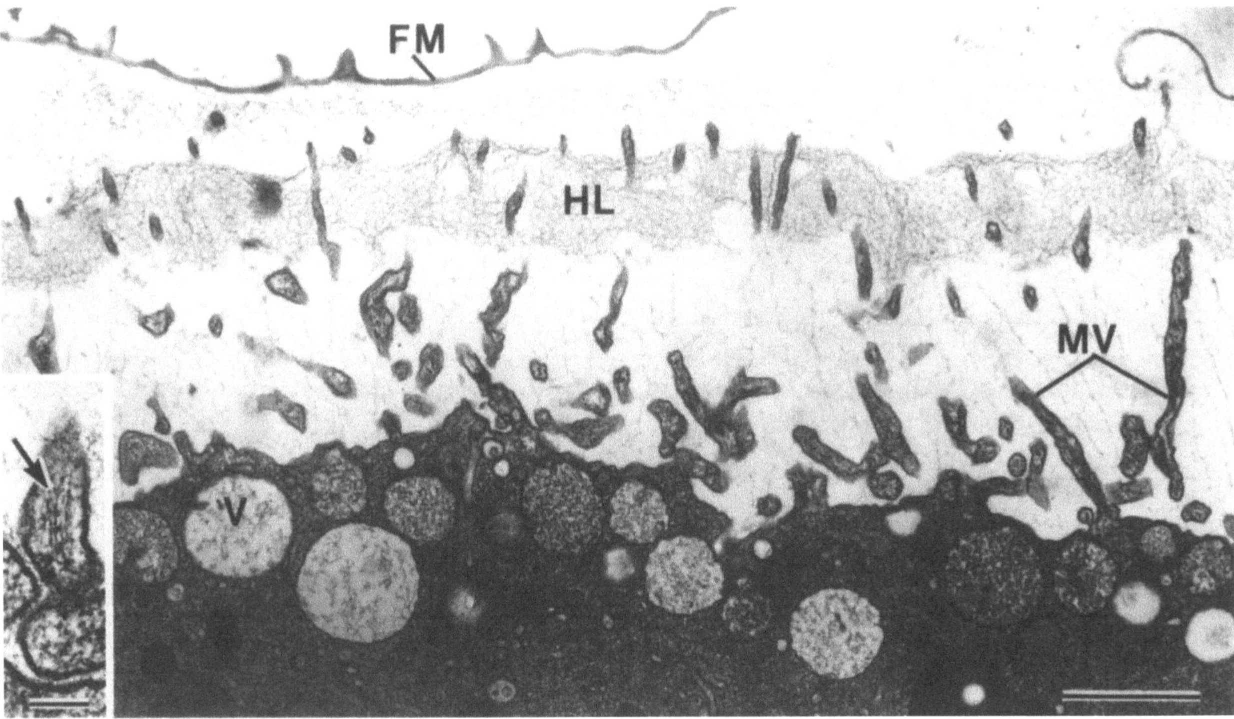


Fig. 1. A freeze-substituted *H. pulcherrimus* egg at the streak stage exhibiting a layer of vacuoles (*V*) close to the plasma membrane. *FM* Fertilization membrane. *HL* Hyaline layer. Bar, 1 µm. × 18,300. Inset: Arrow indicates actin filament bundles within microvilli (*MV*). Bar, 0.1 µm. × 81,000

the cortical region of the egg at the streak stage rapid-frozen and then freeze-substituted. Actin filament bundles were seen in microvilli (Fig. 1, inset). These filaments seemed to be cross-linked by some materials to form the parallel array. Other cytoskeletal structure could not be seen clearly in the cortex. There was a layer of electron-translucent vacuoles about 1 µm in diameter close to the plasma membrane. At least two kinds of such vacuoles were observed; one contained sparse meshwork structures and the other contained denser materials. However, we do not distinguish these vacuoles hereafter since we did not notice any difference in the location of these vacuoles after fertilization up to this stage. The average distance between the vacuoles and the plasma membrane was 0.10 µm (s.d. = 0.15 µm, n = 49). Cortices isolated at the same stage by homogenizing method was shown in Fig. 2. When the cortex was isolated under a hypotonic condition using glucose-free CIM (Fig. 2 *a*), distorted vacuoles were situated about 1 µm apart from the plasma membrane in the cortex. Most of the microvilli seemed to have swallen. A layer of a meshwork of actin filaments, about 1 µm thick, was observed beneath the plasma membrane. When the cortex was isolated under an isotonic condition using 0.8 M glucose-containing CIM

(Fig. 2 *b*), the vacuoles seemed to be intact. The average distance between the vacuoles and the plasma membrane was 0.14 µm (s.d. = 0.18 µm, n = 23). A meshwork of actin filaments, into which the rootlets of the microvillar actin bundles penetrated, was seen in the gap between the vacuoles and the plasma membrane. Microvilli seemed to be normal in their shape and contained core actin bundles.

In the cortex isolated on a substratum under a hypotonic condition (Fig. 3 *a*), few vacuoles remained. No layer of the meshwork of actin filaments was observed although core actin bundles were detected in microvilli. When the cortex was isolated on a substratum under an isotonic condition (Fig. 3 *b*), intact vacuoles were seen close to the plasma membrane. The average distance between the vacuoles and the plasma membrane was 0.11 µm (s.d. = 0.14 µm, n = 36). Microvilli seemed to be normal. Some of core actin filament bundles of microvilli extended down into the cortex and attached to the vacuole by their side. Actin filaments were observed in the gap between the vacuoles, but they did not seem to form a well-developed meshwork (see also Fig. 3 *b*, inset). Considering that there was a layer of the vacuoles close to the plasma membrane in the freeze-substituted image (Fig. 1), the cortical struc-

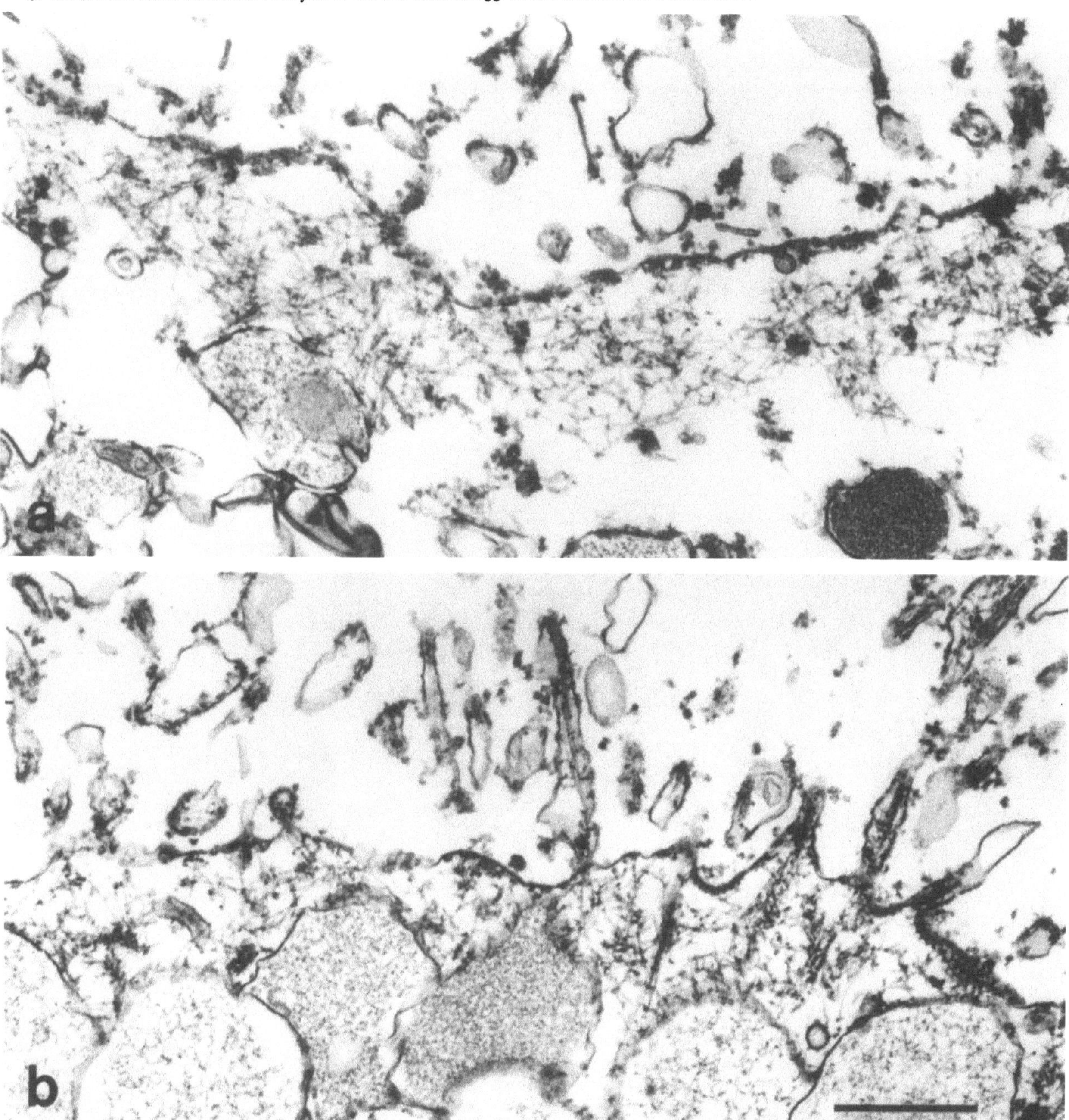

Fig. 2. Cortices of *H. pulcherrimus* eggs isolated at the streak stage by the homogenizing method. *a* A cortex isolated with a hypotonic CIM. The space between the plasma membrane and the layer of distorted vacuoles is occupied with a meshwork of actin filaments. *b* A cortex isolated with an isotonic CIM. The vacuoles seem to be intact. Some vacuoles are apart from the plasma membrane making a gap occupied with a meshwork of actin filaments. Bar, 0.5 µm. × 47,000

ture is preserved best when the cortex was isolated on a substratum under an isotonic condition. Therefore, we used the substratum method to investigate the cortical cytoskeleton of the sea urchin egg.

### 3.2. Cortical Cytoskeleton of Unfertilized Eggs

The cortical region of an unfertilized eggs observed by the freeze-substitution technique is shown in Fig. 4.

Short microvilli and cortical granules were the characteristic structure of this region. Filamentous materials which are not well organized were seen in the microvilli.

The cortex of an unfertilized egg isolated on a substratum is shown in Fig. 5 *a*. At a higher magnification (Fig. 5 *b*), several short microfilaments were seen in microvilli. Some small granules were attached on the fil-

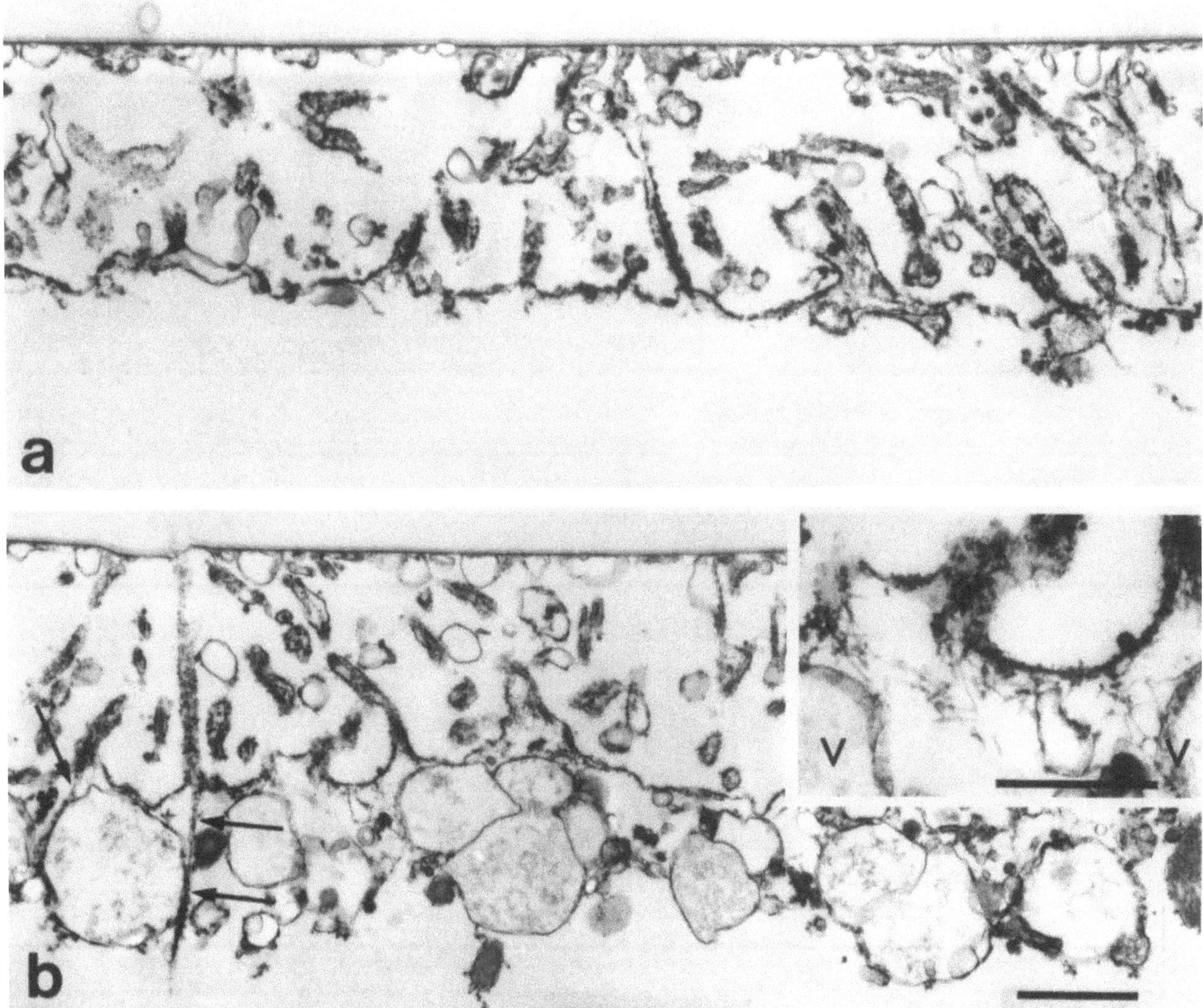

Fig. 3. Cortices of *H. pulcherrimus* eggs isolated at the streak stage by the substratum method. *a* A cortex isolated with the hypotonic CIM. Few vacuoles remain in the cortex. There is no layer of a meshwork of actin filaments. *b* A cortex isolated with the isotonic CIM. The vacuoles seem to be intact and are situated close to the plasma membrane. No continuous layer of a meshwork of actin filaments is observed. Core actin filament bundles (arrow) of microvilli project into the cortex and attach laterally to the vacuole. Bar, 1 μm. × 17,000. Inset: Relatively small number of actin filaments are seen to form a small meshwork in the gap between the vacuoles (*V*). Bar, 0.5 μm. × 46,000

ament and the filaments seemed to be cross-linked with each other by thin strands. Microfilaments slightly extended out of microvilli. There were a few microfilaments around the cortical granules. There was no electron-translucent vacuoles which were seen in the cortex isolated from the egg at the streak stage. By myosin S-1 treatment, a typical arrowhead structure was formed on each filament (Fig. 5 *c*), demonstrating that these microfilaments were actin filaments. These filaments showed the polarity with arrowheads pointing away from the plasma membrane indicating that they attached to the plasma membrane at their barbed ends.

Isolated cortices of unfertilized eggs obtained by homogenizing method were rapid-frozen, deeply etched and rotary-shadowed without fixation since it was difficult to rapid-freeze the cortices isolated on a substratum. It should be noted that there was almost no structural difference between the cortex of the unfertilized egg isolated by homogenizing method and the cortex isolated by substratum method (data not shown). The replica image showed that some actin filaments attached to the cortical granule (Fig. 5 *d*). Actin filaments were hardly identified in microvilli probably because that microvilli were filled with some materials.

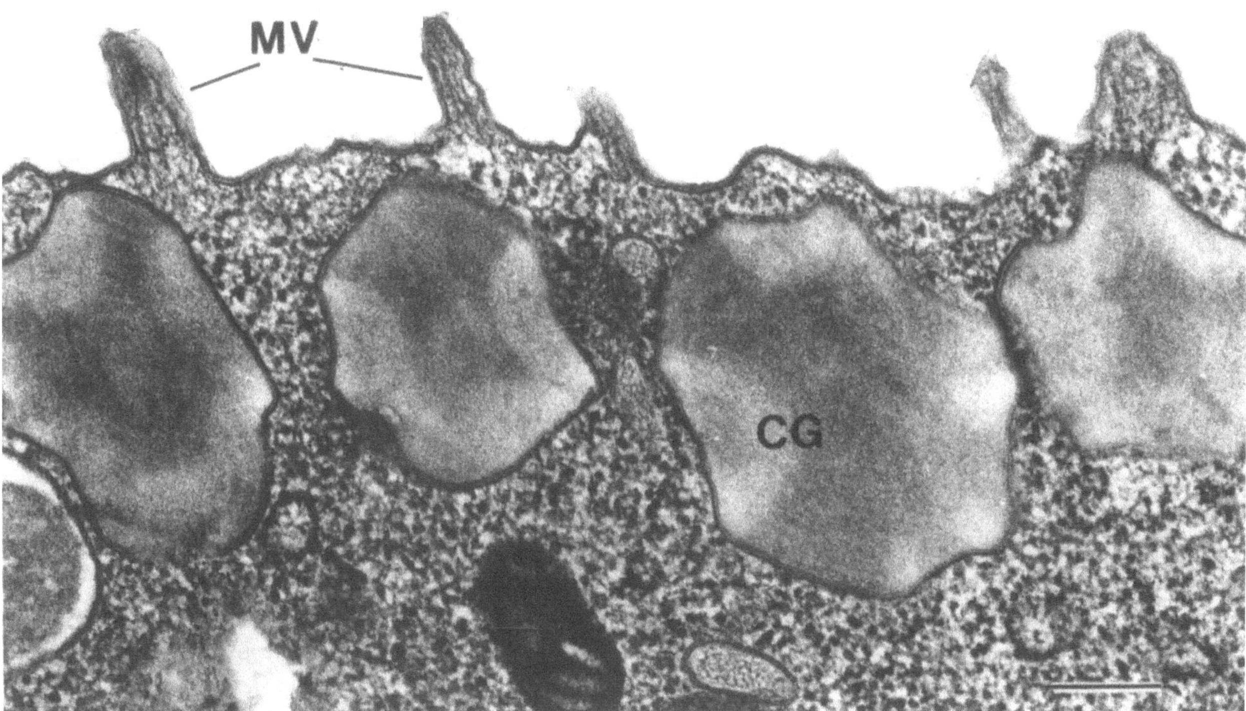


Fig. 4. A freeze-substituted unfertilized *H. pulcherrimus* egg. Filamentous materials which are not well organized are seen in the microvilli (*MV*). *CG* Cortical granule, Bar, 0.2 μm. × 73,500

### 3.3. Cortical Cytoskeleton of Eggs 2.5 min and 12 min After Insemination

The cortical region of an egg 2.5 min after insemination as observed by the freeze-substitution technique is shown in Fig. 6 a. Cortical granules had discharged. Microvilli were seen to be elongating and were not straight. In the microvilli, filamentous materials were recognized: some of them seemed to form bundles, while others were curly and less organized. The electron-translucent vacuoles were located about 1 μm away from the plasma membrane. Many small vesicles (about 0.1 μm in diameter) and some coated pits were seen in the region between them.

Next, cortex isolated from an egg at 2.5 min after insemination was examined (Fig. 6 b). The region between the plasma membrane and the vacuoles, which was about 1–1.5 μm, was filled with actin filaments. Coated pits were also the prominent structures in this region. Actin filaments seemed to gather around the coated pit (Fig. 6, inset). The shape of microvilli was severely altered. This may be an artifact induced by the attachment of the plasma membrane to the protamine-coated substratum.

At 12 min after insemination, microvilli elongated and became straight in the freeze-substitution image (Fig. 7 a). Their appearance was almost the same as that at the streak stage (see Fig. 1); the core actin filaments seemed to be cross-linked to form distinct bundles. The space between the plasma membrane and the electron-translucent vacuoles reduced to about 0.5 μm thick. In cortices isolated at this stage, the appearance of microvilli was not significantly altered (Fig. 7 b). Both the actin filaments and the coated pits in the region between the plasma membrane and the electron-translucent vacuoles reduced in number.

### 4. Discussion

We compared the ultrastructure of actin cytoskeleton in the cortex of the sea urchin egg isolated by two methods: homogenizing method and substratum method. As a result, the cortical ultrastructure was well preserved in the latter method using the isotonic isolation medium. So far, the vacuoles were seen to be situated close to the plasma membrane when the cortical region of the intact egg at every developmental stage was examined by electron microscopy except some period after fertilization (Hᴀʀʀɪs 1968, Uᴇᴍᴜʀᴀ and Eɴᴅᴏ 1976, Sᴄʜʀᴏᴇᴅᴇʀ 1981, Usᴜɪ and Yᴏɴᴇᴅᴀ

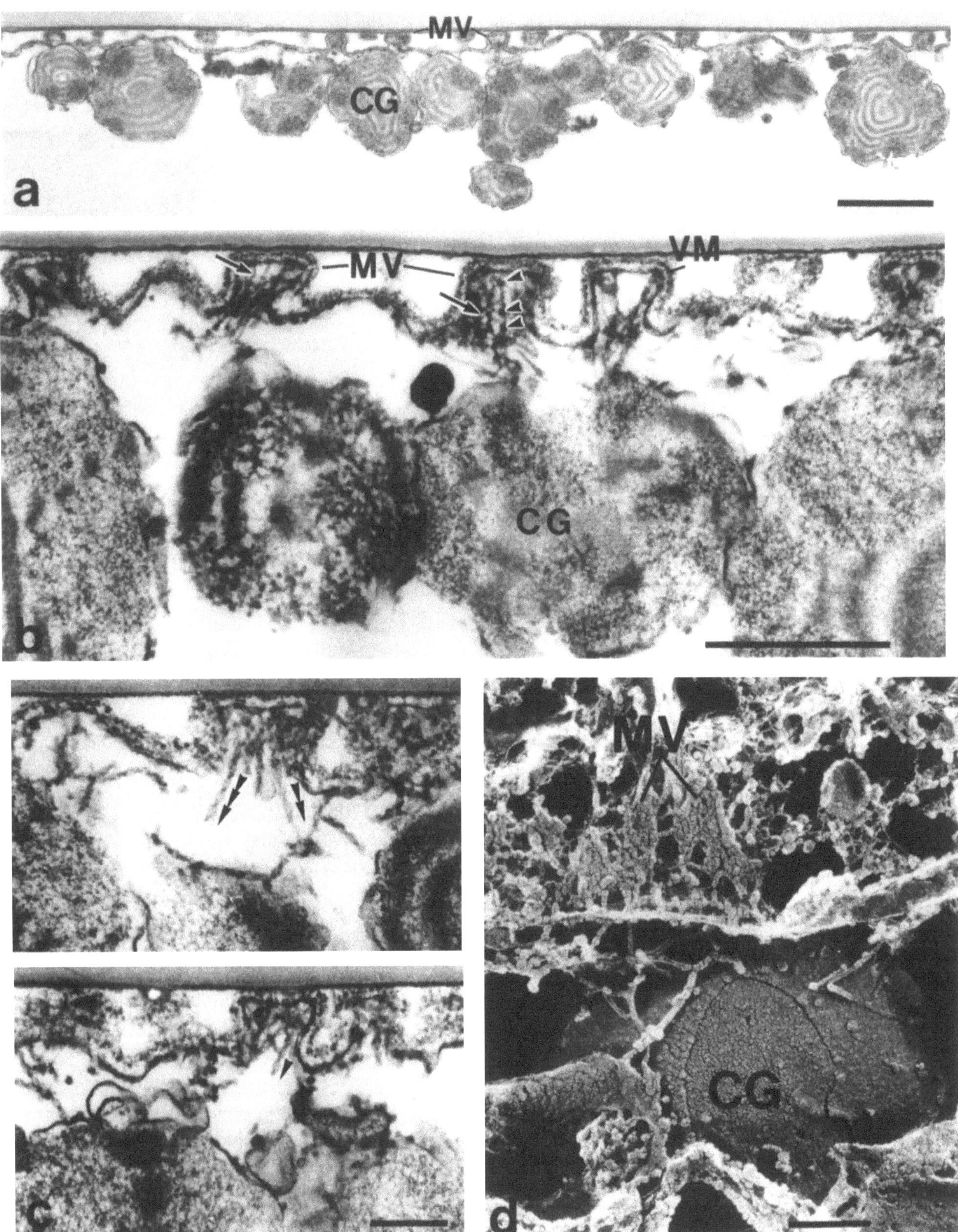

Fig. 5. Cortices isolated from *H. pulcherrimus* unfertilized eggs. *a–c* Cortices isolated by the substratum method. *a* A low magnification view. *MV* Microvilli; *CG* cortical granules. Bar, 1 µm. × 17,000. *b* A high magnification view. Note short microfilaments within short microvilli (*MV*). Some granules (arrowheads) attach on the filaments and the filaments are seemed to be cross-linked with thin threads (arrows). *CG* Cortical granule; *VM* vitelline membrane. Bar, 0.5 µm. × 67,000. *c* Cortices treated with myosin S-1. Short microfilaments bind S-1 to form arrowhead complex pointing away from the plasma membrane (arrowheads). Bar, 0.2 µm. × 69,000. *d* Freeze-etch replica image of an unfertilized egg cortex. Short actin filaments are seen to attach to the cortical granule (*CG*). *MV* Microvilli. Bar, 0.2 µm. × 71,000

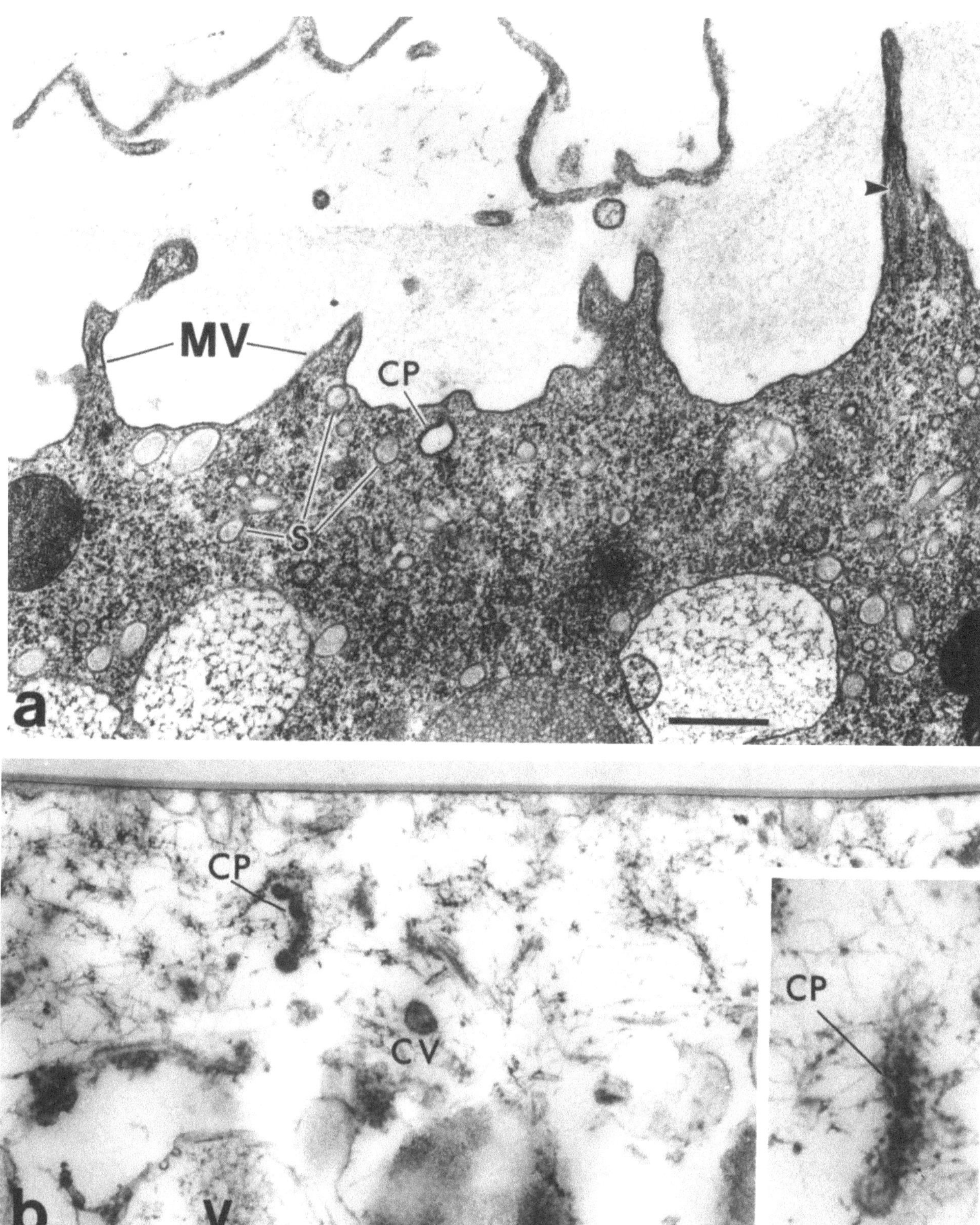

Fig. 6. A freeze-substituted egg (*a*) and an isolated cortex (*b*) at 2.5 min after insemination. *a* Microvilli (*MV*) are elongating and are not straight. Within microvilli, filamentous materials are recognized: some of them seem to form bundles (arrowheads), while the others are curly and less organized. There are electron-translucent vacuoles about 1 µm away from the plasma membrane. *CP* Coated pit; *S* small vesicles about 0.1 µm in diameter. Bar, 0.5 µm. ×36,300. *b* A layer rich in actin filaments appears. Note that the coated pit (*CP*) is associated with actin filaments (inset). *CV* Coated vesicle; *V* vacuole. Bar, 0.5 µm. ×33,000. Inset: A higher magnification of a coated pit. Bar, 0.2 µm. ×65,000

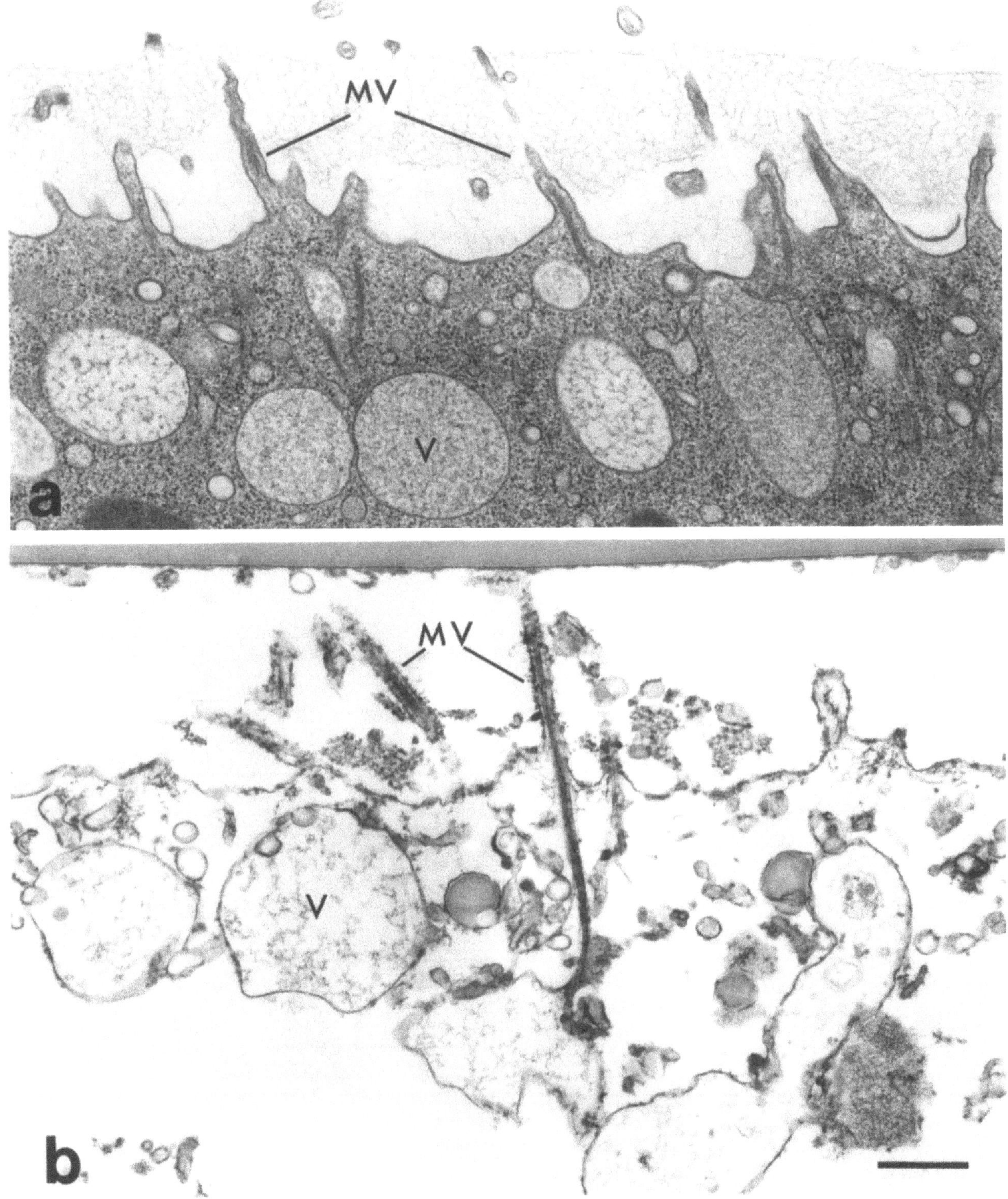



Fig. 7. A freeze-substituted egg (*a*) and an isolated cortex (*b*) at 12 min after insemination. Microvilli (*MV*) elongate and seem to have become rigid. Bar, 0.5 µm. ×30,000

1982). On the other hand, the vacuoles, if any, are situated apart from the plasma membrane in the cortex isolated by homogenizing method (Begg and Rebhun 1979, Kidd and Mazia 1980, Mabuchi *et al.* 1980, Carron and Longo 1982, Fig. 2 *a*). In such cortex, the space between the plasma membrane and the layer of vacuoles is occupied by a well-developed meshwork of actin filaments. We did not see such a well-developed meshwork in the cortex isolated by the substratum method, but saw such a meshwork in the cortex isolated by homogenizing method especially with the hypotonic isolation medium. Therefore, the well-developed meshwork may have been formed artificially.

In the cortical region of the intact egg, at least one end of the actin filament is likely to attach to the plasma membrane. The association of the actin filament and the plasma membrane or that between actin filaments may be broken by osmotic shock or shearing force during isolation by homogenization. Then, free ends may be exposed and actin polymerization may occur at these ends. There is a store of nonpolymeric actin in the egg cytoplasm (Mabuchi and Spudich 1980), the polymerization of which may be inhibited by action of actin-modulating proteins such as depactin and profilin (Mabuchi 1981, Mabuchi and Hosoya 1982). The mode of interaction of actin and these proteins may be altered in the course of the cortex isolation when the cytoplasm and the cortex are exposed to an artificial environment. Thus, it is plausible that actin polymerization occurs in the cortex during isolation by the homogenizing method. On the other hand, egg cytoplasm and soluble components in the cortex will be quickly washed out within several seconds by the substratum method, which will induce no artificial actin polymerization.

The present observation on the freeze-substituted egg supported previous ones that the electron-translucent vacuoles are situated close to the plasma membrane and that there is little room for a continuous layer of the actin filament meshwork. The ultrastructure of the cortex isolated by the substratum method was revealed to be similar to the one *in situ*. Therefore, we reexamined the structure of the cortices from eggs at different stages by the use of this substratum method.

In the unfertilized egg cortex, short actin filaments were clearly shown within short microvilli. This had also been demonstrated as numerous fluorescent spots in the isolated cortex stained with rhodamine-phalloidin (Cline *et al.* 1983, Yonemura and Mabuchi 1987), but had not been shown very clearly by electron microscopy (Chandler and Heuser 1981, Sardet 1984).

These short actin filaments were shown to attach to the plasma membrane at their barbed ends. It was also demonstrated for the first time that these filaments were cross-linked by thin strands and also associated with some granules. These short actin filaments may become nuclei of the actin polymerization which occurs after fertilization. Furthermore, the existence of actin filaments attaching to the cortical granules were confirmed by a freeze-etch replica image.

A layer consisting of numerous actin filaments were formed in the cortical region 2.5 min after fertilization. Formation of this layer corresponds to the dramatic increase in actin content in the cortex after fertilization (Spudich and Spudich 1979, Begg and Rebhun 1979, Wang and Taylor 1979, Mabuchi *et al.* 1980, Vacquier and Moy 1980, Yonemura and Kinoshita 1986, Yonemura and Mabuchi 1987). This filamentous layer also corresponds to the extra-granular zone which is formed after fertilization. This zone is subsequently occupied by a layer of electron-translucent vacuoles which migrate from the inner cytoplasm to the egg surface (Uemura and Endo 1976, Fisher and Rebhun 1983). Therefore, it is not likely that this filamentous layer was artificially induced during preparation. Microinjection of fluorescently labeled actin or alpha-actinin revealed that these proteins accumulate in the cortical layer maximally during several minutes after fertilization and then reduced in amount to reach a steady state (Hamaguchi and Mabuchi 1987, 1988). The role of this layer has not been known at all. Association of actin filaments to the coated pits may suggest the involvement of these filaments in pinocytosis at this stage (Fisher and Rebhun 1983).

Later, the filamentous layer reduced in thickness and finally confined to the region between vacuoles. At the streak stage, microvillar actin filament bundles were the prominent structure of the cortical cytoskeleton. The bundles laterally attached to the vacuole. It is plausible that the bundle link the vacuole to the plasma membrane.

The shape of the elongating microvilli at 2.5 min after insemination seemed to be altered by the attachment to the substratum comparing with the image obtained from freeze-substituted eggs, while those at 12 min after insemination and at the streak stage seemed to be less altered. It has also been reported with other sea urchin eggs that microvilli elongating after the cortical exocytosis are irregular in shape and contain less organized actin filaments (Tilney and Jaffe 1980, Chandler and Heuser 1981, Carron and Longo 1982, Begg *et al.* 1982, Fisher and Rebhun 1983). These suggest

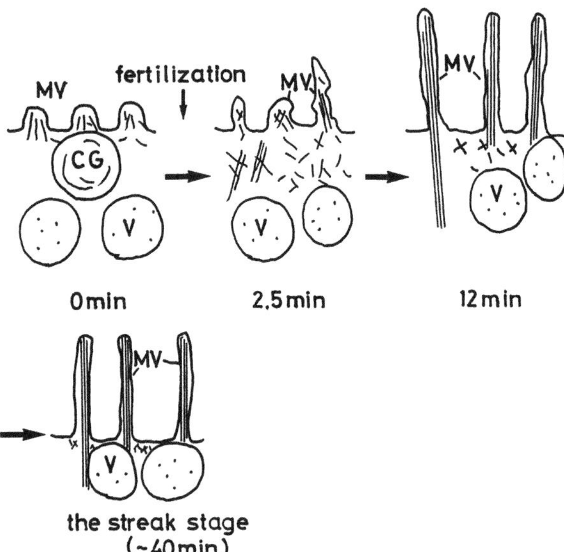


Fig. 8. A model of the sequence of the changes in cortical structures after fertilization. *CG* Cortical granule; *MV* microvilli; *V* electron-translucent vacuole

that the elongating microvilli are not rigid compared to those which accomplished elongation after fertilization, the actin bundles of which may be stabilized by cross-linkers, probably fascin (Otto *et al.* 1979). The sequence of the changes in the cortical structures deduced from the present study is illustrated in Fig. 8. It was shown that the substratum method for the isolation of cortex from the sea urchin egg is an excellent method to investigate cytoskeletal structure in the egg. Further analysis in detail will reveal the mechanism of the cortical cytoskeleton organization and the role of the cortical cytoskeleton in the egg in the course of development. By this method, cleavage furrow including the contractile ring can also be isolated. The structural analysis of the contractile ring will be described elsewhere.

## Acknowledgements

This work was supported in part by research grant from the Ministry of Education, Science and Culture. S.Y. was supported by a fellowship of the Japan Society for the Promotion of Science for Japanese Junior Scientists.

## References

Begg DA, Rebhun LI (1979) pH regulates the polymerization of actin in the sea urchin egg cortex. J Cell Biol 83: 241–248

Begg DA, Rebhun LI, Hyatt H (1982) Structural organization of actin in the sea urchin egg cortex: microvillar elongation in the absence of actin filament bundle formation. J Cell Biol 93: 24–32

Carron CP, Longo F (1982) Relation of cytoplasmic alkalinizaion to microvillar elongation and microfilament formation in the sea urchin egg. Dev Biol 89: 128–137

Chandler DE, Heuser J (1981) Postfertilization growth of microvilli in the sea urchin egg: new view from egg that have been quick-frozen, freeze-fractured, and deeply etched. Dev Biol 82: 393–400

Fisher GW, Rebhun L (1983) Sea urchin egg cortical granule exocytosis is followed by a burst of membrane retrieval via uptake into coated vesicles. Dev Biol 99: 456–472

Hamaguchi Y, Mabuchi I (1986) Alpha-actinin accumulation in the cortex of echinoderm eggs during fertilization. Cell Motil Cytoskeleton 6: 549–559

— — (1988) Accumulation of fluorescently labeled actin in the cortical layer in sea urchin eggs after fertilization. Cell Motil Cytoskeleton 9: 153–163

Harris P (1968) Cortical fibers in fertilized eggs of the sea urchin *Strongylocentrotus purpuratus*. Exp Cell Res 52: 677–681

Heuser JE, Reese TS, Dennis MJ, Jan Y, Jan L, Evans L (1979) Synaptic vesicle exocytosis captured by quick freezing and correlated with quantal transmitter release. J Cell Biol 81: 275–300

Kidd P, Mazia D (1980) The ultrastructure of surface layers isolated from fertilized and chemically stimulated sea urchin eggs. J Ultrastruct Res 70: 58–69

Mabuchi I (1981) Purification from starfish eggs of a protein that depolymerizes actin. J Biochem 89: 1341–1344

— (1986) Biochemical aspects of cytokinesis. Int Rev Cytol 101: 175–213

— Hosoya H (1982) Actin-modulating proteins in the sea urchin egg. II. Sea urchin egg profilin. Biomed Res 3: 465–476

— — Sakai H (1980) Actin in the cortical layer of the sea urchin egg. Changes in its content during and after fertilization. Biomed Res 1: 417–426

— Spudich JA (1980) Purification and properties of soluble actin from sea urchin eggs. J Biochem 87: 785–802

Otto JJ, Kane RE, Bryan J (1979) Redistribution of actin and fascin in sea urchin eggs after fertilization. Cell 17: 285–293

Sardet C (1984) The ultrastructure of the sea urchin egg cortex isolated before and after fertilization. Dev Biol 105: 196–210

Schroeder TE (1975) Dynamics of the contractile ring. In: Inoue S, Stephens RE (eds) Molecules and cell movement. Raven Press, New York, pp 305–334

— (1981) Interaction between the cell surface and the cytoskeleton in cleaving sea urchin eggs. In: Poste G, Nicolson GL (eds) Cytoskeletal elements and plasma membrane organization. Elsevier/North Holland, Amsterdam, pp 170–216

Spudich A, Spudich JA (1979) Actin in triton-treated cortical preparations of unfertilized and fertilized sea urchin eggs. J Cell Biol 82: 212–226

Tsukita S, Usukura J, Tsukita S, Ishikawa H (1982) The cytoskeleton in myelinated axons: a freeze-etch replica study. Neuroscience 7: 2135–2147

— Yano M (1985) Actomyosin structure in contracting muscle detected by rapid freezing. Nature 317: 182–184

Uemura I, Endo Y (1976) Electron microscopic observations in the extra-granular zone of the embryo of the sea urchin, *Hemicentrotus pulcherrimus*. Dev Growth Differ 18: 399–406

Usui N, Yoneda M (1982) Ultrastructural basis of the tension increase in sea urchin eggs prior to cytokinesis. Dev Growth Differ 24: 453–465

Vacquier VD (1981) Dynamic changes of the egg cortex. Dev Biol 84: 1–26

— Moy GW (1980) The cytolitic isolation of the cortex of the sea urchin egg. Dev Biol 77: 178–190

Wang YL, Taylor DL (1979) Distribution of fluorescently labeled actin in living sea urchin eggs during early development. J Cell Biol 82: 672–679

Yonemura S, Kinoshita S (1986) Actin filament organization in the sand dollar egg cortex. Dev Biol 117: 171–183

— Mabuchi I (1987) Wave of cortical actin polymerization in the sea urchin egg. Cell Motil Cytoskeleton 7: 46–53

Protoplasma (1988) [Suppl. 2]: 116–130



# The Actin Cytoskeleton and Polar Water Permeability in Characean Cells

RANDY WAYNE** and M. TAZAWA*

Department of Botany, Faculty of Science, University of Tokyo

Received February 23, 1988
Accepted May 7, 1988

Dedicated to Professor Dr. NOBURO KAMIYA on the occasion of his 75th birthday

## Summary

The hydraulic resistance was measured on internodal cells of *Chara corallina* and *Nitellopsis obtusa* using the method of transcellular osmosis described by KAMIYA and TAZAWA in 1956. The transcellular hydraulic resistances of *N. obtusa* and *C. corallina* are 2.63 and 1.11 pm$^{-1}$ s Pa, resepectively. These values correspond to osmotic permeability coefficients ($P_{os}$) of 102.6 and 243.1 µm s$^{-1}$, respectively. Cytochalasin A, B, and E (1–30 µg ml$^{-1}$) increase the hydraulic resistance in a concentration-dependent manner. The order of effectiveness is: CE > CA = CB. CE increases the activation energy of water transport from 16.4 kJ mol$^{-1}$ to 32.5 kJ mol$^{-1}$ indicating that it increases the hydraulic resistance by eliminating a low resistance pathway. Cytochalasin B and E specifically increase the hydraulic resistance to endoosmosis; even when the driving force for transcellular osmosis is as low as 0.06 MPa. The effect of the cytochalasins is independent of their effect on cytoplasmic streaming since stopping streaming with N-ethyl maleimide or electrical stimulation has no effect on hydraulic conductivity. The possibility is discussed that a cortical actin cytoskeleton interacts with the plasma membrane in order to regulate the transport of water.

*Keywords:* Actin cytoskeleton; *Chara corallina*; Cytochalasin; *Nitellopsis obtusa*; Polarity; Water transport.

*Abbreviations:* APW artificial pond water; CA cytochalasin A; CB cytochalasin B; CE cytochalasin E; DMSO dimethylsulfoxide; NEM N-ethyl maleimide; K transcellular osmotic constant (pico m$^3$ s$^{-1}$ Pa$^{-1}$); k' transcellular hydraulic permeability coefficient (pm s$^{-1}$ Pa$^{-1}$); R transcellular hydraulic resistance (pico m$^{-3}$ s Pa); $r_{tot}$ total transcellular hydraulic resistance (pm$^{-1}$ s Pa); $L_p$ hydraulic conductivity (pm s$^{-1}$ Pa$^{-1}$); $L_{pen}$ endoosmotic hydraulic conductivitiy (pm s$^{-1}$ Pa$^{-1}$); $L_{pex}$ exoosmotic hydraulic conductivity (pm s$^{-1}$ Pa$^{-1}$); $L_{pen}^{-1}$ endoosmotic hydraulic resistance (pm$^{-1}$ s Pa); $L_{pex}^{-1}$ exoosmotic hydraulic resistance (pm$^{-1}$ s Pa); $J_v$ rate of water flow (pico m$^3$ s$^{-1}$); $\bar{V}_w$ partial molar volume of water (pico m$^3$ mol$^{-1}$).

* Correspondence and Reprints: Department of Botany, Faculty of Science, University of Tokyo, Hongo, Tokyo 113, Japan.

** Present address: Section of Plant Biology, Cornell University, Ithaca, NY 14853, U.S.A.

## 1. Introduction

In 1956 KAMIYA and TAZAWA used the method of transcellular osmosis to describe quantitatively the movement of water through a single internodal cell of *Nitella*. The method consists simply of dividing a cell between two compartments and replacing the water in one compartment with a solution of nonelectrolytes like sucrose. The solution establishes a difference in the water potential gradient across the plasma membrane between the water and solution sides, and thus drives water out of one cell end (the exoosmotic end) and draws water into the other end of the cell (the endoosmotic end). Water thus moves transcellularly (OSTERHOUT 1949 a, b). The transcellular osmosis induced by subjecting the exoosmotic cell half to a solution is called forward transcellular osmosis and the transcellular osmosis induced by replacing the solution again with water is called backward transcellular osmosis.

The forward transcellular osmosis takes place in two phases; an initial rapid exponential phase and a subsequent slower linear phase which lasts indefinitely. KAMIYA and TAZAWA (1956) explained that the slowing down of water movement is a result of the establishment of a polar distribution of intracellular solutes, *i.e.*, dilution of solutes on the endoosmotic side and concentration of solutes on the exoosmotic side. But the transcellular water movement never stops as a consequence of the depolarization of these solutes by an active cytoplasmic streaming. In order to test this hypothesis we added cytochalasin B to cells of *Chara corallina* and *Nitellopsis obtusa* to inhibit streaming and then observed what happened to the transcellular transport of water. Contrary to our expectation, we found that CB markedly inhibits the initial rapid water movement but has a lesser effect on the continuous, linear transcellular transport of water.

Kamiya and Tazawa (1956) observed that transcellular water transport in *Nitella flexilis* occurs in a polar fashion. That is the hydraulic resistance to exoosmosis is greater than that to endoosmosis. This polarity is also observed in higher plant cells in that the rate constant for deplasmolysis is greater than the rate constant for plasmolysis (Höfler 1930, Levitt *et al.* 1936). There has been much controversy over whether the observed polarity is real or only apparent as a result of the unstirred layers because the outflowing water on the exoosmosis side will sweep away the osmoticum and consequently reduce the driving force on the exoosmotic side (Dainty and Hope 1959, Dainty 1963 a, b, Kiyosawa and Tazawa 1973). However, a polarity still exists, albeit somewhat reduced even after the unstirred layers are taken into consideration (Dainty 1963 a, b, Dainty and Ginzburg 1964 a, Tazawa and Kamiya 1965, 1966).

There still remained the question of whether the observed polarity is an intrinsic character of the living cell or is a result of a nonspecific dehydration by the osmoticum of the membrane on the exoosmotic side (Dainty and Ginzburg 1964 b, Kiyosawa and Tazawa 1972, 1973, Tazawa and Kiyosawa 1973). Using the pressure probe technique, Steudle and Zimmermann (1974) induced a water potential gradient across the membrane of *Nitella flexilis* hydrostatically thus eliminating any effect of dehydration on the membrane by an osmoticum. With this protocol they observed a polarity in the transmembrane transport of water. Furthermore the resistance to outward water movement was higher than the resistance to inward water movement. However, using smaller hydrostatic gradients Steudle and Tyerman (1983) were unable to observe any polarity in *Chara corallina*, indicating perhaps that a minimum flow rate is required for the manifestation of a polarity.

Here we present our observations on transcellular water movement in *Chara* and *Nitellopsis* and confirm the suggestion made by Kamiya and Tazawa (1956) that the polarity is an intrinsic character of the living cell and show that actin microfilaments participate in the establishment of the cellular polarity which reduces the hydraulic resistance on the endoosmotic side.

## 2. Materials and Methods

### 2.1. Plant Materials

*Chara corallina* Klein ex Willd., em. R.D.W. (= *Chara australis* R. Brown) and *Nitellopsis obtusa* (Desv. in Lois.) J. Gr. were grown in a soil water mixture in large plastic buckets at 25 ± 2 C with 15 hL: 9 hD photoperiod. Internodal cells of *Chara corallina* or *Nitellopsis obtusa* were isolated and placed in artificial pond water (APW: 0.1 mM NaCl, 0.1 mM CaCl$_2$ and 0.1 mM KCl) buffered by 10 mM Mes titrated with Tris to pH 5.5. The cells were then placed on a shaker for 3 h to remove the CaCO$_3$ deposited on the wall. The cells were then transferred to APW buffered by 2 mM Hepes-NaOH (pH 7.3). The cells remained in this medium at least overnight before use.

### 2.2. Measurement of Transcellular Osmosis

The rate of transcellular osmosis was measured in the apparatus shown in Fig. 1. This apparatus is modeled after the apparatus designed by Tazawa and Kamiya (1966). Internodal cells were first treated in 1% DMSO (0.136 Osm = 0.33 MPa at 20 °C) in a covered chamber for 35 minutes and subsequently rinsed in water for 1 min. The cell was then placed in the apparatus such that the length of the cell part in chamber B was two times the length of the cell part in chamber A. Chamber A and chamber B were physically separated by a 4–7 mm (depending on the chamber) silicone seal (HVG, Toray Silicone, Tokyo, Japan) and the cell parts in both chambers A (endoosmotic side) and B (exoosmotic side) were bathed in water. Forward transcellular osmosis was initiated by replacing the solution in chamber B with 100 mM sorbitol (0.244 MPa at 20 °C) and the quantity of water movement was recorded every 10 seconds for 40–70 seconds. The magnitude of water movement was measured by following the rate of movement of a column of water in a glass capillary with an Olympus CH microscope equipped with a × 4 objective lens in the case of *Chara* or a × 10 objective lens in the case of *Nitellopsis*. The sensitivity of the apparatus is about 5 nl, which was approximately 1% of the water volume transported within the first 60 seconds. The water in the capillary was continuous with the solution in chamber A. The solution in chamber B was replaced with water

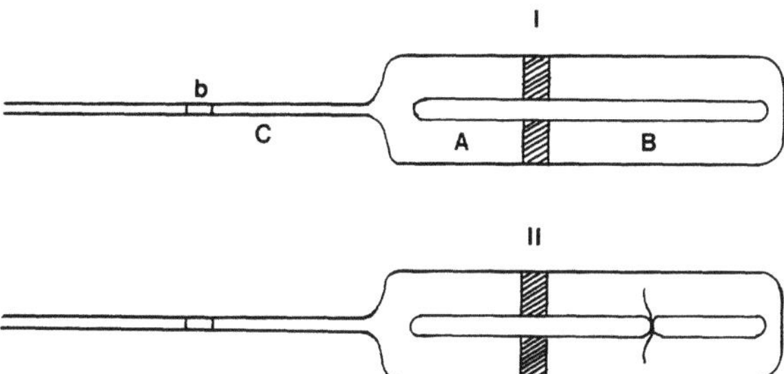



Fig. 1. A diagram (top view) of the apparatus used for transcellular osmosis. The initial condition where the cell part in chamber *A* is one-half the length of the cell part in chamber *B* is shown in *I*. After ligation, the cell lengths in chambers *A* and *B* that can contribute to transcellular osmosis are equal (*II*). The volumes of chambers *A* and *B* were 750 and 2250 μl, respectively. The volume of the capillary (*c*) was 615.3 nl mm$^{-1}$. The flux of water was determined by following the movement of an air bubble (*b*) in the capillary. When electrical stimulation was given to the cells, silver wires were placed along the length of chambers *A* and *B*

to initiate backward transcellular osmosis and the cells were allowed to reach their original equilibrium conditions during the next 10 minutes except when stated otherwise. Subsequently, the cell part in chamber B was ligated (Kamiya and Kuroda 1956) at its middle so that the length of the cell in chamber B that could participate in transcellular osmosis was equal to the length of the cell part in chamber A. Again transcellular osmosis was initiated with 100 mM sorbitol and was recorded every 10 seconds for 40–70 seconds. At the end of the experiment, the cell end distal to the ligation was cut off to ensure that the ligation was complete.

The cell could also be placed asymmetrically in the apparatus in a way that the length of the cell part in chamber A was equal to one-half of the length of the cell part in chamber B and the water flow was measured. Then the cell was removed and reset so that the length of the cell part in chamber B was equal to one-half the length of the cell part in chamber A and again water flow was measured. Experiments done in this manner yielded values of $L_{pen}^{-1}$, $L_{pex}^{-1}$ and $r_{tot}$ similar to those obtained with the ligation method (data not shown). In order to eliminate any interference from the small volume change that occurs during the onset of transcellular osmosis (0–5 sec) (Tazawa and Kamiya 1966) or the buildup of unstirred layers (more than 40 seconds after the onset of transcellular osmosis) the amount of water movement between 10 and 40 seconds and 10 and 30 seconds was used to calculate the rate of transcellular osmosis in *Chara* and *Nitellopsis*, respectively. Experiments were performed at room temperature (20–26 °C) unless stated otherwise.

### 2.3. Determination of the Total Hydraulic Resistance Using Transcellular Osmosis

The rate of transcellular osmosis was measured in the apparatus described above. Internodal cells were first treated in 1% DMSO (0.136 Osm = 0.32 MPa at 20 °C) in a covered chamber for 35 minutes and subsequently rinsed in water for 1 min. The cells were then placed symmetrically in water in the apparatus. Forward transcellular osmosis was initiated by replacing the solution in chamber B with 100 mM sorbitol (0.100 Osm = 0.237 MPOa at 20 °C) and the quantity of water movement was recorded every 10 seconds for 40–70 seconds by measuring the rate of movement of an air bubble as detailed above. The total transcellular hydraulic resistance was determined from the following equation (Kiyosawa and Tazawa 1972, Tazawa 1980):

$$r_{tot} = \frac{A \Delta \Pi}{J_v}$$

where $J_v$ is the observed flow of water movement in pico $m^3/s$; A is the surface area (in pico $m^2$) of the cell part in each chamber; and $\Delta \Pi$ is the osmotic gradient (in Pa) used to drive transcellular osmosis. The total hydraulic resistance is given in $pm^{-1}$ s Pa.

### 2.4. Analysis of Results

The hydraulic conductivities for endoosmosis ($L_{pen}$) and exoosmosis ($L_{pex}$) were calculated using the following equations. The general equations can be found in Kamiya and Tazawa (1956) and they are based on the following assumptions: (1) The membranes form the main resistance to water permeation; and (2) The resistance to endoosmosis and the resistance to exoosmosis are independent, both work in series and can be summed.

The initial rate of water flow ($J_v$, in pico $m^3 s^{-1}$) is proportional to the osmotic gradient ($\Delta \Pi$, in Pa) used to initiate transcellular osmosis.

$$J_v = K \Delta \Pi \tag{1}$$

The transcellular osmosis constant (K) (Tazawa and Kamiya 1965) was calculated by dividing the initial flux of transcellular water movement by the magnitude of the osmotic gradient.

$$K = J_v / \Delta \Pi \tag{2}$$

K (pico $m^3$ $s^{-1}$ $Pa^{-1}$) is a constant that is given as $A_{en}A_{ex}L_{pen}L_{pex}/(A_{en}L_{pen} + A_{ex}L_{pex})$. The transcellular osmotic constant (K) is equal to the transcellular hydraulic permeability coefficient (k′, in pm $s^{-1}$ $Pa^{-1}$) times the surface area of half of a cell (A, in pico $m^2$) when the cell is equally partitioned.

$$K = k' A \tag{3}$$

where k′ is given as $L_{pen}L_{pex}/(L_{pen} + L_{pex})$.

In the initial transcellular osmosis before ligation $A_{en} = A_{ex}/2 = A$ and the transcellular osmotic constant ($K_1$) equals:

$$K_1 = \frac{2 A L_{pen} L_{pex}}{(L_{pen} + 2 L_{pex})} \tag{4}$$

After ligation, $A_{en} = A_{ex} = A$ and the transcellular osmotic constant ($K_2$) equals:

$$K_2 = \frac{A L_{pen} L_{pex}}{(L_{pen} + L_{pex})} = k' A \tag{5}$$

The polarity of water movement ($\alpha$) is defined as the ratio of $L_{pen}$ to $L_{pex}$. That is:

$$\alpha = L_{pen}/L_{pex} \tag{6}$$

We can substitute $L_{pen}$ in Eqs. (4) and (5) with $\alpha L_{pex}$ to get the following equations:

$$K_1 = \frac{2 \alpha L_{pex}}{\alpha + 2} \tag{7}$$

$$K_2 = \frac{\alpha A L_{pex}}{\alpha + 1} \tag{8}$$

From Eqs. (7) and (8) we get:

$$\alpha = \frac{2(K_1 - K_2)}{2 K_2 - K_1} \tag{9}$$

Rearranging Eq. (8) we can solve for $L_{pex}$

$$L_{pex} = \frac{K_2(\alpha + 1)}{\alpha A} \tag{10}$$

We can now calculate $L_{pen}$ from Eq. (6) as $\alpha L_{pex}$

$$L_{pen} = \alpha L_{pex} = K_2(\alpha + 1)/A \tag{11}$$

The reciprocal of $L_{pen}$ and $L_{pex}$ are the transcellular hydraulic resistances for endoosmosis and exoosmosis, respectively. When the cell is partitioned in equal halves ($A_{en} = A_{ex} = A$),

$$R = (L_{pen}^{-1} + L_{pex}^{-1}) A^{-1} \tag{12}$$

Then we can introduce the total transcellular hydraulic resistance ($r_{tot}$) in units of $pm^{-1}$ s Pa.

$$r_{tot} = L_{pen}^{-1} + L_{pex}^{-1} \tag{13}$$

Therefore

$$r_{tot} = AR = A/K = 1/k' \tag{14}$$

where $r_{tot}$ is in units of $pm^{-1}$ s Pa, A is in units of pico $m^2$, R is in units of pico $m^{-3}$ s Pa and K is in units of pico $m^3$ $s^{-1}$ $Pa^{-1}$. Assuming that there is no polarity, or that $L_{pen} = L_{pex} = L_p$, we get from Eq. (13)

$$L_p = 2 r_{tot}^{-1} \tag{15}$$

where $L_p$ is the hydraulic conductivity across a single membrane.

### 2.5. Measurement of Cellular Osmotic Pressure

Cellular osmotic pressure was measured by the turgor balance method (Tazawa 1957). Solution osmotic pressures were periodically tested by the turgor balance method and the vapor pressure osometer (Wescor 5100C) making sure to pay attention to the ill effect of DMSO on the osmometer chamber.

### 2.6. Measurement of Membrane Potential ($E_m$), Membrane Resistance ($R_m$), and Membrane Excitability During Transcellular Osmosis

Figure 2 shows the experimental setup for measuring $E_m$ and $R_m$ during transcellular osmosis. The setup was similar to that described by Hayama et al. (1979). The cells were equally partitioned into two 1 cm diameter pools that were separated by a 1 cm vaseline seal. The cells were bathed in APW containing 1% DMSO or APW containing 1% DMSO and 30 µg ml$^{-1}$ CE. A glass Ag−AgCl microcapillary electrode filled with 3 M KCl was inserted into the vacuole of the cell half in chamber B. An agar bridge containing 100 mM KCl was used as the reference electrode. The electrical resistance of the membrane was determined by measuring the change in the membrane potential after applying rectangular 0.2 µA current pulses through Ag−AgCl wires placed in the two pools. The resistance due to components other than membranes was small (Hayama et al. 1979) and was neglected. $E_m$ and $R_m$ of the endoosmotic side during forward osmosis was measured by replacing the solution in chamber A with APW containing 0.4 M sorbitol. $E_m$ and $R_m$ during backward osmosis of this same side was measured by replacing the sorbitol solution with the original solution. $E_m$ and $R_m$ of the exoosmotic side during forward osmosis was measured by replacing the solution in chamber B with APW containing 0.4 M sorbitol. $E_m$ and $R_m$ of this same side during backward osmosis was measured by replacing the sorbitol solution with the original solution.

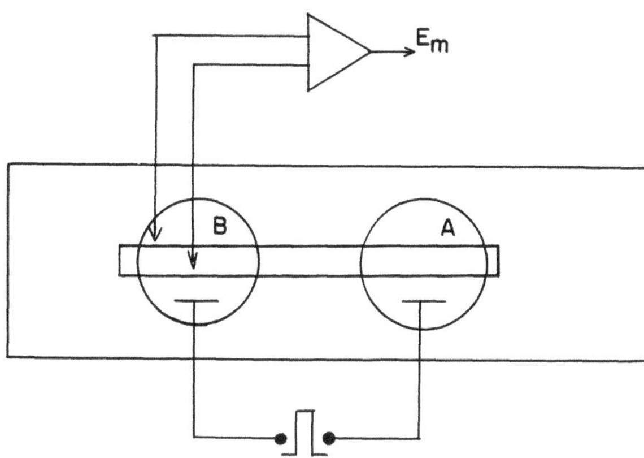



Fig. 2. A diagram of the experimental setup used for measuring membrane potential ($E_m$) and resistance ($R_m$) during transcellular osmosis. See text for details

### 2.7. Extraction and Measurement of Intracellular ATP Content

Cells were rapidly frozen in liquid nitrogen after a 35 minutes treatment in either 1% DMSO or 30 µg ml$^{-1}$ CE. ATP was extracted in a boiling buffer which included 25 mM K$^+$-Hepes, 10 mM K$^+$-EDTA and 0.3% H$_2$O$_2$ (pH 7.4). The extract was analysed by the firefly-flash method with an ATP photometer (Chemglow photometer J4-7441; Aminco, Silver Spring, Md., U.S.A.) following the method of Mimura et al. (1984).

### 2.8. Chemicals

In all experiments DMSO was used as a carrier. Cytochalasin A, B, and E, N-ethyl maleimide (NEM) and colchicine were all dissolved in DMSO (so that the final concentration of DMSO always was one percent) and the cells were treated with these solutions for 35 minutes. Preliminary experiments showed that the rates of osmosis were similar whether the drugs were present or absent in the chambers during transcellular osmosis. As an economic measure, therefore, we eliminated the drugs from the chambers during transcellular osmosis.

Cytochalasins A, B, and E, and colchicine were obtained from Sigma. Cytochalasin B was also obtained from Aldrich. NEM was obtained from Nakarai Chemical Co.

## 3. Results

### 3.1. Hydraulic Resistance

Internodal cells of *Nitellopsis* and *Chara* have a total transcellular hydraulic resistance ($r_{tot}$) of 2.63 and 1.11 pm$^{-1}$ s Pa, respectively. This corresponds to a hydraulic conductivity ($L_p$) of 0.76 pm s$^{-1}$ Pa$^{-1}$ for *Nitellopsis* and 1.80 pm s$^{-1}$ Pa$^{-1}$ for *Chara*. If the volume flow is expressed as the flow of water molecules, then the units of $L_p$ can be converted to the units of the osmotic permeability coefficient ($P_{os}$) by multiplying $L_p$ with RT/$\bar{V}_w$ (Dainty 1963 b). $P_{os}$ for *Nitellopsis* is 102.6 µm s$^{-1}$ and 243.1 µm s$^{-1}$ for *Chara* (Table 1). Although these values are similar to values

Table 1. *The total transcellular hydraulic resistances, hydraulic conductivities and osmotic permeability coefficients of Nitellopsis obtusa and Chara corallina*

| | $r_{tot}$ | $L_p$ | $P_{os}$ |
|---|---|---|---|
| | pm$^{-1}$ s Pa | pm s$^{-1}$ Pa$^{-1}$ | µm s$^{-1}$ |
| Species | | | |
| *N. obtusa* (n = 47) | 2.63 ± 0.11 | 0.76 | 102.6 |
| *C. corallina* (n = 51) | 1.11 ± 0.07 | 1.80 | 240.54 |

The total transcellular hydraulic resistances ($r_{tot}$) were determined for the cells which were firstly treated with 1% DMSO for 35 min and then rinsed with distilled water for 1 min. After temperature equilibration cells were subjected to transcellular osmosis using a driving force of 100 mM sorbital. The hydraulic conductivities ($L_p$) and the osmotic permeability coefficient ($P_{os}$) were calculated from the following formulas: $L_p = 2(1/r_{tot})$ and $P_{os} = L_p RT/\bar{V}_w$ (where $L_p$ is the hydraulic conductivity in pm s$^{-1}$ Pa$^{-1}$, R is the gas constant), T is the absolute temperature in K (set to 298 K) and $\bar{V}_w$ is the partial molar volume of water (Hansson Mild and Løvtrup 1985). Please note that the total transcellular hydraulic resistance represents the total resistance to both endoosmosis and exoosmosis; therefore $L_p$ represents the average hydraulic conductivity. The value for the total transcellular hydraulic resistance represents the mean ± SEM for measurements made from May 1987 to February 1988

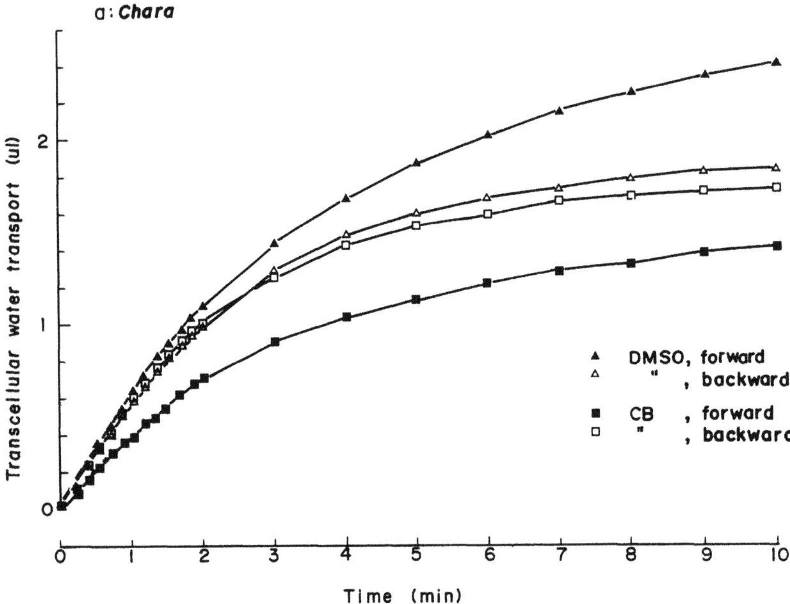


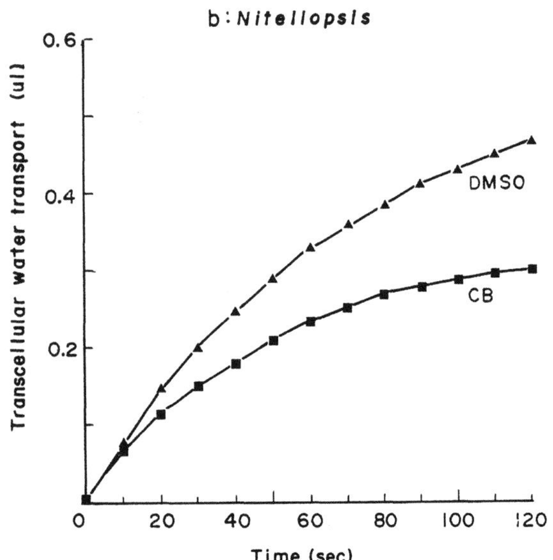


Fig. 3. The time course of transcellular osmosis in *Chara* (*a*) and *Nitellopsis* (*b*). In the case of *Chara*, the cell was treated with 1% DMSO for 35 min and placed symmetrically in a chamber that had 1% DMSO in both chambers *A* and *B*. Forward transcellular osmosis was initiated by the addition of 1% DMSO plus 100 mOsm sorbitol to chamber *B*. After 15 minutes, the sorbitol solution was replaced with 1% DMSO and backward transcellular osmosis was observed for 15 minutes. Subsequently, the same cell was taken out and treated with 100 µg ml$^{-1}$ CB (in 1% DMSO) for 35 minutes and subjected to forward and backwards transcellular osmosis as described above except CB was included in all the solutions. In the case of *Nitellopsis*, the cell was treated first with 1% DMSO for 35 min, rinsed in water and symmetrically placed in the apparatus which had distilled water on both sides. Subsequently, the water in chamber *B* was replaced with 100 mOsm sorbitol and forward transcellular osmosis was followed for 2 min. Then the same cell was treated with 30 µg ml$^{-1}$ CB (in 1% DMSO) for 35 minutes, rinsed and subjected to forward transcellular osmosis. Each cell is a representative experiment. More than 10 cells of each species were used. Notice the different scales of the two abscissae

obtained with characean cells by other workers they are approximately 20–300 times higher than the values obtained with other plant cells (Kamiya and Tazawa 1956). The difference in the hydraulic resistances of *Chara* and *Nitellopsis* are reproducible throughout the year.

### 3.2. Effects of Cytochalasins A, B, and E on the Total Transcellular Hydraulic Resistance

Cytochalasin B inhibits transcellular water movement in the internodal cells of *Chara* and *Nitellopsis* (Fig. 3). Cytochalasin B specifically inhibits forward transcel-

lular osmosis, but has almost no effect on backward transcellular osmosis (Fig. 3 *a*). Cytochalasin B inhibits the initial rate of forward transcellular osmosis during the exponential phase (0–40 sec). However, the slow linear phase (5–10 min) still continues indefinitely albeit at a reduced rate. The inhibition of water movement results from the ability of CB to increase the hydraulic resistance to transcellular water movement. Low concentrations of CB (3–30 µg ml$^{-1}$ or 6–60 µM) increase $r_{tot}$ in a concentration dependent manner (Figs. 4 and 6 *b*). After washing away CB, the effects of 30 µg ml$^{-1}$ CB are always completely reversible (Fig. 5); the

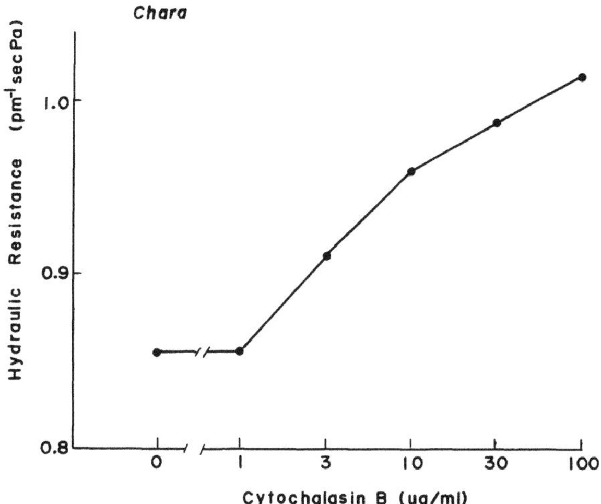



Fig. 4. The effect of cytochalasin B on the total transcellular hydraulic resistance ($r_{tot}$) in *Chara*. One and the same cell was treated with increasing concentrations of CB. The cell was placed symmetrically in the apparatus and water was in chambers *A* and *B*. The water in chamber *B* was replaced with 100 mOsm sorbitol to initiate forward transcellular osmosis. After a 4.5 hr wash in water, the hydraulic resistance returned to the control value (o)

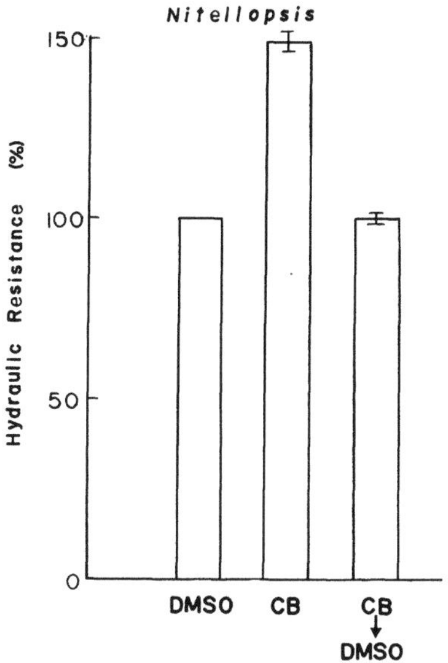



Fig. 5. The reversibility of the cytochalasin B induced increase in the total transcellular hydraulic resistance ($r_{tot}$) in *Nitellopsis*. Cells were sequentially treated with 1% DMSO, 30 µg ml⁻¹ CB and 1% DMSO for 35 min each and after each treatment subjected to transcellular osmosis. Each bar represents the mean ± SEM for 3 cells. The 100% value is equal to 2.02 ± 0.24 pm⁻¹ s Pa

effects of 100 µg ml⁻¹ are sometimes reversible (Fig. 4). By contrast, Steudle and Tyerman (1983) found that CB had no effect on $L_p$ in *Chara corallina*. The reason for the discrepancy remains unknown although it may

be related to the fact that their control cells show no polarity.

Cytochalasin A (CA) and cytochalasin E (CE) also increase $r_{tot}$ in both *Chara* and *Nitellopsis* (Fig. 6 *a* and *b*). In both genera, CE is the most effective of the three tested; 1 µg ml⁻¹ gives a half-maximal response. The effects of CE are irreversible. Cytochalasin A is slightly less effective than CB in *Chara* and slightly more effective than CB in *Nitellopsis*.

### 3.3. The Effect of Cytochalasin B on the Cellular Osmotic Pressure

The cellular osmotic pressure has an effect on transcellular water movement (Kiyosawa and Tazawa 1972, 1973). In order to test whether CB inhibits transcellular osmosis by altering the cellular osmotic pressure, we determined the cellular osmotic pressure of the living cell by the turgor balance method. Five cells of each species were measured. Compared to the untreated controls, 1% DMSO increases the cellular osmotic pressure from 0.253 Osm (0.617 MPa at 20 °C) to 0.292 Osm (0.712 MPa at 20 °C) in *Chara* and from 0.296 Osm (0.702 MPa at 20 °C) to 0.338 Osm (0.824 MPa at 20 °C) in *Nitellopsis*. CB does not have any effect on the cellular osmotic pressures when compared to the DMSO control. In 100 µg ml⁻¹ CB, the cellular osmotic pressure is 0.292 Osm in *Chara* and 0.338 Osm in *Nitellopsis*. These data also show that CB has no noticable effect on the membrane permeability to sorbitol.

### 3.4. The Effect of Cytochalasin E on the Activation Energy of Transcellular Osmosis

Cells of *Nitellopsis* were treated with or without 30 µg ml⁻¹ CE for 35 min and subsequently washed in water for 10 minutes while being shaken at 40 rev min⁻¹ to remove all the DMSO. Then the cells were subjected to transcellular osmosis at various temperatures ranging from 5–25 °C. The activation energy ($E_a$) for transcellular osmosis was calculated form Arrhenius plots. The $E_a$ for transcellular osmosis of the DMSO control in *Nitellopsis* is 16.38 kJ mol⁻¹ (3.9 kcal mol⁻¹). This is similar to the values found by Kiyosawa (1975) for *Chara* and Tazawa and Kamiya (1966) for *Nitella flexilis* but lower than the values found by Dainty and Ginsburg (1964 a) for *Nitella translucens*. However in *Nitellopsis*, after treatment with CE, the activation energy increases to 32.5 kJ mol⁻¹ (7.76 kcal mol⁻¹) indicating that CE causes the elimination of a low energy pathway for water (Table 2).

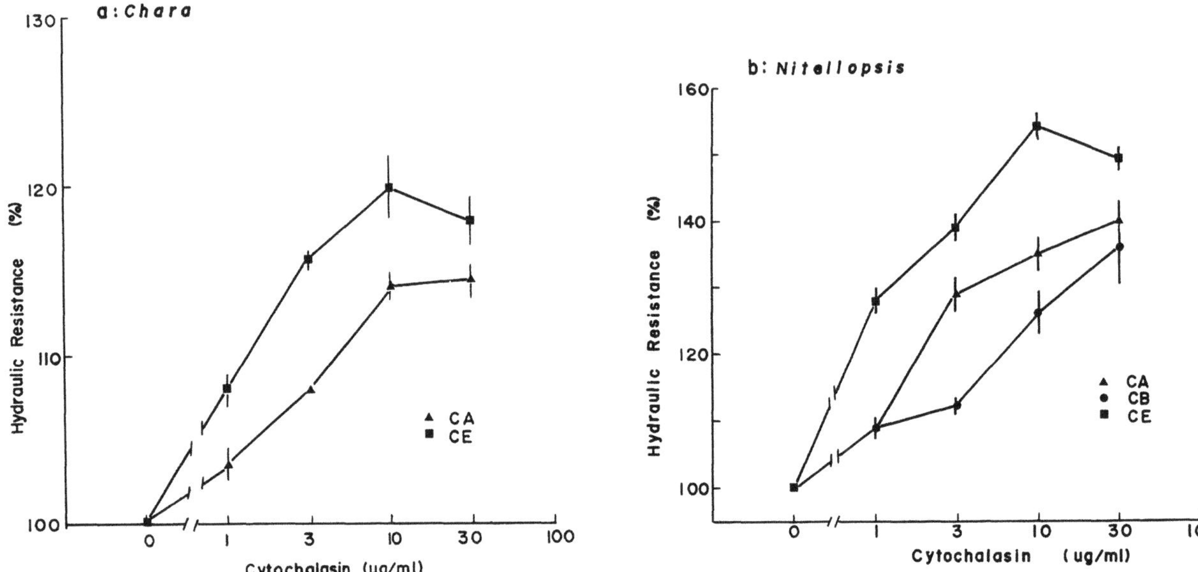


Fig. 6. The effect of cytochalasins on the total transcellular hydraulic resistance ($r_{tot}$) of *Chara* (*a*) and *Nitellopsis* (*b*). The abscissae represent the concentrations of CA, CB or CE. The ordinates represent the increase in the hydraulic resistance relative to the DMSO control. In the case of 6 a: The controls for CA and CE were 1.1 ± 0.21 and 0.97 ± 0.19 pm$^{-1}$ s Pa, respectively. When the cells were repeatedly exposed to 1% DMSO (5 times), the initial control value was 100% = 0.97 ± 0.21 pm$^{-1}$ s Pa and the values varied between 96.75 and 100% (data not shown). Each point represents the average ± standard error of the mean for 3 cells. In the case of 6 b: The controls for CA, CB, and CE are: 100% = 3.05 ± 0.59, 2.36 ± 0.38 and 3.06 ± 0.43 pm$^{-1}$ s Pa, respectively. When cells were repeatedly exposed to 1% DMSO (5 times), the initial control value was 100% = 2.48 ± 0.27 pm$^{-1}$ s Pa and the values varied between 96 and 102.3% (data not shown). Each point represents the mean ± SEM for 3 cells

## 3.5. Differential Effects of the Cytochalasins on Endoosmotic and Exoosmotic Hydraulic Resistances in Nitellopsis

We devised a new method to determine the endoosmotic and exoosmotic hydraulic resistances using cel-

Table 2. *The effect of 30 μg ml$^{-1}$ CE on the activation energy ($E_a$) for transcellular osmosis in Nitellopsis*

| Cell no. | Activation Energy | | | |
|---|---|---|---|---|
| | DMSO | | Cytochalasin E | |
| | kJ mol$^{-1}$ | (kcal mol$^{-1}$) | kJ mol$^{-1}$ | (kcal mol$^{-1}$) |
| 1 | 13.83 | (3.30) | 32.30 | (7.72) |
| 2 | 15.68 | (3.75) | 32.60 | (7.79) |
| 3 | 19.62 | (4.69) | — | — |
| 4 | — | — | 32.73 | (7.82) |
| $\bar{X}$ | 16.38 | (3.91) | 32.50 | (7.78) |
| S. E. | 0.99 | (0.24) | 0.13 | (0.03) |

Cells were pretreated in either 1% DMSO or 30 μg/ml CE for 35 min, washed in distilled water for 10 min and then subjected to transcellular osmosis at various temperatures ranging from 5–25 C. The difference between the DMSO control and the CE-treated cells is significant at the 0.05 level

lular ligation to make either endoosmosis or exoosmosis limiting. We used this method to determine whether CB increases the endoosmotic resistance or the exoosmotic resistance. Figure 7 shows clearly that CB and CE specifically increases the endoosmotic hydraulic resistance which results in an increase in $r_{tot}$ and a loss or reversal in the polarity. Similar results are observed with *Chara* (data not shown).

When we perform transcellular osmoses under various osmotic pressure gradients, we find that both $r_{tot}$ and $L_{pex}^{-1}$ increase as the transcellular osmotic gradient increases from 0.06 MPa to 0.50 MPa at 23 °C. By contrast $L_{pen}^{-1}$ decreases as the osmotic gradient increases (Fig. 8). When transcellular osmosis is induced by a gradient as small as 0.06 MPa (25 mOsm), $L_{pex}^{-1}$ is still larger than $L_{pen}^{-1}$ indicating that a polarity still exists under physiological conditions. Although a linear extrapolation to zero indicates that a polarity exists at a zero flow rate, there is no reason to assume that a linear extrapolation is correct. We suggest that the polarity is not a function of the membrane alone but results from a physiological interaction between the membrane, the cytoplasm and the external milieu (see Discussion).

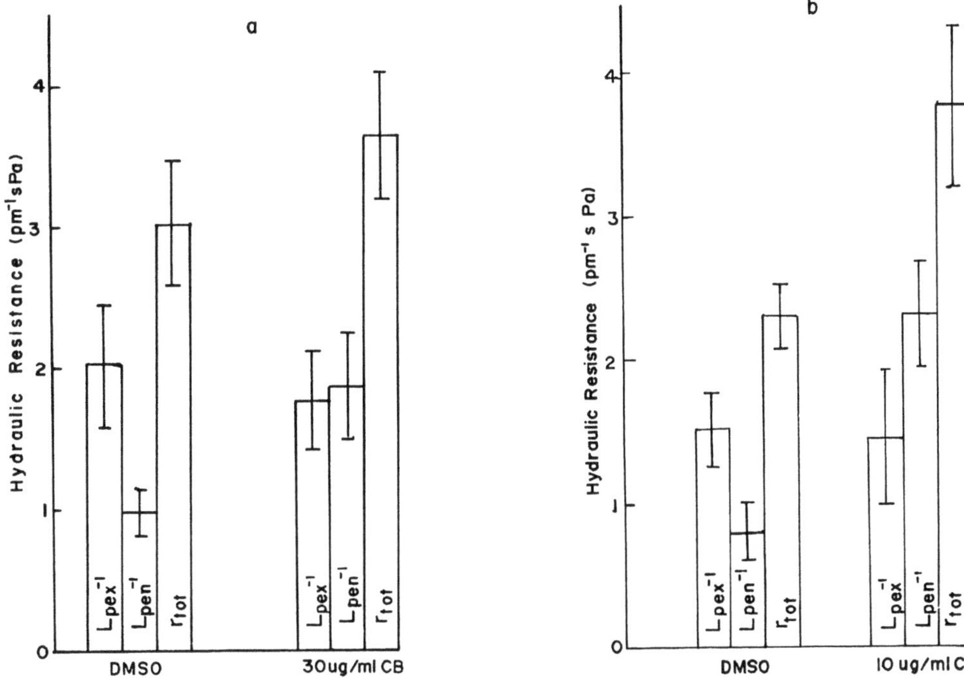

Fig. 7. The effect of cytochalasin B and E on the endoosmotic ($L_{pen}^{-1}$), exoosmotic ($L_{pex}^{-1}$) and total transcellular ($r_{tot}$) hydraulic resistance in *Nitellopsis*. Cells were treated in either *a* 1% DMSO or 30 µg ml$^{-1}$ CB for 35 min or *b* 2% DMSO or 10 µg ml$^{-1}$ CE for 35 min and then placed asymmetrically in the apparatus and subjected to the ligation method of transcellular osmosis. See Materials and Methods for details. In the case of the CB experiments, cells were temperature-equilibrated for only 5 minutes to minimize the leakage of CB. Each bars represents the mean ± SEM for 5 cells. Levels of significance obtained from t-tests are p < 0.2, non significant, and p < 0.05 for $r_{tot}$, $L_{pex}^{-1}$ and $L_{pen}^{-1}$, respectively, of CB treated cells relative to control cells. In the case of CE treated cells, the levels of significance obtained from t-tests are p < 0.05, non significant, and p < 0.05 for $r_{tot}$, $L_{pex}^{-1}$ and $L_{pen}^{-1}$, respectively

A particularly novel observation is that the polarity observed when the osmotic gradient is as small as 0.06 MPa is still sensitive to CE. CE eliminates or partially reverses the polarity (Fig. 9), thus providing strong support for the hypothesis that the polarity is an intrinsic but physiologically variable characteristic of the living cell (Fig. 8).

Cytochalasin B has a dramatic effect on cytoplasmic streaming (NAGAI and KAMIYA 1977). Therefore we tested whether or not CB increases the endoosmotic hydraulic resistance by interferring with streaming. We inhibited the streaming electrically by inducing an action potential with a 1.1 µA electric current reciprocally between two cell ends or chemically by treating the cell with 0.2 mM NEM in 1% DMSO (CHEN and KAMIYA 1975). This low concentration of NEM, when applied in DMSO stops streaming in 5 minutes. Two mM NEM in DMSO stops streaming instantly (data not shown). Inhibiting streaming by either electrical or chemical means does not effect the endoosmotic resistance in *Chara* and *Nitellopsis* (data not shown).

Colchicine (5 mM, 60 min) has little effect on the initial rate of water movement indicating that microtubules do not participate in the regulation of hydraulic resistance (data not shown).

### 3.6. The Effect of Cytochalasin E on the Electrical Response Induced by Transcellular Osmosis

Transcellular osmosis induces a transcellular potential difference and differential ion movements in *Nitella flexilis* (KATAOKA *et al.* 1979, NISHIZAKI 1955, TAZAWA and NISHIZAKI 1956) indicating that other membrane properties besides hydraulic resistance become polarized. HAYAMA *et al.* (1979) showed that the potential difference results from a transcellular osmosis-induced depolarization on the endoosmotic side and a hyperpolarization on the exoosmotic side. We also find that transcellular osmosis induces a difference in the electrical properties of the endo- and exoosmotic sides in *Nitellopsis*. Transcellular osmosis induces an action potential on the endoosmotic side and a very small slow depolarization on the exoosmotic side in cells treated with APW or APW plus 1% DMSO (Fig. 10 *a* and *b*).

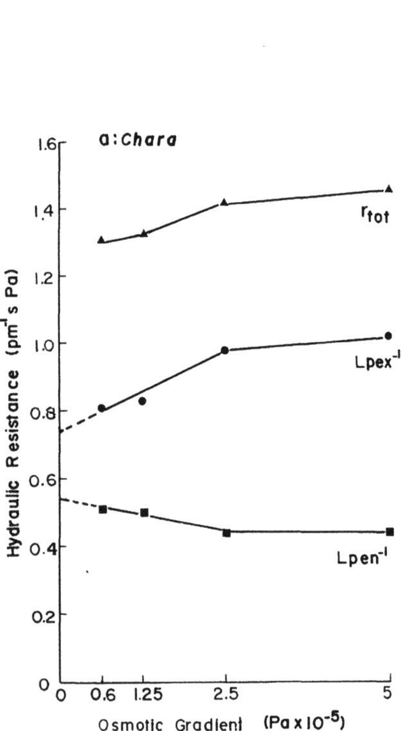


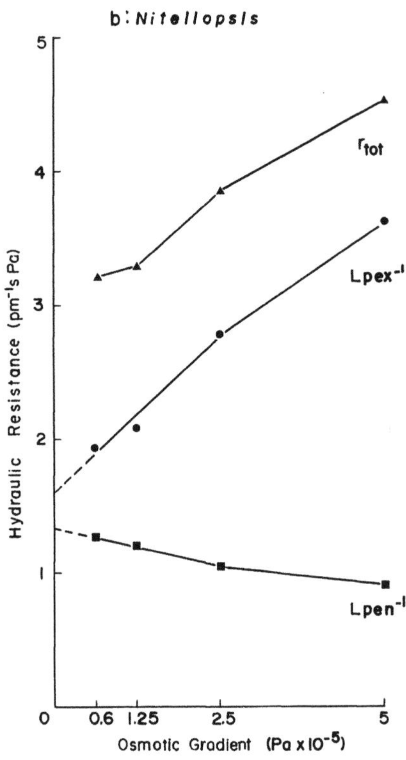


Fig. 8. The effect of the magnitude of the osmotic gradient on the endoosmotic ($L_{pen}^{-1}$), exoosmotic ($L_{pex}^{-1}$) and total transcellular ($r_{tot}$) hydraulic resistance in *Chara* (*a*) and *Nitellopsis* (*b*). Each graph shows one representative cell for each species. In the case of *Chara*, 9 cells were tested. In the case of *Nitellopsis*, 3 cells were tested

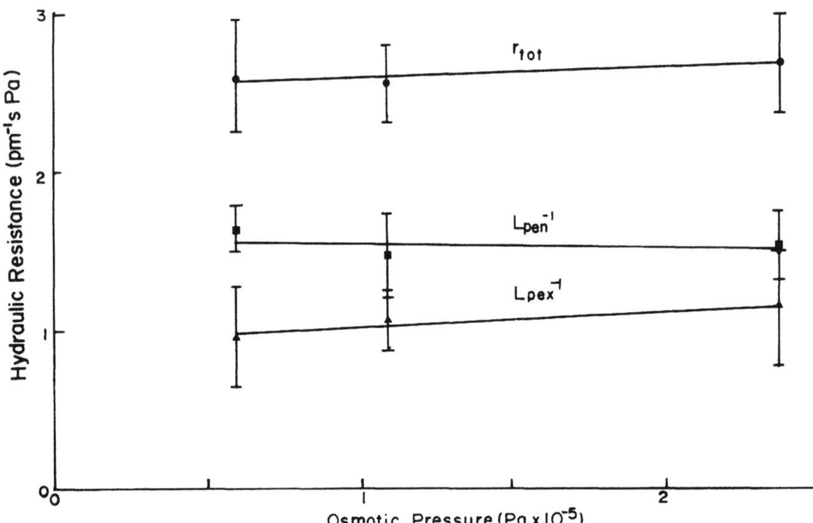


Fig. 9. The effect of the osmotic gradient on the endoosmotic ($L_{pen}^{-1}$), exoosmotic ($L_{pex}^{-1}$) and total transcellular ($r_{tot}$) hydraulic resistance in cells of *Nitellopsis* treated with cytochalasin E. Three cells were treated with $30\,\mu g\,ml^{-1}$ CE for 35 min and then washed with water on a shaker ($44\,rev\,s^{-1}$) to remove the CE and DMSO. The average of three individual experiments is shown with SEM

In contrast to forward transcellular osmosis, backward transcellular osmosis does not induce any electrical changes on either side. However in 7 out of 10 cells treated with $30\,\mu g\,ml^{-1}$ CE for 35 minutes, forward transcellular osmosis was unable to induce an action potential on the endoosmotic side although the cells were still capable of generating an action potential in response to an electric current (Fig. 10 *c*). CE did not have any effect on backward transcellular osmosis. CE does not have any effect on the membrane resistance, the membrane potential or the ability of the cell to generate an action potential in response to an electrical

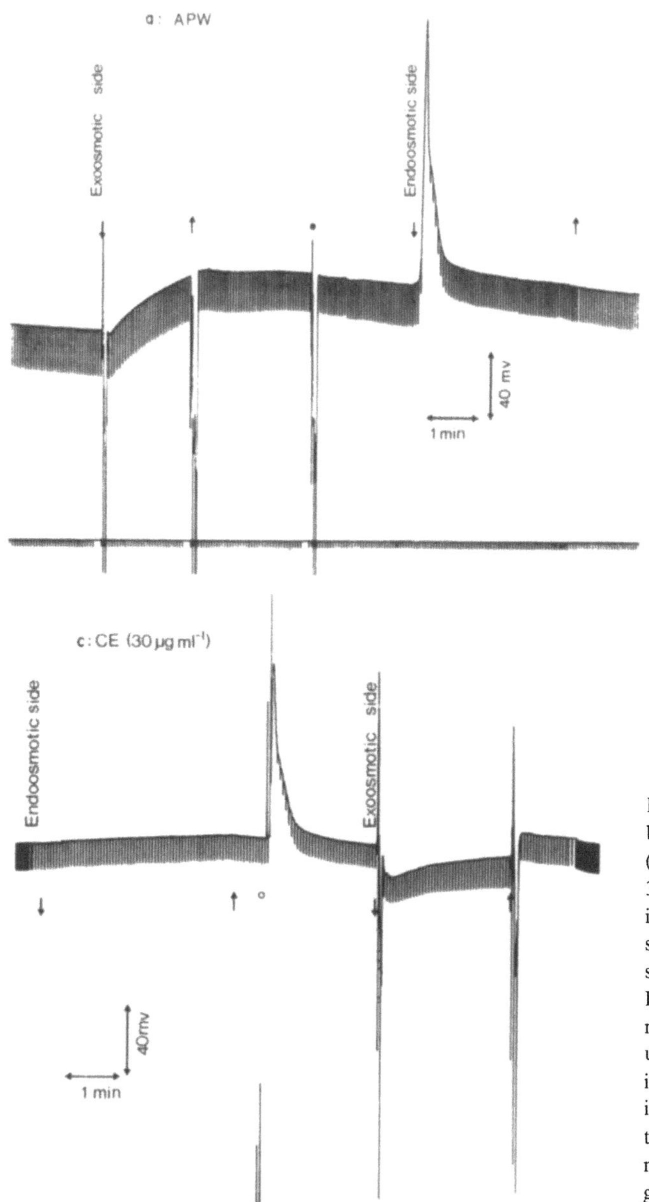
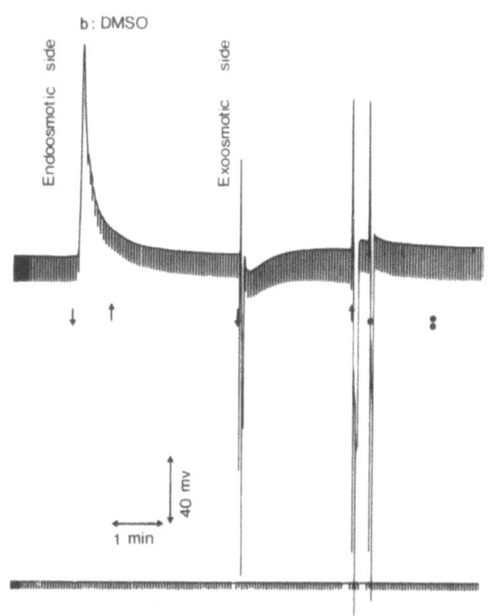

Fig. 10. The effect of cytochalasin E on the electrical response induced by transcellular osmosis in *Nitellopsis*. Cells were put in *a* APW (pH 5.6), *b* APW plus 1% DMSO or *c* APW, 1% DMSO and 30 µg ml$^{-1}$ CE in an open chamber with two compartments as shown in Fig. 2. Transcellular osmosis was induced with 400 mM aqueous sorbitol placed into either chamber $A$ or $B$. The diameter of the cell shown was 0.58 mm and the diameter of each chamber was 10 mm. Downward arrow indicates the onset of forward transcellular osmosis, the upward arrow indicates the onset of backward transcellular osmosis. A single black dot ($\cdot$) represents an isoosmotic change in chamber $A$, a double black dot ($\colon$) indicates an isoosmotic change in chamber $B$, and an open dot (o) represents an electrical stimulus that was given to the cell. The upper trace represents the change in membrane potential and the lower trace represents the current pulses given to the cell

stimulus (Table 3). CE does not stop streaming or transcellular water movement by depleting the intracellular ATP levels. The concentrations of ATP found in the DMSO control cells and in cells treated with 30 µg ml$^{-1}$ CE for 35 min are $3.10 \pm 0.69$ mM (4) and $3.96 \pm 0.72$ mM (4), respectively. Actually the treatment of cells with CE results in an increase in the intracellular ATP content, presumably by preventing ATP hydrolysis by the actomyosin system.

## 4. Discussion

Cells of *Chara corallina* and *Nitellopsis obtusa* exhibit a flow rectification or polarity in their hydraulic conductivity ($L_p$) which was first found in characean cells by Kamiya and Tazawa (1956). The exoosmotic resistance ($L_{pex}^{-1}$) is greater than the endoosmotic resistance ($L_{pen}^{-1}$). The polarity is enhanced by increasing the transcellular osmotic gradient. This was interpreted to be a consequence of a nonspecific dehydration of the membrane caused by the osmotic solution on the exoosmotic side as suggested by Dainty and Ginzburg (1964a) and Kiyosawa and Tazawa (1972, 1973) for characean cells and by Rich *et al.* (1968) and by Blum and Forster (1970) for red blood cells (see Tazawa 1972 for a review). However, the general dehydration effect of osmotic solutions on wa-

Table 3. *The effect of cytochalasin E on membrane resistance ($R_m$), membrane potential ($E_m$) and excitability (Excit.) in Nitellopsis*

| Treatment | Electrical properties of the membrane | | |
|---|---|---|---|
| | $R_m$ ($\Omega$ m$^2$) | $E_m$ (mV) | Excit. (%) |
| 1% DMSO | 1.36 ± 0.13 (32) | −180.44 ± 0.44 (32) | 80 (25) |
| 30 µg ml$^{-1}$ CE | 1.38 ± 0.19 (19) | −185.60 ± 1.79 (15) | 80 (10) |

Cells were pretreated in 1% DMSO-APW or 30 µg/ml CE-APW for 35 min and the membrane potential was measured by the conventional microelectrode technique. The resistance was measured by applying 0.2 µA square current pulses and measuring the change in membrane potential. Excitability was determined by applying an electrical current of approximately 1 µA to the cells. Each value represents the mean ± SEM. Numbers in parentheses represent the number of experiments

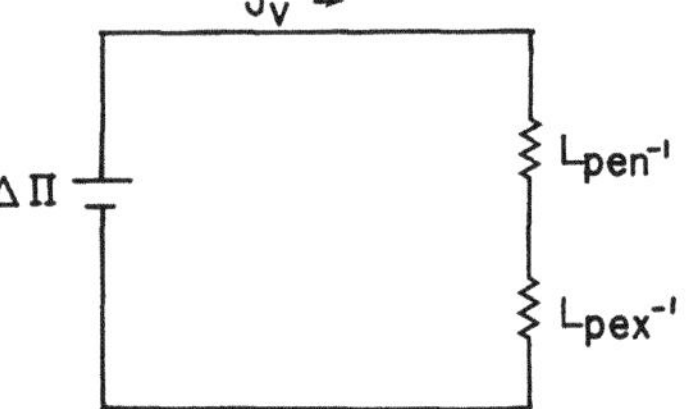


Fig. 11. The equivalent circuit for transcellular osmosis. $\Delta$ $\Pi$: osmotic pressure of the sorbitol solution on the exoosmotic side. $L_{pen}^{-1}$ and $L_{pex}^{-1}$: hydraulic resistances on the endoosmotic and exoosmotic side, respectively. $J_v$: rate of water flow

ter movement has been questioned. Using a method that minimizes the perturbation to the cell, FARMER and MACEY (1970) found that the hydraulic conductivity is independent of the osmolarity of the medium and further showed that the data of RICH *et al.* (1968) are consistent with this interpretation if we assume that a rectification of osmotic flow exists. Later, using an NMR method (PIRKLE *et al.* 1979), CHIEN and MACEY (1977) showed that the diffusional water permeability of red blood cells is independent of the osmolarity and concluded that the apparent dependence of the hydraulic permeability on osmolarity is a consequence of an intrinsic cellular polarity.

The polarity of water movement in characean cells also seems to be intrinsic since a polarity exists when the osmotic gradient used to drive transcellular osmosis is only 0.06 MPa. Furthermore, the observations of STEUDLE and ZIMMERMANN (1974) show that a polarity exists in characean cells even when the water movement is controlled by hydrostatic pressure and not by osmotic pressure. The most striking data against the idea that the polarity is a result of a general dehydration of the membrane are the reversal of polarity due to cytochalasins since it seems unlikely that the cytochalasins can act to prevent a membrane dehydration on the exoosmotic side.

The analysis of our results is based on Ohm's Law using an equivalent electrical circuit as an analog (Fig. 11). We have assumed throughout the analysis that the flow of water ($J_v$) is linearly related to the pressure gradient ($\Delta\Pi$) through the constant of proportionality 1/R. Experimentally we have demons-

trated that this is true for osmotic gradients as small as 0.06 MPa. However, the influx of water on the endoosmotic side may remove ions etc. from the protoplasmic surface of the plasma membrane and cause a buildup at the protoplasmic surface of the plasma membrane on the exoosmotic side. A flow of water thus creates a cytoplasmic polarity that in turn may act upon the membrane and differentially create or gate channels on the two sides of the cell. If this be true, $L_{pen}^{-1}$ and $L_{pex}^{-1}$ may act as non-linear resistors. There is evidence that the endoosmotic resistance acts like a non linear-resistor and decreases as a consequence of an increased water flow (Fig. 8)(HAYAMA and TAZAWA 1978, KIYOSAWA and TAZAWA 1973, STEUDLE and ZIMMERMANN 1974). Therefore a linear extrapolation of the data presented in Fig. 9 to zero is unwarranted and we suggest that the polarity in water movement is a consequence of differential cytoplasmic-membrane protein interactions on the endoosmotic- and exoosmotic sides and not the presence of rectifying channels per se in the membrane.

It is generally assumed that water moves through the lipid bilayer by a solubility-diffusion mechanism (ALBERTS *et al.* 1983, FINKLESTEIN 1984). This is supported by observations that water moves through plasma membranes at a rate comparable to the rate of water movement through lipid bilayers (ZIMMERMANN and STEUDLE 1978). The water permeability coefficients of lipid membranes range from 0.2–100 µm s$^{-1}$ (CASS and FINKLESTEIN 1967, FETTIPLACE and HAYDON 1980, FINKLESTEIN 1984, OSCHMAN *et al.* 1974); a range sufficiently large to account for the various rates of water movement in almost all cell types measured. Secondly, the activation energy of water movement across the plasma membrane (42–63 kJ mol$^{-1}$) is usually similar to the energy of activation of water movement across lipid bilayers (FETTIPLACE and HAYDON 1980, HANSSON MILD and LØVTRUP 1985, TOMOS *et al.* 1981).

Lastly, in *Valonia*, the diffusional water permeability is equal to the osmotic water permeability ($2.4\,\mu\text{m s}^{-1}$) and there seems to be no apparent solvent-solute interactions (Gutknecht 1967, 1968), a characteristic of lipid membranes without pores (Cass and Finklestein 1967). If water moves primarily through a lipid pathway by a solubility-diffusion mechanism in characean cells, it is difficult to understand how water can move in a polar fashion. However if we consider that water can also move through the protein components of the membrane, we can invoke a mechanism for the polar transport of water.

What is the evidence that water can move through channels created by membrane proteins in characean cells? The hydraulic conductivity of characean cells is markedly higher than those of most other plant cells (Dainty *et al.* 1974, Kamiya and Tazawa 1956, Bennet-Clark 1959, Zimmermann and Steudle 1978). In fact it is comparable to the rate of water flow in red blood cells and epithelial cells (Dick 1966). In red blood cells and epithelial cells the osmotic permeability coefficient is substantially larger than the diffusion permeability coefficient indicating that water moves through aqueous pores (Finklestein 1987). Furthermore in characean internodal, red blood and epithelial cells, the activation energy for transmembrane water movement is comparable to the activation energy for the self-diffusion of water (Dainty and Ginzburg 1964a, Kiyosawa 1975, Tazawa and Kamiya 1965, Pirkle *et al.* 1979, Whittembury *et al.* 1984), which is smaller than that for water movement across lipid bilayer membranes. This is further evidence that water moves through water-filled channels. Additional evidence comes from the observations of solvent-solute interactions in *Chara* (Steudle and Tyerman 1983). Lastly, inhibitors of protein function have been shown to increase the hydraulic resistance in red blood, endothelial and epithelial cells (Benga *et al.* 1983, Fischbarg *et al.* 1987, Naccache and Sha'afi 1974, Brown *et al.* 1975, Lukacovic *et al.* 1984, Whittembury *et al.* 1984). These studies indicate that the anion channel (band 3) and the glucose transporter may serve as water channels. Although the lipid bilayer may be the sole pathway for water transport in the majority of cells, it is likely that intrinsic membrane proteins as well as the lipid bilayer serve as pathways for water movement in cells with an inherently high hydraulic conductivity.

The intriguing possibility exists that the high hydraulic conductivities are a result of the aggregation of membrane proteins. A kinetic analysis of band 3 proteins, which were purified from red blood cells, and then inserted into lipid bilayers shows that the formation of aqueous channels is a consequence of the formation of tetramers of band 3 proteins (Benz *et al.* 1984). Indeed the possibility that aggregates of proteins may serve as the water channel was first proposed by Pinto da Silva (1973) when he observed during freeze-etch experiments that sublimation of water at $-100\,^{\circ}\text{C}$ occurs primarily through protein aggregates. Further support for the hypothesis that water moves through membrane protein aggregates comes from studies on amphibian bladders. Treatment of toad bladders with vasopressin, which decreases their hydraulic resistance to osmotic water flow, induces an increase in the number of protein aggregates observed by freeze fracture electron microscopy (Kachadorian *et al.* 1975, Parisi *et al.* 1985). In bladders, cytochalasin B inhibits both the formation of protein aggregates in the membrane and the hormone-induced decrease in the hydraulic resistance (Taylor *et al.* 1973, Davis *et al.* 1974, Parisi *et al.* 1985) indicating that the actin cytoskeleton is involved in the regulation of the hydraulic resistance. The volume of mouse C3H-2K cells is also slightly influenced by CB (Iida and Yahara 1986). The actin cytoskeleton is involved in other membrane events including the polarization of $Ca^{2+}$-channels in *Fucus* and *Funaria* (Brawley and Robinson 1985, Saunders 1986). Cytochalasin B however also binds to the glucose transporter (Lin and Spudich 1974) thus making its site of action obscure. However, CE, which does not bind to the glucose transporter (Jung and Rampal 1977, Rampal *et al.* 1980, Yahara *et al.* 1982), increases hydraulic resistance in characean cells, indicating that CE as well as CB acts on the actin microfilaments. The greater effectiveness of CE compared with either CA or CB in increasing the hydraulic resistance correlates with its greater effectiveness in inhibiting actin-mediated processes (Yahara *et al.* 1982). The idea that contractile proteins may influence the hydraulic resistance of the membrane was first proposed by Goldacre in 1952. Indeed actin may be involved in the water regulating mechanisms of contractile vacuoles and pinocytosis (Dick 1966) and intracellular water transport (Allen and Francis 1965). However it is important to note that CB may not always modulate hydraulic conductivity by interacting with microfilaments. For example CB inhibits water transport in corneal endothelial cells (Fischbarg *et al.* 1987) by directly acting on the glucose transporter.

In characean cells we have demonstrated that cytochalasins A, B, and E increase the endoosmotic hydraulic resistance during forward transcellular osmosis.

Apparently, the cytochalasin-sensitive pathway does not participate in backward transcellular osmosis. The increase in the endoosmotic resistance during forward transcellular osmosis leads to a loss or reversal of polarity in cytochalasin-treated cells. The loss of the hydraulic polarity occurs in parallel with a loss in the electrical polarity. The sensitivity of the polarity to cytochalasins indicates to us that the polarity must be an intrinsic character of the living cell.

The actin bundles on the ectoplasm/endoplasm interface, which participate in force generation for streaming (KAMIYA 1959, 1981, KURODA and KAMIYA 1956, PALEVITZ and HEPLER 1975) influence membrane transport processes, including $OH^-$ efflux (LUCAS and DAINTY 1977, LUCAS and SHIMMEN 1981) and intercellular rubidium transport (DING and TAZAWA unpublished). However, these bundles seem not to be important in the regulation of hydraulic resistance since stopping streaming by either electrical stimulation or NEM has no effect on the hydraulic resistance. The system of transverse filaments within the cortical cytoskeleton which bind anti-actin or NBD-phallacidin (NOTHNAGEL *et al.* 1981, WILLIAMSON 1985, WILLIAMSON *et al.* 1986) may contribute to the regulation of membrane activities, including water permeability and some electrical phenomena (Fig. 10). WILLIAMSON *et al.* (1986) observed that these filaments were stabilized in the presence of CB. Perhaps a dynamic action or the possibility of fragmentation is necessary for these filaments to regulate membrane properties.

Since an action potential is induced only on the endoosmotic side, and only during forward transcellular osmosis, it seems that something occurs in this localized area of the cell only during forward transcellular osmosis. The cessation of streaming, depolarization, or action potential (HAYAMA and TAZAWA 1978, HAYAMA *et al.* 1979) that occur only on the endoosmotic side and only during forward transcellular osmosis point to an increase in the intracellular $[Ca^{2+}]$ in the endoosmotic side during forward transcellular osmosis, since it is known that the cytoplasmic free $[Ca^{2+}]$ is an important regulatory factor of cytoplasmic streaming (TAZAWA and SHIMMEN 1987, TAZAWA *et al.* 1987). Perhaps a rapid influx of water washes away $Ca^{2+}$ that is bound to the internal surface of the plasma membrane on the endoosmotic side and this results in an increase in the cytoplasmic $[Ca^{2+}]$, which subsequently acts on the microfilaments. A change in the microfilament organization in turn may induce or allow an aggregation of membrane proteins, which results in a decrease in the hydraulic resistance on the endoosmotic

side only during forward transcellular osmosis. A change in the hydraulic resistance may also result from a localized change in the intracellular pH, osmolarity or ionic strength on the endoosmotic side.

The membrane proteins with which the cortical actin microfilaments presumably interact still remain unknown. KIYOSAWA and OGATA (1987) provide evidence for the presence of water channels. They showed that the electrical resistance of the cell membrane remains unchanged when the osmotic pressure of the medium is raised even though the hydraulic resistance increases. Although this data indicates that ions can not pass through at least some of the channels that water can pass through, it is still possible that water can move through ion channels (KUKITA and YAMAGISHI 1983). It seem likely that various proteins or aggregates of proteins are able to serve as water channels and open or close upon transcellular osmosis depending on the various ionic activities and osmotic pressure differences that occur in the endoosmotic end exoosmotic cell halves during forward and backward transcellular osmosis (*e.g.*, FINKLESTEIN 1987, ZIMMERBERG and PARSEGIAN 1986). In short, the flux of water, as well as ions, is subject to biological regulation as well as the laws of physics.

## Acknowledgements

This work was supported in part by grants from the Ministry of Education Culture and Science of Japan and the Japanese Society for the Promotion of Science. We warmly thank S. AMINO, D.-Q. DING, M. KATSUHARA, T. KOHNO, Y. OKAZAKI, T. SHIINA, T. SHIMMEN, K. TAKESHIGE and Y. TOMINAGA of the Laboratory of Plant Physiology and M. WATANABE of the National Institute of Basic Biology who have contributed to every aspect of this research from repeating critical experiments to correcting grammatical errors in the final manuscript. Their valuable discussions and technical help made this paper possible. We also warmly thank T. and K. KITABATAKE who generously created a real home for R.W. in Japan.

## References

ALBERTS B, BRAY D, LEWIS J, RAFF M, ROBERTS K, WATSON JD (1983) Molecular biology of the cell. Garland, New York London, 1146 pp

ALLEN RD, FRANICS DW (1965) Cytoplasmic contraction and the distribution of water in *Amoeba*. Symp Soc Expl Biol 19: 259–271

BENGA G, POP VI, POPESCU O, IONESCU M, MIHELE V (1983) Water exchange through erythrocyte membranes: nuclear magnetic resonance studies on the effects of inhibitors and of chemical modification of human membranes. J Membr Biol 76: 129–137

BENNET-CLARK TA (1959) Water relations of cells. In: STEWART FC (ed) Plant physiology, vol 2. Academic Press, New York, pp 104–191

BENZ R, TOSTESON MT, SCHUBERT D (1984) Formation and properties of tetramers of band 3 protein from human erythrocyte membranes in planer lipid bilayers. Biochim Biophys Acta 775: 347–355

Blum RM, Forster RE (1970) The water permeability of erythrocytes. Biochim Biophys Acta 203: 410–423

Brawley SH, Robinson KR (1985) Cytochalasin treatment disrupts the endogeneous currents associated with cell polarization in fucoid zygotes: studies on the role of F-actin in embryogenesis. J Cell Biol 100: 1173–1184

Brown PA, Feinstein MB, Sha'afi RI (1975) Membrane proteins related to water transport in human erythrocytes. Nature 254: 523–525

Cass A, Finklestein A (1967) Water permeability of thin lipid membranes. J Gen Physiol 50: 1765–1784

Chen JCW, Kamiya N (1975) Localization of myosin in the internodal cell of Nitella as suggested by differential treatment with N-ethylmaleimide. Cell Struct Funct 1: 1–9

Chien DY, Macey RI (1977) Diffusional water permeability of red cells Independence on osmolarity. Biochim Biophys Acta 464: 45–52

Dainty J (1963 a) The polar permeability of plant cell membranes to water. Protoplasma 57: 220–228

— (1963 b) Water relations of plant cells. Adv Bot Res 1: 279–326

— Ginzburg BZ (1964 a) The measurement of hydraulic conductivity (osmotic permeability to water) of internodal characean cells by means of transcellular osmosis. Biochim Biophys Acta 79: 102–111

— — (1964 b) The permeability of the cell membrane of Nitella translucens to urea, and the effect of high concentrations of sucrose on this permeability. Biochim Biophys Acta 79: 112–121

— Hope AB (1959) The water permeability of cells of Chara australis R Br. Aust J Biol Sci 12: 136–145

— Vinters H, Tyree M (1974) A study of transcellular osmosis and the kinetics of swelling and shrinking in cells of Chara corallina. In: Zimmermann U, Dainty J (eds) Membrane transport in plants. Springer, Berlin Heidelberg New York, pp 59–63

Davis WL, Goodman DBP, Schuster RJ, Rasmussen H, Martin JH (1974) Effects of cytochalasin B on the response of toad urinary bladder to vasopressin. J Cell Biol 63: 986–997

Dick DAT (1966) Cell water. Butterworths, London, 153 pp

Farmer REL, Macey RI (1970) Perturbation of red cell volume: Rectification of osmotic flow. Biochim Biophys Acta 196: 53–65

Fettiplace R, Haydon DA (1980) Water permeability of lipid membranes. Physiol Rev 60: 510–550

Finklestein A (1984) Water movement through membrane channels. Curr Top Membr Transport 21: 295–308

— (1987) Water movement through lipid bilayers, pores, and plasma membranes. Theory and reality. John Wiley and Sons, New York (Distinguished lecture series of the Society of General Physiologists, vol 4)

Fischbarg J, Liebovitch LS, Koniarek JP (1987) Inhibition of transepithelial osmotic water flow by blockers of the glucose transporter. Biochim Biophys Acta 898: 266–274

Goldacre RJ (1952) The folding and unfolding of protein molecules as a basis of osmotic work. Int Rev Cytol 1: 135–164

Gutknecht J (1967) Membranes of Valonia ventricosa: apparent absence of water-filled pores. Science 158: 787–788

— (1968) Permeability of Valonia to water and solutes: apparent absence of aqueous membrane pores. Biochim Biophys Acta 163: 20–29

Hansson Mild K, Løvtrup S (1985) Movement and structure of water in animal cells. Ideas and experiments. Biochim Biophys Acta 822: 155–167

Hayama T, Nakagawa S, Tazawa M (1979) Membrane depolarization induced by transcellular osmosis in internodal cells of Nitella flexilis. Protoplasma 98: 73–90

— Tazawa M (1978) Cessation of cytoplasmic streaming accompanying osmosis-induced membrane depolarization in Nitella flexilis. Cell Struct Funct 3: 47–60

Höfler K (1930) Eintritts- und Rückgangsgeschwindigkeit der Plasmolyse. Jb Wiss Bot 73: 300–350

Iida K, Yahara I (1986) Reversible induction of actin rods in mouse CH3-2K cells by incubation in salt buffers and by treatment with non-ionic detergents. Exp Cell Res 164: 492–506

Jung CY, Rampal AL (1977) Cytochalasin B binding sites and glucose transport carrier in human erythrocyte ghosts. J Biol Chem 252: 5456–5463

Kachadorian WA, Wade JB, DiScala VA (1975) Vasopressin: induced structural change in toad bladder luminal membrane. Science 190: 67–69

Kamiya N (1959) Protoplasmic streaming. In: Heilbrunn LV, Weber F et al (eds) Protoplasmatologia, vol 8. Springer, Wien, part 3 a

— (1981) Physical and chemical basis of cytoplasmic streaming. Ann Rev Plant Physiol 32: 205–236

— Kuroda K (1956) Artificial modification of the osmotic pressure of the plant cell. Protoplasma 46: 423–436

— Tazawa M (1956) Studies on water permeability of a single plant cell by means of transcellular osmosis. Protoplasma 46: 394–422

Kataoka H, Nakagawa S, Hayama T, Tazawa M (1979) Ion movements induced by transcellular osmosis in Nitella flexilis. Protoplasma 99: 179–187

Kiyosawa K (1975) Studies on the effects of alcohols on membrane water permeability on Nitella. Protoplasma 86: 243–252

— Ogata K (1987) Influence of external osmotic pressure on water permeability and electrical conductance of Chara cell membrane. Plant Cell Physiol 28: 1013–1022

— Tazawa M (1972) Influence of intracellular and extracellular tonicities on water permeability in characean cells. Protoplasma 74: 257–270

— — (1973) Rectification characteristics of Nitella membranes in resepect to water permeability. Protoplasma 78: 203–214

— — (1977) Hydraulic conductivity of tonoplast-free Chara cells. J Membr Biol 37: 157–166

Kukita F, Yamagishi S (1983) Effects of an outward water flow on potassium currents in a squid giant axon. J Membr Biol 75: 33–44

Kuroda K, Kamiya N (1956) Velocity distribution of the protoplasmic streaming in Nitella cells. Bot Mag (Tokyo) 69: 544–554

Levitt J, Scarth GW, Gibbs RD (1936) Water permeability of isolated protoplasts in relation to volume change. Protoplasma 26: 237–248

Lin S, Spudich JA (1974) Biochemical studies on the mode of action of cytochalasin B. J Biol Chem 249: 5778–5783

Lucas WJ, Dainty J (1977) Spatial distribution of functional $OH^-$ carriers along a characean internodal cell: determined by the effect of cytochalasin B on $H^{14}CO_3^-$ assimilation. J Membr Biol 32: 75–92

— Shimmen T (1981) Intracellular perfusion and cell centrifugation studies on plasmalemma transport processes in Chara corallina. J Membr Biol 58: 227–237

Lukacovic MF, Toon M, Solomon AK (1984) Site of cation leak induced by mercurials sulfhydryl reagents. Biochim Biophys Acta 772: 313–320

Mimura T, Shimmen T, Tazawa M (1984) Adenine-nucleotide levels and metabolism-dependent membrane potential in cells of *Nitellopsis obtusa* Groves. Planta 162: 77–84

Naccache P, Sha'afi RI (1974) Effect of PCMBS on water transfer across biological membranes. J Cell Physiol 83: 449–456

Nagai R, Kamiya N (1977) Differential treatment of *Chara* cells with cytochalasin B with special reference to its effect on cytoplasmic streaming. Exp Cell Res 108: 231–237

Nishizaki Y (1955) Bioelectric phenomena accompanying osmosis in a single plant cell. Cytologia 20: 32–40

Nothnagel EA, Barak LS, Sanger JW, Webb WW (1981) Fluorescence studies on modes of cytochalasin B and phallotoxin action on cytoplasmic streaming in *Chara*. J Cell Biol 88: 364–372

Oschman JL, Wall BJ, Gupta BL (1974) Cellular basis of water transport. Symp Soc Exp Biol 28: 305–350

Osterhout WJV (1949 a) Movement of water in cells of *Nitella*. J Gen Physiol 32: 553–557

— (1949 b) Transport of water from concentrated to dilute solutions in cells of *Nitella*. J Gen Physiol 32: 559–565

Palevitz BA, Hepler PK (1975) Identification of actin *in situ* at the ectoplasm-endoplasm interface of *Nitella*. J Cell Biol 65: 29–38

Parisi M, Pisam M, Merot J, Chevalier J, Bourguet J (1985) The role of microtubules and microfilaments in the hydroosmotic response to antidiuretic hormone. Biochim Biophys Acta 817: 333–342

Pinto da Silva P (1973) Membrane intercalated particles in human erythrocyte ghosts: sites of preferred passage of water molecules at low temperature. Proc Natl Acad Sci USA 70: 1339–1343

Pirkle JL, Ashley DL, Goldstein JH (1979) Pulse nuclear magnetic resonance measurements of water exchange across the erythrocyte membrane employing a low Mn concentration. Biophys J 25: 389–406

Rampal AL, Pinkofsky H, Jung CY (1980) Structure of cytochalasins and cytochalasin B binding sites in human erythrocyte membranes. Biochemistry 19: 679–683

Rich GT, Sha'afi RI, Romualdez A, Solomon AK (1968) Effect of osmolarity on the hydraulic permeability coefficient of red cells. J Gen Physiol 52: 941–954

Saunders MJ (1986) Cytokinin activation and redistribution of plasma-membrane ion channels in *Funaria*. Planta 167: 402–409

Steudle E, Tyerman SD (1983) Determination of permeability coefficients, reflection coefficients, and hydraulic conductivity of *Chara corallina* using the pressure probe: effects of solute concentration. J Membr Biol 75: 85–96

— Zimmermann U (1974) Determination of the hydraulic conductivity and of reflection coefficients in *Nitella flexilis* by means of

direct cell-turgor pressure measurement. Biochim Biophys Acta 332: 399–412

Taylor A, Mamelak M, Reaven E, Maffly R (1973) Vasopressin: Possible role of microtubules and microfilaments in its action. Science 181: 347–350

Tazawa M (1957) Neue Methode zur Messung des osmotischen Wertes einer Zelle. Protoplasma 48: 342–359

— (1972) Membrane characteristics as revealed by water and ionic relations of algal cells. Protoplasma 75: 427–460

— (1980) Cytoplasmic streaming and membrane phenomena in cells of *Characeae*. In: Grantt E (ed) Handbook of phycological methods. Developmental and cytological methods. Cambridge University Press, Cambridge, pp 179–193

— Kamiya N (1965) Water relations of characean internodal cell. Ann Rep Biol Works Fac Sci Osaka Univ 13: 123–419

— — (1966) Water permeability of a characean internodal cell with special reference to its polarity. Aust J Biol Sci 19: 399–419

— Kiyosawa K (1973) Analysis of transcellular water movement in *Nitella*: A new procedure to determine the inward and outward water permeabilities of membranes. Protoplasma 78: 349–364

— Nishizaki Y (1956) Simultaneous measurement of transcellular osmosis and the accompanying potential difference. Jap J Bot 15: 227–238

— Shimmen T (1987) Cell motility and ionic relations in characean cells as revealed by internal perfusion and cell models. Int Rev Cytol 109: 259–312

— — Mimura T (1987) Membrane control in the Characeae. Ann Rev Plant Physiol 38: 95–117

Tomos AD, Steudle E, Zimmermann U, Schulze E-D (1981) Water relations of leaf epidermal cells of *Tradescantia virginiana*. Plant Physiol 68: 1135–1143

Whittembury G, Carpi-Medina P, Gonzalez E, Linares H (1984) Effect of para-chloromercuribenzenesulfonic acid and temperature on cell water osmotic permeability of proximal straight tubules. Biochim Biophys Acta 775: 365–373

Williamson RE (1985) Immobilization of organelles and actin bundles in the cortical cytoplasm of the alga *Chara corallina* Klein ex. Willd. Planta 163: 1–8

— Perkin JL, McCurdy DW, Craig S, Hurley UA (1986) Production and use of monoclonal antibodies to study the cytoskeleton and other components of the cortical cytoplasm of *Chara*. Eur J Cell Biol 41: 1–8

Yahara I, Harada F, Sekita S, Yoshihira K, Natori S (1982) Correlation between effects of 24 different cytochalasins on cellular structures and cellular events and those on actin *in vitro*. J Cell Biol 92: 69–78

Zimmerberg J, Parsegian VA (1986) Polymer inaccessible volume changes during opening and closing of a voltage-dependent ionic channel. Nature 323: 36–39

Zimmermann U, Steudle E (1978) Physical aspects of water relations of plant cells. Adv Bot Res 6: 45–117

Protoplasma (1988) [Suppl. 2]: 131–136


# A New Class of Photoactivated Fixatives for Immunocytochemistry

ELENA MCBEATH* and KEIGI FUJIWARA

Department of Structural Analysis, National Cardiovascular Center Research Institute, Fujishiro-dai, Suita, Osaka

Received March 11, 1988
Accepted May 31, 1988

Dedicated to Professor Dr. NOBURO KAMIYA on the occasion of his 75th birthday

## Summary

Loss of antigenicity upon fixation has been a problem in immunocytochemistry. We previously reported [J Cell Biol (1984) 99: 2061–2073] on a new fixation technique using photoactivated 1,3,5-triazido-2,4,6-trinitrobenzene (TTB). This fixative preserves antigenicity in general better than the many conventional fixatives we have tried. We communicate here on the existence and some properties of a whole host of structurally related compounds, many of which can also act as immunocytochemical fixatives upon irradiation. Although TTB still produces the best antigenic preservation of all the compounds tested, certain of the compounds have special properties, such as water solubility and the ability to be rapidly photoactivated, which allow them to be used in a wider variety of situations.

*Keywords:* Immunocytochemistry; Fixative; Photoactivation; Microtubule; Intermediate filament.

*Abbreviations:* TTB 1,3,5-triazido-2,4,6-trinitrobenzene; IgG immunoglobulin G; TNB 1,3,5-trinitrobenzene.

## 1. Introduction

Light microscopic observation of biological material using the more intrusive techniques, such as immunofluorescence microscopy, usually requires fixation of the biological material. A fixation method for immunocytochemistry must not only preserve the biological structure but also keep antibody binding sites on an antigen intact and accessible to antibodies. It has been the common experience, however, that antigenicity is lost as the extent of morphological preservation is improved (NAKANE 1975, CANDE *et al.* 1977, FUJIWARA and POLLARD 1980). We have recently developed a new

fixation method for immunofluorescence microscopy using light activated 1,3,5-triazido-2,4,6-trinitrobenzene (TTB) (MCBEATH and FUJIWARA 1981, 1984). Cells fixed by this method retain excellent antibody binding ability as well as good cellular morphology at the light microscopic level.

We originally reasoned that TTB might act as a fixative because it contains three azido groups, each of which can be photoactivated to form a nitrene (JI 1979), a free radical that can crosslink biological structures. Such a free radical is usually short lived and its reaction is known to be essentially temperature independent (JI 1979). However, we found that fixation with TTB was temperature dependent and that photoactivated TTB stored at 0 °C for many hours is still capable of fixing cells (MCBEATH and FUJIWARA 1984). These facts suggest that the nitrenes formed by irradiating TTB are not directly responsible for its ability to fix and that irradiated analogues of TTB without azide groups might also act as fixatives.

To obtain insights into the mechanism of fixation by photoactivated TTB, a systematic study was initiated on the effectiveness of various irradiated TTB analogues as fixatives. Using antitubulin immunofluorescence staining, we investigated the morphology and staining intensities of microtubules in tissue culture cells fixed with one of 32 compounds structurally related to TTB.

## 2. Materials and Methods

*2.1. Chemicals*

In addition to TTB (ICN Pharmaceuticals, Inc., Plainview, NY and Polysciences, Inc. Warrington, PA), the following chemicals were

---

* Correspondence and Reprints: Department of Structural Analysis, National Cardiovascular Center Research Institute, Fujishiro-dai, Suita, Osaka, Japan.

used: benzotrifuroxan (Aldrich, Milwaukee, WI), 1,3,5-trinitrobenzene (Eastman Kodak Co., Rochester, NY), 2,4-dinitrofluorobenzene (Sigma, St. Louis, MO), phenol (Fisher, Fair Lawn, NJ), 1,3-dihydroxybenzene (Fisher), tannic acid (Mallinckrodt, Paris, KY), 2,4-dinitrophenol (Aldrich), 1,3-dinitrobenzene (Fisher), 1,3,5-trihydroxybenzene (Mallinckrodt), 4-fluoronitrobenzene (Eastman Kodak), nitrobenzene (Fisher), 2-fluorobenzoic acid (ICN), 2-hydroxynaphthalene (Eastman Kodak), 1-hydroxynaphthalene (J. T. Baker Chemical Co., Phillipsburg, NJ), toluene (Fisher), benzene (J. T. Baker), naphthalene (MCB Manufacturing Chemists, Inc., Gibbstown, NJ), 3,5-dihydroxytoluene (Fisher), methoxybenzene (Fisher), 1,4-benzoquinone (Fisher), benzoic acid (Fisher), 2,4,6-trinitrophenol (MCB), 4-nitrophenol (Aldrich), 1,4-diaminobenzene (Fisher), 3,3',4,4'-tetraaminobiphenyl (Sigma), 1,5-diaminonaphthalene (Aldrich), 2-nitrophenol (Aldrich), 3-nitrophenol (Aldrich), 2,4-dihydroxy-1,3,5-trinitrobenzene (Eastman Kodak), benzofuroxan (Aldrich), 3-fluorobenzoic acid (ICN), benzenephosphonic acid (ICN), 4-aminobenzenesulfonic acid (J. T. Baker), NaCl (Fisher), Triton X-100 (Sigma), beta-mercaptoethanol (Sigma), acetic acid (VWR Scientific Inc., San Fransico, CA), HCl (Fisher), and NaOH (VWR).

### 2.2. Fixation and Staining of Cells

One ml of a 1% solution of each TTB analogue was made in methanol and irradiated with a 1,000 watt Hg-Xe lamp for 30 minutes, either on ice or dry ice, as previously reported for TTB (MCBEATH and FUJIWARA 1984). Rat kangaroo kidney epithelial cells (PtK-2) were grown on glass coverslips, dehydrated with methanol and fixed with these solutions as described for TTB (MCBEATH and FUJIWARA 1984). Cells were treated first with either rabbit antitubulin (FUJIWARA and POLLARD 1978) or rabbit anticytokeratin (MCBEATH and FUJIWARA 1984) and then with rhodamine labelled goat anti-rabbit IgG (Cappel Laboratories, West Chester, PA). Every effort was made to maintain constant conditions for dehydration, fixation and antibody staining. Because the cell density varied from culture to culture, an internal control staining was included in all the experiments, as described in the next section.

To evaluate the effectiveness of fixation by different TTB analogues, we compared the microtubule staining intensity as well as the overall morphology of cells and the microtubule distribution pattern inside cells. We chose microtubules as the test antigen since their immunofluorescence staining pattern is known to be sensitive to various conventional fixatives.

### 2.3. Quantitation of Staining Intensity

Microtubule staining intensity was determined from the photographic exposure time of confluent portions of the coverslip using a Leitz Orthomat automatic camera system. For each coverslip, exposures were taken of six different fields and the average exposure time determined. Typically, a field contained about 25 interphase cells. Since exposure time is inversely related to intensity, the inverse of average exposure time is directly proportional to staining intensity. To compare relative staining intensities between coverslips from different experiments, one coverslip from each experiment was fixed with the same activated analogue, usually 1% 1,3,5-trinitrobenzene (TNB) activated for 1 minute on ice. The relative staining intensity of cells fixed with TNB in this manner was taken as 1 and all other relative intensities were made proportional to this value.

### 2.4. Fluorescence Microscopy

Epifluorescence microscopy was done using Leitz Orthoplan equipped with a Ploemopak 2 containing a Leitz N-2 filter and a Zeiss × 63 planapo phase lens (N.A. 1.4 oil). Fluorescent images were recorded by Leitz Orthomat using 35 mm Tri-X film (Eastman Kodak Co., Rochester, NY) exposed at ASA 1,000 and developed in Acufine (Acufine, Inc., Chicago, IL).

## 3. Results

The TTB analogues were solubilized in methanol and photoirradiated before being used as fixatives for immunofluorescence microscopy. When the morphology and the staining intensities of microtubules in PtK-2 cells fixed with one of 32 analogues were evaluated, four simple conclusions emerged: (i) many TTB analogues were capable of fixing cells after being irradiated by light; (ii) none of the TTB analogues could fix cells without prior photoirradiation; (iii) irradiated methanol (Fig. 1 *a*) did not fix cells any better than methanol without irradiation; and (iv) of the 32 analogues tested, some could not fix cells, even after being irradiated, any better than methanol alone. These analogues were 1,4-diaminobenzene, 3,3',4,4'-tetraaminobiphenyl, 1,5-diaminonaphthalene, 2-nitrophenol, 3-nitrophenol, 2,4-dihydroxy-1,3,5-trinitrobenzene, benzofuroxan, 3-fluorobenzoic acid, and benzenephosphonic acid. Although a large number of the irradiated compounds were able to preserve tubulin antigenicity (Table 1), there were some differences in the extent of microtubule preservation, both between different analogues and sometimes between the same analogue activated at different temperatures.

### 3.1. Analogues Containing Nitro Groups

The activated analogues that gave the best results were the nitro substituted benzene compounds. The more nitro groups present on benzene, the better the analogue acted as a fixative. TTB is one of the most highly nitro substituted benzene compounds, and is the best fixative of all the compounds tested so far. Benzotrifuroxan, into which irradiated TTB most likely degenerates (KORSUNSKII *et al.* 1971), was also an excellent fixative when photoactivated. Since non-irradiated benzotrifuroxan did not fix cells, fixation by irradiated TTB cannot be due to the presence of benzotrifuroxan. Instead, it must be due to some other product of irradiated TTB and/or benzotrifuroxan. Irradiated benzofuroxan did not fix under any condition. Of the three best fixing compounds (TTB, benzotrifuroxan, and TNB), TNB was the most sensitive to light. Good microtubule (Fig. 1 *b*) and intermediate filament (Fig. 1 *c*)

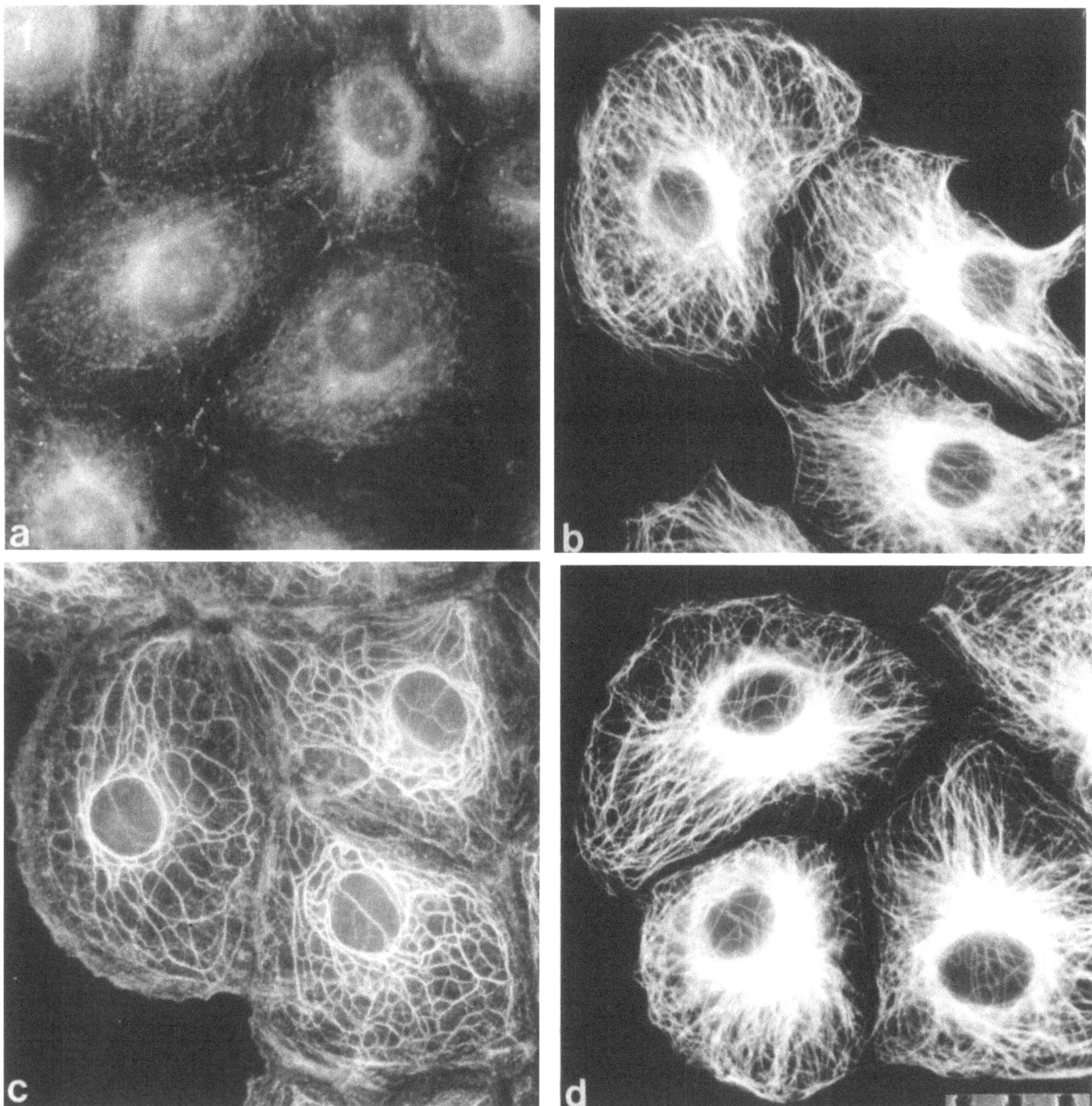

Fig. 1. Immunofluorescence micrographs of PtK-2 cells stained with antitubulin or anticytokeratin intermediate filament. *a* Cells were fixed with irradiated methanol and stained with antitubulin. No distinct microtubules are visible. *b* and *c* Cells were fixed with TNB irradiated using a 1,000 watt Hg-Xe lamp for 1 minute on ice. Microtubules (*b*) stained with antitubulin and cytokeratin intermediate filaments (*c*) stained with anticytokeratin are well preserved. *d* Cells were fixed with photoactivated picric acid in water containing 150 mM NaCl (pH 1) and stained with antitubulin. Microtubules are stained. One micrometer division = 10 µm

staining was seen in cells fixed with TNB activated for 1 minute while TTB, benzotrifuroxan, picric acid, and benzene all required irradiation times of between one to two orders of magnitude longer to produce similar antitubulin staining results. TNB was so sensitive to ultraviolet light that the same results could be obtained with 20 minutes of irradiation from an unfiltered, hand held, 4 watt, short wave ultraviolet lamp as from 1 minute of irradiation from the 1,000 watt lamp. All the nitro substituted benzenes worked better when activated on ice than on dry ice. In several cases, this is probably due to the precipitation of the compound at dry ice temperature.

### 3.2. Analogues Containing Hydroxy Groups

The hydroxy substituted benzene and naphthalene compounds produced the next best fixation. Unlike the nitro benzene compounds, the presence of more (alternating) hydroxyl groups did not appear to improve tubulin antigenic preservation. Furthermore, the hydroxy benzene compounds all fixed cells similarly, whether activated on ice or on dry ice. Activated benzene and naphthalene, although noticeably less effective than their respective hydroxy substituted analogues, still gave reasonable fixation. Both benzene and naphthalene fixed better when activated on ice than on dry ice, again probably because of precipitation at the colder temperature.

### 3.3. Analogues Containing Nitro-hydroxy Combinations or Other Groups

Because the nitro and hydroxy subsituted benzenes work so well, it might be expected that a compound containing both of these groups would be an even better fixative. On the contrary, the presence of hydroxyl and nitro groups on the same benzene ring, usually ortho to each other, often dramatically reduced the compound's ability to become a fixative when irradiated. A reduction or increase in the number of nitro groups relative to 2,4-dinitrophenol abolished their ability to fix, particularly when irradiated on ice. When the hydroxyl group was replaced with fluorine on 2,4-dinitrophenol and on 4-nitrophenol, better fixation was obtained (unless the compound precipitated).

As is evident from the isomers of nitrophenol, both the position of the functional groups relative to one another on the benzene ring and the type of functional group are important in determining whether or not the irradiated analogue can fix. For example, while irradiated 2-nitrophenol and 3-nitrophenol did not, 4-nitrophenol activated on dry ice did preserve tubulin antigenicity. Similarly, fluorine ortho to the carboxylic acid improved activated benzoic acid's ability to fix while fluorine meta to the carboxylic acid destroyed it. Irradiated amine containing TTB analogues did not fix microtubules.

### 3.4. Temperature Sensitivity

The ability of TTB and its analogues to fix, once activated, deteriorated rapidly, indicating that the pho-

toactivated compounds were unstable. Nevertheless, short-term storage was possible. We found that fixatives lasted longer as the storage temperature was lowered and that some compounds were more sensitive to storage temperature than others. For example, after irradiation, benzene and the nitro compounds, TTB and TNB, lost their ability to fix much faster than phenol.

Freezing irradiated TTB in liquid nitrogen appears to greatly extend its life time as a fixative. The microtubules of cells fixed with TTB which had been activated and then stored for one year in liquid nitrogen stained almost as well as those fixed with just activated TTB.

### 3.5. Solvent Effect

We also investigated whether or not some of the compounds could be used in water. TTB is not soluble in water. However, methanol could be removed from activated TTB by vacuum and the resulting powder partially dissolved in the same volume of 150 mM NaCl in water. If the solution was buffered to pH 7 with 10 mM phosphate and used to fix living tissue culture cells, most cells came off the coverslip and the microtubules of the remaining cells (permeabilized with 0.2% Triton X-100 for 2 seconds) did not stain. However, if buffered to pH 4, or even better, pH 2, microtubules stained quite nicely although not nearly as brightly as in methanol. The TTB analogue, 2,4,6-trinitrophenol (picric acid) is soluble in water. When a 1% aqueous solution was activated on ice at low pH in 150 mM NaCl, it could preserve both microtubule (Fig. 1 d) and cytokeratin intermediate filament morphology and antigenicity. Adequate fixation occurs within 30 seconds. Activation at pH 1 or 2 gave much better staining results than pH 4 while no tubulin antigenicity was preserved at pH 7. Picric acid activated at pH 1 or 2 and then neutralized could not fix microtubules, and this pH effect was not reversible since the ability to fix was not recovered by lowering the pH from 7 back to 1 or 2. It is possible that the activated picric acid is so reactive at pH 7 that it cannot last after irradiation. However, microtubules of cells placed in the solution during irradiation were not preserved. Nonirradiated picric acid at low pH, or just low pH phosphate buffered saline without picric acid did not preserve microtubule antigenicity. It is interesting to note that picric acid becomes a fixative in low pH water when activated on ice but the same compound in methanol does not work if irradiated on ice.

Both benzotrifuroxan and 2-fluorobenzoic acid are

only slightly soluble in water and when a saturated solution at pH 4 and pH 2 respectively were activated 30 minutes on ice and used as fixatives, microtubules stained only faintly. The faintness of the image is most likely due to the low solubility of these compounds in water. Many other water soluble TTB analogues were also tested: 1% benzene at pH 1, 1% phenol at pH 1 or 7–8, 10% 1,3-dihydroxybenzene at pH 3–4, 0.5% 1,3,5-trihydroxybenzene at pH 2 or 7, 1% tannic acid at pH 3, 1% 1,4-benzoquinone at pH 2, and 1% 4-aminobenzenesulfonic acid at pH 2–3. These anlogues were dissolved in 150 mM NaCl in water, irradiated for 30 minutes on ice and used to fix cells. None were capable of preserving tubulin antigenicity. It appears, therefore, that the nitro groups on picric acid, not the hydroxy group, are important for activation in water. One trivial possibility for the mechanism of fixation by activated TTB analogues is that irradiation of the analogues is converting methanol to formaldehyde or formic acid which then fixes cells. However, this is not the case since microtubules cannot be fixed by either compound in methanol any better than by methanol alone. Furthermore, irradiated formaldehyde and formic acid in methanol also failed to fix microtubules.

### 3.6. Effect of Amino Acid Analogues

To obtain insights into the mechanism of fixation by nitro benzene compounds, we used 1% TNB in methanol irradiated for 1 minute on ice as a model system. Various types of molecules were solubilized or suspended in irradiated TNB in methanol before using it as a fixative. The added compound was usually in molar excess to TNB. Addition of the following compounds abolished or greatly reduced microtubule staining intensity: beta-mercaptoethanol (1%), 1,4-diaminobenzene (1%), and imidazole (3%). When cells were treated for 5 minutes with these compounds either before or after fixation with irradiated TNB, no effect on microtubule staining intensity was observed. Because beta-mercaptoethanol, 1,4-diaminobenzene and imidazole are analogues of cysteine, lysine and histidine respectively, the results suggest that irradiated TNB reacts with these amino acids. Carboxyl groups do not react with irradiated TNB since acetic acid (0.85%) did not interfere with its fixation.

### 3.7. Effect of Acid and Base

The use of imidazole in the experiment described above led us to investigate the pH effect on irradiated TNB. When imidazole was added to irradiated TNB, the solution turned orange. TNB at 0.1 to 0.5% is used as a pH indicator in ethanol, turning from colorless to orange at pH 12–14 (CRC Handbook of Chemistry and Physics, 1979). Since photoactivated TTB and picric acid were pH sensitive in water, the imidazole effect on the activated component of irradiated TNB might also be due to the basic nature of imidazole. When a 3 times molar excess of HCl was added to irradiated TNB and then neutralized with an equivalent amount of NaOH, microtubules, although fainter, were preserved. When the same amount of acid and base were added to irradiated TNB but in the reverse order, no tubulin staining was seen. No microtubules were seen in cells fixed with irradiated TNB containing a 10 to 1 or 100 to 1 molar ratio of irradiated TNB to NaOH, although spindle staining was just detectable in cells fixed with irradiated TNB in the presence of the smaller amount of NaOH. These observations suggest that the activated component of irradiated TNB is destroyed by hydroxide ions.

## 4. Discussion

We have previously reported a procedure for improved antigenic and morphological preservation at the light microscope level using photoactivated TTB (McBeath and Fujiwara 1981, 1984). We present here our finding that a whole set of structurally related compounds can be photoactivated to produce fixatives capable of preserving antigenicity and cellular structures. The mechanism of fixation by these compounds is not yet clear. However, since so many different types of benzene derivatives are capable of becoming fixatives upon irradiation, there may be some underlying photochemical reaction common to them all. For example, benzene is known to form photoisomers upon irradiation with a strong ultraviolet light (Bryce-Smith and Gilbert 1976, 1977). The formation of many of the isomers is temperature dependant, an unusual characteristic for a photoprocess. Furthermore, most of these highly stressed isomers are thermally unstable at room temperature, reverting back to benzene or to some other form, and can react with other compounds. Since the photoproduced fixatives have similar characteristics to these isomers, photoisomerization may be a possible mechanism by which the fixative is formed. The effectiveness of each compound as a fixative is strongly influenced by the type and/or position of its substituent groups. These groups may modify the underlying photochemical reaction or may help to stabilize the product(s) of the reaction, as is the case with the photoisomers of benzene derivatives.

Some TTB analogues were not converted to fixatives after being irradiated. According to the proposed hypothesis, it is possible that these analogues, due to their substituent groups, may not undergo the postulated photochemical reaction. It is also possible that some analogues, although they do undergo the photochemical reaction, may not form products capable of reacting with cells to fix them for immunocytochemistry. The products may instead be compounds that do not last long enough to react with the antigens or that react with the antigens in ways that do not preserve their antigenicity.

Although irradiated TTB still gives the best antigenic preservation of tubulin among the many compounds tested, several of the TTB analogues have unusual properties that make them more useful for other purposes. For example, TNB is extremely light sensitive and can be activated in a reasonable amount of time using an inexpensive, hand held, short wave ultraviolet lamp. Picric acid can be activated in a low pH water-based solution in which membranes (and presumably membrane antigens) can be preserved. Once the mechanism of fixative formation and reaction is better understood, the proper TTB analogue can be selected so that many other applications of this unique class of fixatives will be possible.

## Acknowledgement

We thank Dr. GILBERT CHIN for providing us with many of the benzene derivatives and for his helpful discussions. The study reported here was conducted at the Department of Anatomy and Cellular Biology, Harvard Medical School, Boston, MA. The work was supported by NIH predoctoral training grant GM 07226 to E.M., NSF grant PCM 81-19171 to K.F. and by a grant for Cardiovascular Disease (62A-3) from the Ministry of Health and Welfare (Japan).

## References

BRYCE-SMITH D, GILBERT A (1976) The organic photochemistry of benzene—I. Tetrahedron 32: 1309–1326

— — (1977) The organic photochemistry of benzene—II. Tetrahedron 33: 2459–2490

CANDE WZ, LAZARIDES EL, MCINTOSH JR (1977) A comparison of the distribution of actin and tubulin in the mammalian mitotic spindle as seen by indirect immunofluorescence. J Cell Biol 72: 552–567

FUJIWARA K, POLLARD TD (1978) Simultaneous localization of myosin and tubulin in human tissue culture cells by double antibody staining. J Cell Biol 77: 182–195

— — (1980) Techniques for localizing contractile proteins with fluorescent antibodies. In: FRIEDLANDER M (ed) Current topics in developmental biology, vol 14, part 2, immunological approaches to embryonic development and differentiation. Academic Press, New York, pp 271–296

JI TH (1979) The application of chemical crosslinking for studies on cell membranes and the identification of surface receptors. Biochim Biophys Acta 559: 39–69

KORSUNSKII BL, APINA TA (1971) Kinetics of the thermal decomposition of 1,3,5-triazido-2,4,6-trinitrobenzene in solution. Bull Acad Sci USSR Div Chem Sci 21: 1971–1973

MCBEATH ER, FUJIWARA K (1981) An improved fixation method for light level antigen localization using a photoactivatable cross-linker. J Cell Biol 91: 339 a

— — (1984) Improved fixation for immunofluorescence microscopy using light-activated 1,3,5-triazido-2,4,6-trinitrobenzene (TTB). J Cell Biol 99: 2062–2073

NAKANE P (1975) Recent progress in the peroxidase-labeled antibody method. Ann NY Acad Sci 254: 203–210

WEAST RC, ASTLE MJ (eds) (1979) CRC Handbook of chemistry and physics, 60th edn. CRC Press, Boca Raton, FL. pp D 150–151

Protoplasma (1988) [Suppl. 2]: 137–144


# An Insoluble Matrix of the Nerve Cytoskeleton

A. G. Loewy[1,*], N. Dollahon[2], P. Klainer[1], and K. Wolfe[1]

[1] Department of Biology, Haverford College, Haverford, and [2] Department of Biology, Villanova University, Villanova, Pennsylvania

Received April 8, 1988
Accepted July 1, 1988

Dedicated to Professor Dr. Noburo Kamiya on the occasion of his 75th birthday

## Summary

When the abdominal nerve cord of the crayfish is extracted with a 6 M guanidine HCl solvent and 5% β-mercaptoethanol a matrix remains which retains some of the distinctive morphological characteristics of the unextracted nerve cord. When the nerve cord sheath is removed by digestion with collagenase, and the nerve cord ghost washed with the guanidine HCl solvent, the remaining structure continues to resemble the unextracted nerve cord. Further extraction with a 6 M urea, 5% sodium dodecylsulfate (SDS) solvent, and 5% β-mercaptoethanol leaves a highly swollen, transparent nerve cord ghost which in scanning electron microscopy contains a matrix of interconnected linear elements. In transmission electron microscopy the ghosts of frog sciatic and peripheral nerves contain linear elements, networks of interconnecting thin filaments and numerous vesicular ghosts.

*Keywords:* Insoluble cytomatrix; Nerve cytoskeleton; Crayfish abdominal nerve cord; $N^\varepsilon$-(γ-glutamic)lysine crosslinks.

*Abbreviations:* HCl hydrochlaric acid; SDS sodium dodecyl sulfate; Tris Tris(hydroxymethyl) aminomethane; EDTA ethylenediaminetetraacetate; SEM scanning electron microscope; TEM transmission electron microscope; PMSF phenylmethyl sulfonylfluoride; NaCl sodium chloride.

## 1. Introduction

It is generally assumed that all the proteins of the cytoskeleton can be solubilized with strong protein unfolding reagents such as guanidine HCl, urea, and SDS in the presence of disulfide reducing reagents. However, we have shown that when striated muscle fibers are extracted with 6 M guanidine HCl and 5% β-mercaptoethanol a component of the cytoskeleton remains which consists of less than 0.2% of the original cell

protein mass and possessing a number of morphological characteristics of the unextracted cells (Loewy *et al.* 1984). This insoluble matrix is made of evenly distributed linear elements which help retain the original shape of the cell and the periodicity of the original striations (Loewy *et al.* 1985). The extracted muscle fibers even show the indentations left by the extracted nuclei (Loewy *et al.* 1986). A urea and SDS-insoluble component of the cytoplasm has also been observed in the cytoplasm of the plasmodial slime mold *Physarum polycephalum* (Gassner *et al.* 1983, 1985). In this instance transmission electron microscopy showed the presence of a pervasive network of superthin (2–3 nm) filaments. It is therefore of interst to determine if networks of superthin filaments, insoluble in strong protein solvents, are present in other cell types such as those of nerve fibers. In this study we report on work performed with nerve cells which, like muscle fibers, are composed of linear elements actively engaged in the transduction of chemical energy into work. An important feature of nerve axons is that the mechanism of cytoplasmic motility includes the movement of vesicles. As we shall see, the insoluble matrices of nerve tissue retain some of the distinctive morphology of the unextracted nerve cord and consist of interconnected linear, reticular and vesicular elements. For purposes of low-resolution general morphology we performed scanning electron microscopic (SEM) studies of the abdominal nerve cord of the crayfish because of its characteristic structure consisting of 5 ganglia, a terminal and 3 pairs of lateral roots per ganglion (Fig. 1). For transmission electron microscopic (TEM)

---

* Corrspondence and Reprints: Department of Biology, Haverford College, Haverford, PA 19041, U.S.A.

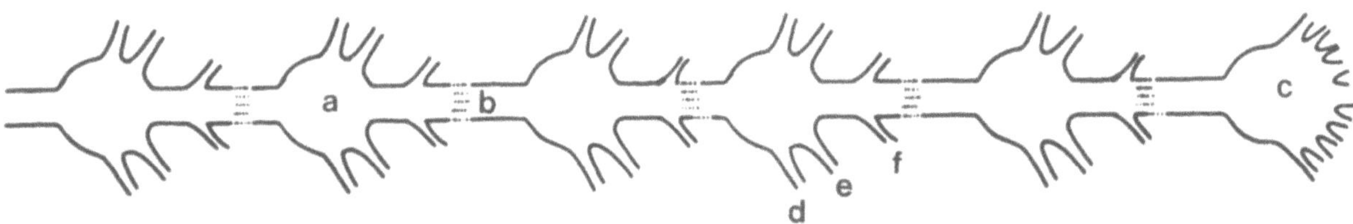


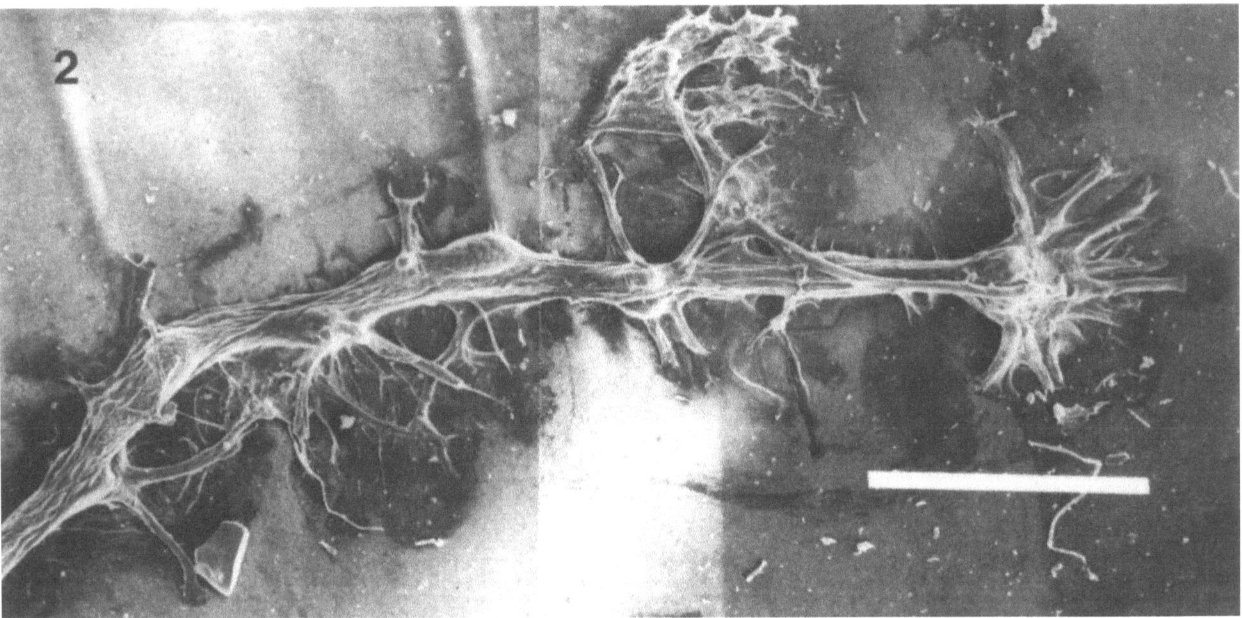


Fig. 1. Diagram of the abdominal nerve cord of the crayfish (*Procambarus clarkii*) showing: ganglia (*a*), connective (*b*), terminal (*c*), first (*d*), second (*e*), and third lateral roots (*f*)

Fig. 2. Abdominal nerve cord extracted with 6 M guanidine HCl solvent and washed with acetone on a Nucleopore filter. Ganglia, terminal and lateral roots can be clearly distinguished. Nerve cord has shrunk to 20–25% of its original length. Bar: 1.5 mm

studies we selected sciatic and peripheral frog nerves which gave us good high resolution images of linear elements, networks of interconnecting fine filaments and vesicular ghosts. Although the SEM and TEM studies appear to complement each other it should be remembered that they were performed on crayfish and frog materials respectively.

## 2. Materials and Methods

Three solvents designed to solubilize, unfold, and extract proteins from nerve tissue were: (1) guanidine HCl solvent [6 M guanidine HCl, 0.1 M sodium carbonate-bicarbonate buffer, 20 mM EDTA, 20 mM phenylmethyl sulfonylfluoride (PMSF), pH 8.5]; (2) urea solvent [6 M urea, 20 mM PMSF, 20 mM EDTA]; and (3) urea-SDS solvent [6 M urea, 5% SDS, 20 mM PMSF, 20 mM EDTA, 20 mM Tris buffer, pH 8.0]. We shall refer to these solvents as guanidine

HCl, urea and urea-SDS solvents respectively. To all three solvents β-mercaptoethanol was added just prior to use to a final concentration of 5% v/v.

For SEM studies the abdominal nerve cords of the crayfish (*Procambarus clarkii*) were excised from the animal and subjected to three successive levels of extraction as follows:

(1) Excised nerve cords were immersed in 5 ml per nerve cord of guanidine HCl solvent and extracted by gentle shaking at 37 °C for 48 hours.

(2) The extracted nerve cords were washed 3 times with 2 ml per nerve cord of NaCl-Tris buffer (0.15 M NaCl, 50 mM Tris, 5 mM EDTA, pH 7.5) and then treated with 1 ml per nerve cord of high purity collagenase solution [0.5 mg/ml, type VII collagenase (Sigma Chemical Co.), 10 mM Tris buffer, 5 mM calcium chloride, 5 mM N-ethylmaleimide, 0.01% Thimerosal (Sigma chemical Co.)] at 37 °C for 24 hours. These nerve cord ghosts were then extracted for a second time with 5 ml per nerve cord of guanidine HCl solvent at 37 °C for 24 hours.

(3) The nerve cord ghosts were then washed 3 times with 2 ml per nerve cord urea solvent and then extracted with 5 ml per nerve cord with urea-SDS solvent at 37 °C per 24 hours.

Nerve cord ghosts extracted as described above were prepared for scanning electron microscopy by washing on Nucleopore filters (pore size 1 µm) with successive acetone solutions starting at 10% and increasing in 10% steps to 100%. The dried ghosts attached to Nulcopore filters were then stored in a dessicator over $P_2O_5$. The filters were mounted on SEM stubs with double face tape, coated with a thin layer (10 nm) of sputtered gold in a Denton Desk-1 apparatus and examined with a Hitachi S-570 scanning electron microscope at 20 kV accelerating voltage. For transmission electron microscopy of frog sciatic and peripheral nerves the sequence of the washing procedure was reversed so as to conclude with a 6 M guanidine HCl wash to allow glutaraldehyde fixation. Thus the sequence consisted of urea-SDS solvent, urea, guanidine HCl solvent, NaCl-Tris buffer, collagenase treatment, guanidine HCl solvent without β-mercaptoethanol. The nerve cord ghosts were then fixed in 6 M guanidine HCl containing 2.5% glutaraldehyde, 0.2% formaldehyde, 10% sucrose, 60 mM sodium chloride, 2 mM calcium chloride and 100 mM sodium cacodylate at pH 7.2. The nerve ghosts were fixed for 2 hours at 4 °C and washed for 1 hour in the solution without glutaraldehyde or formaldehyde. The extracted tissue was postfixed for 1 hour in 2% osmium tetroxide, dehydrated in an alcohol series embedded in an epon-araldite mixture, sectioned on an ultramicrotome and examined with a Hitachi H-600 transmission electron microscope.

## 3. Results

### 3.1. SEM Studies of Crayfish Nerves

The abdominal nerve cords of the crayfish (*Procambarus clarkii*) when extracted with the guanidine HCl solvent shrank to 20–25% of their original length. SEM images of these abdominal nerve cords show structures in which the ganglia, connectives, terminals, and 3 pairs of lateral roots per ganglion can still be discerned (Figs. 1 and 2).

The degree to which shrinkage in width due to protein extraction has occurred can be deduced from the extensive folding of the nerve cord sheath shown in Figs. 2 and 3. High magnification images of the folded nerve sheath can be seen in Figs. 3 and 4.

Treatment of the guanidine HCl extracted nerve cords with collagenase followed by another briefer extraction with the guanidine HCl solvent removes the nerve sheath but does not destroy the basic morphology of the nerve cord with the exception of the shape of the roots which shrank and thickened to form small bulbous structures (Figs. 5 and 6). The removal of the nerve sheath with collagenase suggests that this structure contains collagen. The texture of the guanidine HCl insoluble matrix revealed by the removal of the nerve cord sheath appears spongy (Figs. 6 and 7).

The highly shrunken nerve cord ghosts can be made to swell back to 80–90% of the length of the unextracted nerve cord by removing the guanidine HCl solvent with urea solvent and then treating with the urea-SDS solvent. This treatment transforms the nerve cord ghosts into a highly transparent and delicate structure which fragments readily when transferred to the Nucleopore filter. Upon washing by filtration with acetone these soft and delicate objects flatten on the filter allowing the structure of their matrix to be resolved at high magnification. Figure 8 is an image of such a urea-SDS treated nerve cord. Collagenase digestion appears to have been incomplete in this instance because one of the roots still shows what appears to be axons maintaining their integrity. The other root however shows its content flattened against the filter revealing, at higher magnification, linear structures interconnected by fine filaments (Figs. 9 and 10). The appearance of

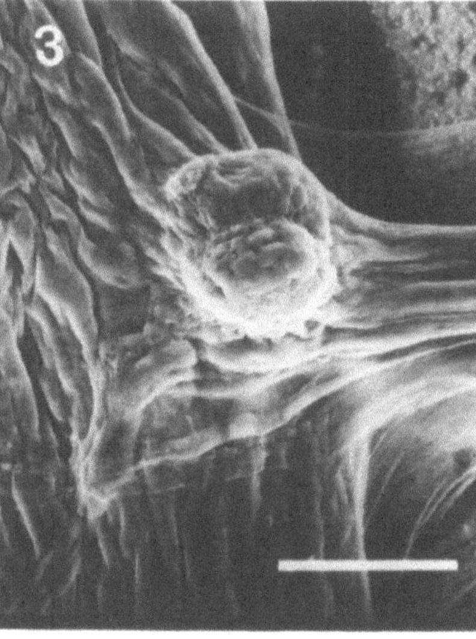
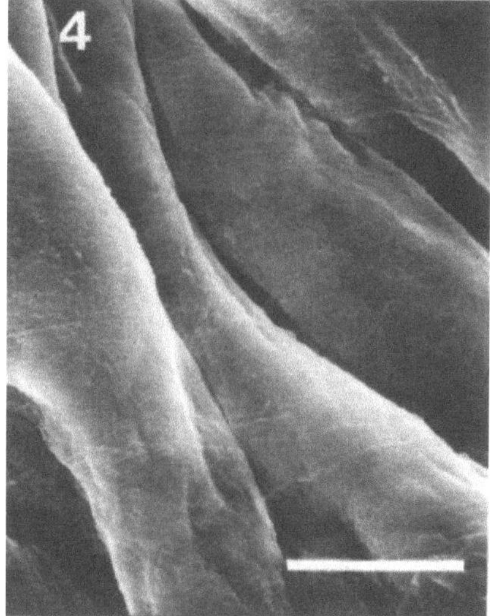

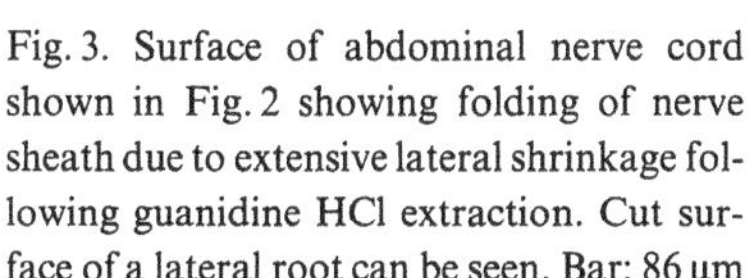

Fig. 3. Surface of abdominal nerve cord shown in Fig. 2 showing folding of nerve sheath due to extensive lateral shrinkage following guanidine HCl extraction. Cut surface of a lateral root can be seen. Bar: 86 µm

Fig. 4. High magnification image of nerve sheath of abdominal nerve cord shown in Fig. 2. Since this structure can be removed with collagenase we conclude that collagen is a major component of the sheath. Bar: 15 µm

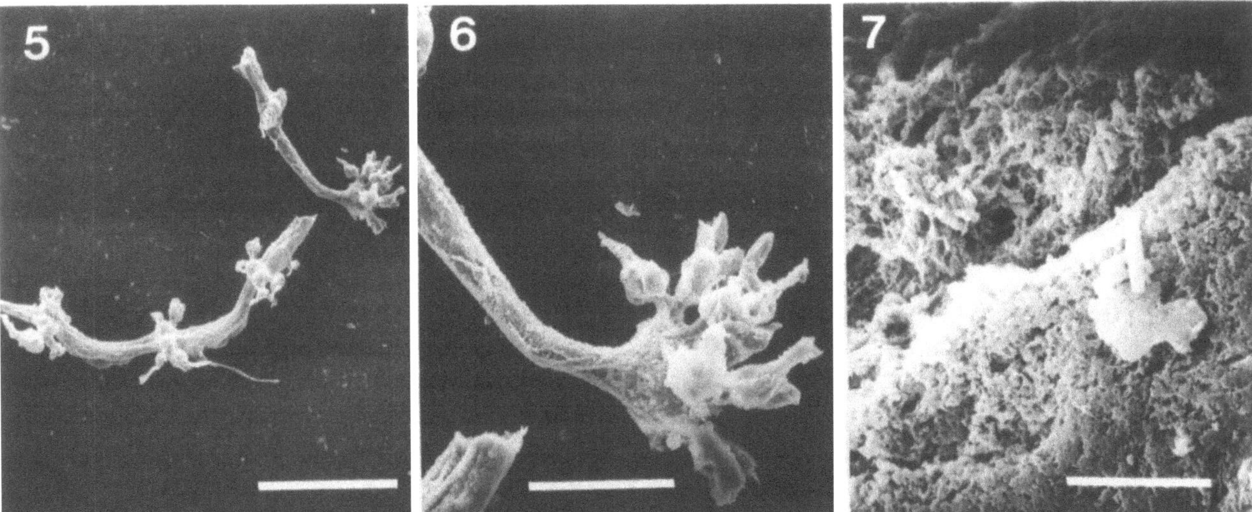

Figs. 5–7. Abdominal nerve cord extracted with guanidine HCl solvent, treated with collagenase and washed with guanidine HCl solvent

Fig. 5. The persistence of the characteristic structure of the abdominal nerve cord except that the roots have shrunk to form a thickened bulbous structure. Bar: 1 mm

Fig. 6. Higher magnification of connective and terminal showing spongy surface of structure revealed by removal of nerve sheath. Bar: 0.3 mm

Fig. 7. High magnification of surface of nerve cord showing spongy, irregular surface. Bar: 7.5 µm

the matrix in the region of the connectives shows a more highly branched structure which however is still oriented longitudinally (Fig. 11). In the region of the ganglion, which contains the cell bodies of the nerves this longitudinal orientation appears to be absent and the insoluble matrix appears to be a network (Figs. 12 and 13). In Fig. 13, the finest filaments seen at high SEM magnification appear to be approximately 25 nm in width.

### 3.2. TEM Studies of Frog Nerves

We found that urea and SDS interferes with embedding and we therefore performed our extraction of frog nerves by reversing our extraction sequence as described above. Transmission electron micrographs of sciatic nerves (Figs. 14–17) showed linear elements of uniform width (45–55 nm) linked to each other by a branched network. Because of the exhaustive treatment of the extracted nerves with collagenase we conclude that these linear elements do not contain collagen. Furthermore, we had shown previously that in the case of striated muscle, collagenase treatment removes all of the hydroxyproline and hydroxylysine from acid hydrolysates (Loewy *et al.* 1984). Abundant vesicular elements, not seen in scanning microscopy were also present. The finest filaments were below 5 nm in width (Fig. 17). Peripheral nerves also showed linear and ves-

icular elements resembling those of sciatic nerves in appearance and texture (Figs. 18 and 19). Globular thickenings at the intersection of some of the branched filaments (Figs. 18 and 19) seem to be characteristic of this type of nerve tissue. We recognize that we are not able in these studies to distinguish between cytoplasm of axons and Schwann cells.

Figures 14–19 were cropped by less than 20% in order to fit them into a single plate.

## 4. Discussion

The concept of the "plasmagel" as a coherent yet dynamic cytomatrix composed of interconnected filaments was propsed many years ago to account for the viscoelastic properties of cytoplasm (Scarth 1927, Chambers and Fell 1931, Frey-Wyssling 1948, Crick and Hughes 1950). Direct microscopic confirmation of this notion has only become possible with high voltage electron microscopy which provides the depth of focus needed to observe the interconnections between the cytoskeletal elements of the cytoplasmic ground substance (Wolosewick and Porter 1979). Porter's concept of the microtrabecular lattice, however, not only points to the interconnectedness between the filaments in the cytoplasm but also raises the possibility that this structure may interact with many cytoplasmic proteins. This latter aspect of the microtra-

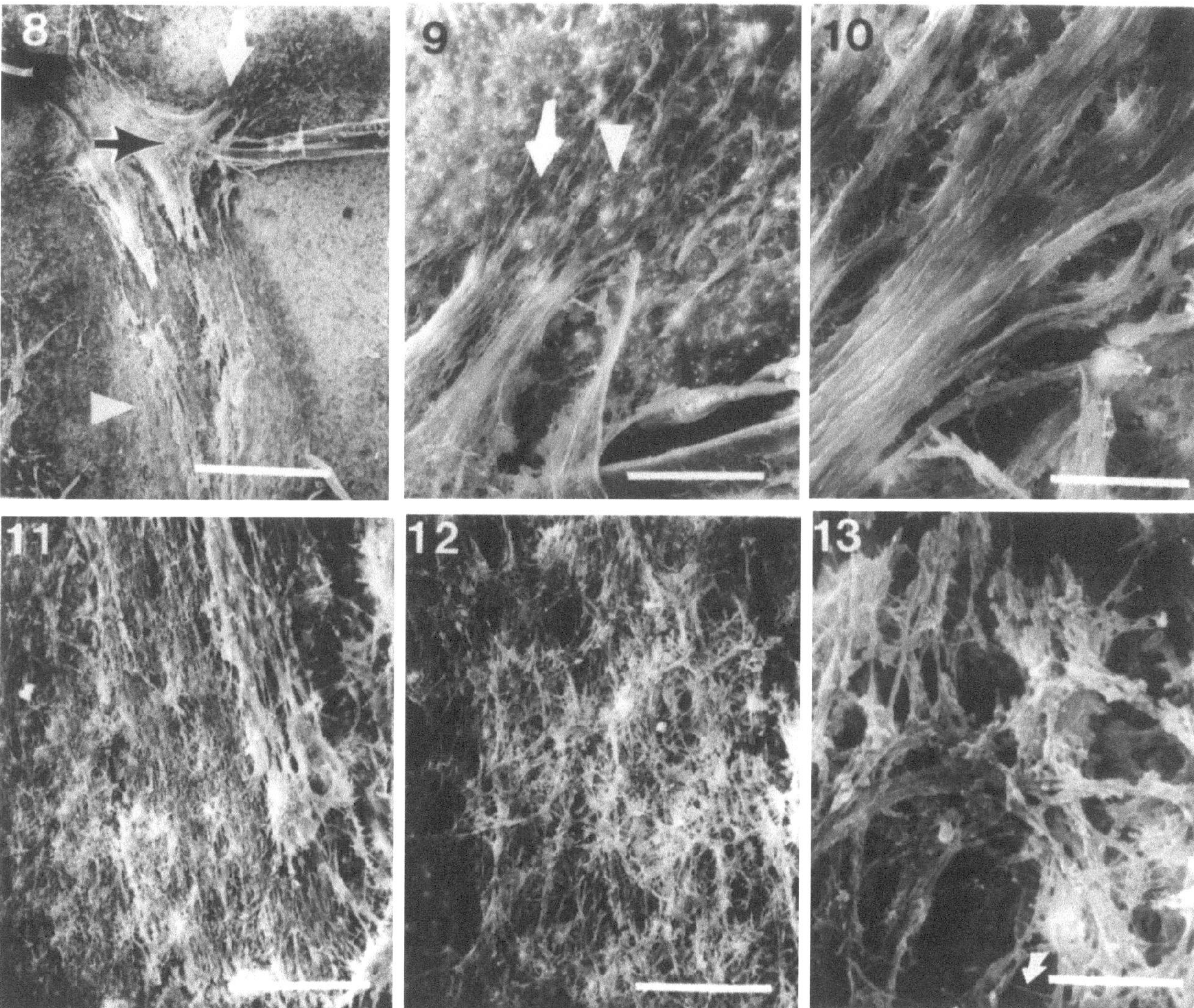

Figs. 8–13. Abdominal nerve cord extracted with guanidine HCl solvent, treated with collagenase, washed with guanidine HCl solvent, washed with urea solvent and then with urea-SDS solvent

Fig. 8. The highly transparent and delicate structure which remains expands back to 80–90% of its original length and flattens on a Nucleopore filter. One root (white arrow) can be seen emanating from the ganglion. The other root shows axons at a stage of incomplete collagenase digestion. Bar: 0.3 mm

Fig. 9. Higher magnification of collagenase digested root shown in Fig. 8. Bar: 60 µm

Fig. 10. High magnification of region identified by arrowhead in Fig. 9. Bar: 12 µm

Fig. 11. Longitudinal orientation of matrix in region of connective identified by arrowhead in Fig. 8. Bar: 86 µm

Fig. 12. Matrix in region of ganglion identified by black arrow in Fig. 8 showing more random orientation of filament structure. Bar: 30 µm

Fig. 13. High magnification image of region shown in Fig. 12. Filament identified by arrow is 25 nm wide. Bar: 3.8 µm

becular lattice has recently received a great deal of attention as independent evidence has accumulated regarding the extent to which the proteins of the "cytosol" are immobilized and specifically localized in the cytoplasm (PAINE 1984). It is important however to recognize that these two independent concepts of interconnectedness and proclivity for protein binding are not easily pursued simultaneously. Indeed, SCHLIWA has demonstrated that by extracting cells with nonionic detergents it is possible to obtain very clear images of microtubules, intermediate filaments and microfilaments as well as of numerous interconnecting superthin (2–3 nm) filaments (SCHLIWA and VAN BLERKOM 1981). It is reasonable to conclude that the clarity of the im-

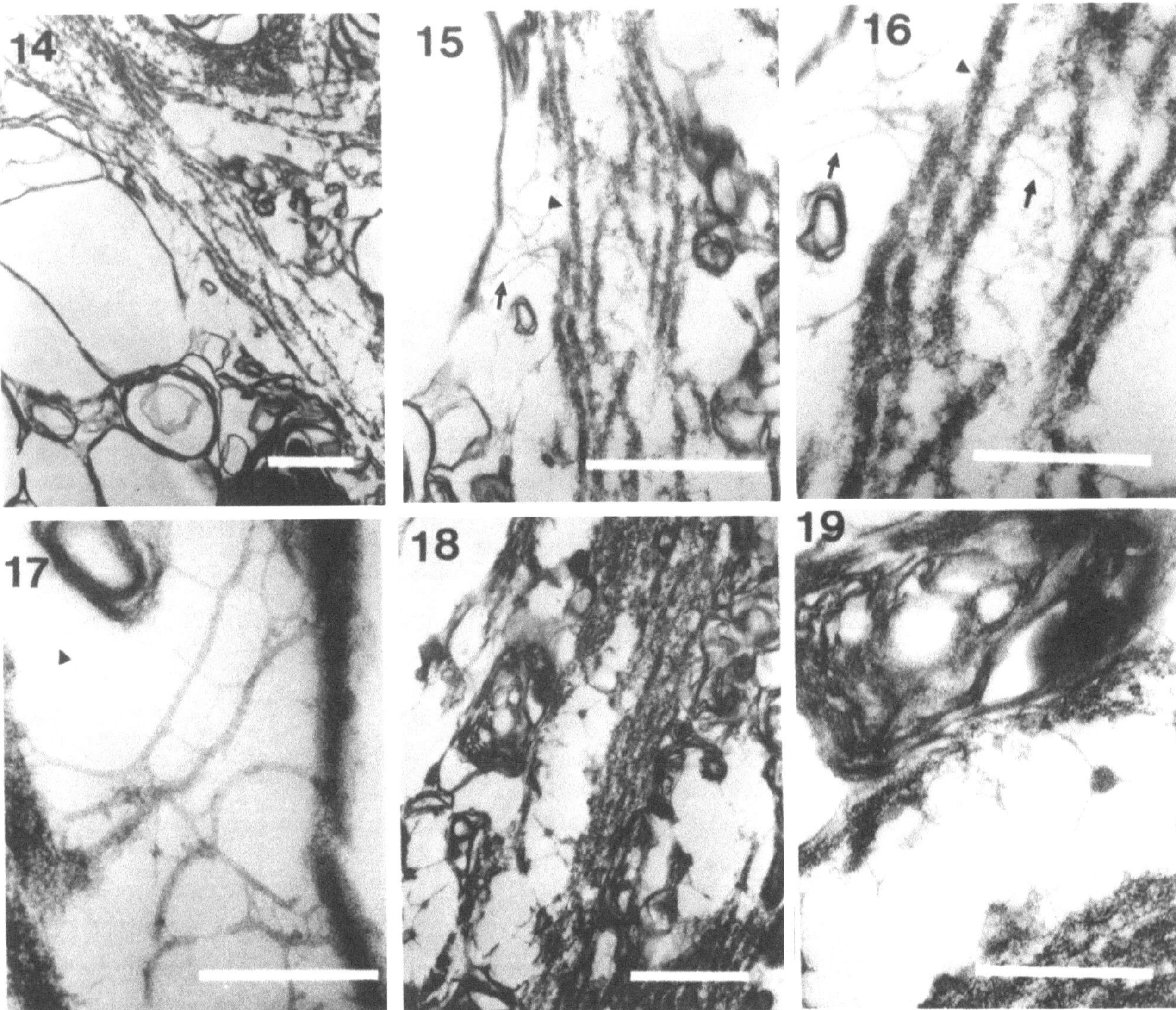

Figs. 14–19. TEM images of frog sciatic (Figs. 14–17) and peripheral (Figs. 18 and 19) nerve ghosts. Nerves were extracted with urea-SDS solvent, urea, guanidine HCl solvent. NaCl-Tris buffer, then treated with collagenase and finally washed with guanidine HCl solvent without β-mercaptoethanol. Bars: Figs. 14, 15, and 18: 1 µm; Figs. 16 and 19: 0.5 µm; Fig. 17: 0.25 µm

Figs. 14–17. Images of sciatic nerve ghosts showing linear elements 45–55 nm in width (arrowheads), interconnecting branched network (small arrows) and vesicles can be seen at various magnifications. Finest filament (Fig. 19, arrowhead) is 3–4 nm in width

Figs. 18 and 19. Images of peripheral nerve ghosts showing linear and vesicular elements and interconnecting branched networks with globular thickenings

ages obtained by SCHLIWA are not only due to the removal of "background" proteins dissolved in the cytoplasm but are also based on the extraction by the detergent of proteins normally bound to the filaments of the cytomatrix. The use of mild solvents such as nonionic detergents did not however turn out to be the final step in the process of stripping soluble proteins from the cytomatrix. MARUYAMA had shown some time ago that skeletal muscle tissue could be extracted with strong protein solvents such as 1 N acetic acid, 0.1 N

NaOH, 6 M urea, and 1% SDS leaving behind an elastic material he named connectin (MARUYAMA *et al.* 1977). It was not clear at the time however whether or not connectin included insoluble components of the extra cellular matrix of muscle tissue. Subsequent work has shown that to distinguish between the extracted remains of the cytoplasm and of the extracellular matrix it is necessary to: (1) remove the extracellular matrix with collagenase (LOEWY *et al.* 1984), or (2) isolate and examine cytoplasm free of contamination with extracellular matrix (Gassner *et al.* 1985), or (3) as we do in this article, examine sectioned material in transmission electron microscopy. All three approaches have now been employed and the surprising outcome of the use of strong protein solvents such as guanidine HCl and urea-SDS (in the presence of disulfide reducing reagents) is that: (1) A covalently-crosslinked matrix composed of ultrathin filaments has been demonstrated in a number of cell types (Loewy *et al.* 1984, Gassner *et al.* 1985); (2) this matrix is distributed throughout the cytoplasm of the cell (Loewy *et al.* 1985); and (3) although the matrix consists of less than 1% of the total cell protein, it retains some of the morphological characteristics of the unextracted cell (Loewy *et al.* 1984, 1985, 1986). These results have been confirmed by our studies of nerve tissue reported in this article. The surface of the central nerve cord of the crayfish has a spongy appearance after extraction with the guanidine HCl solvent and after removal of the nerve sheath with collagenase (Figs. 6 and 7). Since subsequent treatment with the urea-SDS solvent reveals linear elements oriented parallel to the nerve cord, we conclude that the spongy appearance is due to the folding of the linear elements resulting from the extensive shrinkage caused by guanidine HCl extraction.

We have shown previously that the covalently-crosslinked matrix of skeletal muscle contains an unusually high percentage of hydrophobic amino acid residues (Loewy *et al.* 1984). Since the solvent action of guanidine HCl is exerted mainly by interfering with hydrogen bonds and other ionic interactions it is reasonable to conclude that the shrinking observed is due to hydrophobic interactions between remaining components of the matrix which have been stripped by the extraction of more hydrophilic cytoskeletal proteins. According to this view the superfine filaments of the covalently crosslinked cytomatrix are normally at the core, or at least closely associated with filaments, of the cytoskeleton. Schliwa's "connector filaments" would, according to this view, be bare portions of an extensive interconnected cytomatrix and superthin filaments. When nerve cord ghosts are treated with the urea-SDS solvent, the extensive binding of the negatively charged SDS causes the filaments of the matrix to unfold, bringing about an expansion of the nerve cord ghosts. SEM images of this material show an extended and interconnected matrix which, when flattened against the Nucleopore filter, preserves some of its loose and open structure (Figs. 8–13). In the region of the connectives and roots (composed of nerve axons), the cytomatrix consists of filamentous elements mostly oriented parallel to the nerve cord (Figs. 9–11). In the region of the ganglion (composed of cell bodies), the orientation of the filamentous elements is relatively random (Figs. 12 and 13). It is not surprising, however, that most of the filamentous material in the SEM images, is much thicker than 2–3 nm since in the course of the replacement of urea-SDS with acetone, much side to side aggregation is likely to have occurred.

In frog sciatic and peripheral nerves, the initial urea-SDS extraction removes the cytoskeletal proteins such as actin, myosin, tubulin and intermediate filament proteins. We found that, under these conditions, the extracted nerve ghost does not shrink by more than a few percent. When the urea-SDS was exchanged with urea and then with guanidine HCl we also do not see much longitudinal shrinkage, possibly because the nerve sheaths had not been removed with collagenase. Under these conditions one would expect the interaction between the largely hydrophobic filaments of the cytomatrix to cause mainly lateral associations. This indeed is what we see in our TEM images; the parallel orientation of the filaments are dramatically evident. These filaments are much wider (45–55 nm) than any linear structures known in the unextracted axon and we therefore conclude that they must be the result of side to side association of super fine filaments which may have been associated with microtubules (Linck 1985) and laid bare by the extraction process. The filaments interconnecting these linear structures can be observed forming a network of filaments (Figs. 15–17), the finest of which are 3–4 nm in thickness (Fig. 12). One of the surprises of the TEM studies of extracted nerve axons is the numerous vesicle ghosts seen in the preparations. We did not expect that some of the peripheral membrane proteins resist extraction with strong protein solvents such as urea-SDS. It is likely therefore that vesicle ghosts are also composed of covalently crosslinked protein chains.

It is remarkable how the TEM images of the extracted axon with its numerous filaments oriented in parallel, its extensive interconnections, and its vesicles, resembles the images obtained by Hirokawa and his associates (Hirokawa 1982, Hirokawa and Yorifuji 1986), showing microtubules and intermediate filaments interconnected with numerous, branched, superthin filaments. The presence of fine (2–5 nm) filaments in the cytoplasm of many cell types has been reported (Roberts 1987). We do not know which of these fine filaments are covalently crosslinked. The only clear demonstration of an extensive covalently crosslinked matrix of 2–3 nm filaments has been made in the cytoplasm of the plasmodium of *Physarum polycephalum* (GASSNER *et al.* 1985). Low solubility components consisting of 2–3 nm filaments have been found in connection with striated muscle (WANG *et al.* 1979, 1984) and microtubules (LINCK *et al.* 1982, 1985) and their protein subunits have been named titin and tectin respectively. Our methods of extensive extraction with strong protein solvents followed by an examination of the insoluble residue has allowed us to focus directly on the covalently crosslinked components of the cytomatrix.

In summary we have found that: (1) $N^\varepsilon$-($\gamma$-glutamic)lysine crosslinks are present in the insoluble matrices of skeletal muscle (LOEWY and MATACIC 1981, LOEWY *et al.* 1981) and of *Physarum polycephalum* (LOEWY and GASSNER 1988), (2) the glutamic lysine crosslink content of these matrices is lowered by treatment of glycerol extracted cells with $Mg^{2+}$ ATP and raised with $Mg^{++}$ ATP followed by $Ca^{2+}$ (LOEWY and MATACIC 1981, LOEWY *et al.* 1981), (3) these matrices are composed of a network of linear elements (LOEWY *et al.* 1985), and (4) the overall structure of the covalently crosslinked matrices resembles that of the unextracted cell (LOEWY *et al.* 1984, LOEWY *et al.* 1986). In this article we have extended some of these observations to nerve tissue. We have also observed the presence of insoluble matrices in the profiles of vesicles. Whether the insoluble cytomatrix is actively involved in mechanochemical transductions of the cytoplasm must yet be determined, but there is little doubt that, in order to participate in the dynamic activities of cytoplasm, it must be capable of rapid modulation of its crosslinks, a property of cytoplasm which has been recognized by the early students of cell structure and function (SCARTH 1927, CHAMBERS and FELL 1931, FREY-WYSSLING 1948, CRICK and HUGHES 1950).

## Acknowledgements

We wish to acknowledge the stimulation and inspiration which NOBURO KAMIYA has exerted over our thinking during these many years. His love for and his interest in the dynamics of cytoplasm has been an inspiration to all of us. We would like to thank Drs. PHILIP J. STEPHENS and KAREN F. GREIF for the advice they gave us regarding the dissection and handling of the nerve tissues and KARI NADEAU for her review of the manuscript. This work was supported by NIH grant AM 34503 and NSF grant DCB 851112.

## References

CHAMBERS R, FELL HB (1931) Micro operations on cells in tissue culture. Proc Roy Soc Lond 109: 381

CRICK FHC, HUGHES AFW (1950) The physical properties of cytoplasm. A study by means of the magnetic particle method. Exp Cell Res 1: 37–80

FREY-WYSSLING A (1948) Submicroscopic morphology of protoplasm and its derivatives, 2nd edn. Elsevier, New York

GASSNER D, SHRAIDEH Z, WOHLFARTH-BOTTERMANN KE (1983) An extraction resistant tensile protein in the protoplasmic matrix of *Physarum*. Cell Biol Int Rep 7: 905–913

— — — (1985) A giant titin-like protein in *Physarum polycephalum*: evidence for its candidacy as a major component of an elastic cytoskeletal superthin filament lattice. Eur J Cell Biol 37: 44–62

HIROKAWA N (1982) Cross-linked system between neurofilaments, microtubules and membranous organelles in frog axons revealed by the quick-freeze, deep-etching method. J Cell Biol 94: 129–142

— YORIFUJI H (1986) Cytoskeletal architecture of reactivated crayfish axons with special reference to crossbridges among microtubules and between microtubules and membrane organelles. Cell Motil Cytoskeleton 6: 458–468

LINCK RW, ALBERTINI DF, KENNEDY DM, LANGEVIN GL (1982) Tectin filaments: chemically unique filaments of sperm flagellar microtubules. Cell Motil [Suppl 1]: 127–132

LINCK RW, AMOS LA, AMOS WB (1985) Localization of tectin filaments in microtubules of sea urchin sperm flagella by immunoelectron microscopy. J Cell Biol 100: 126–135

LOEWY AG, GASSNER D (1988) Personal Communication

— MATACIC SS (1981) Modulation of the $N^\varepsilon$-($\gamma$-glutamic)lysine crosslink in cellular proteins. I. *In vivo* and *in vitro* studies. Biochim Biophys Acta 668: 167–176

— — RICE P, STERN J (1981) Modulation of the $N^\varepsilon$-($\gamma$-glutamic)lysine crosslink in cellular proteins. II. Fractionation studies. Biochem Biophys Acta 668: 117–185

— SHUPP-BYRNE DE, STORM CA (1986) Covalently-crosslinked matrices in muscle, nerve, and other cells. Acta Biol Hung 37: 94

— WILSON FJ, TAGGART NM, GREENE EA, FRASCA P, KAUFMAN HS, SORRELL MJ (1984) A covalently cross-linked matrix in sekeletal muscle fibers. Cell Motil 3: 463–483

— — — KAUFMAN HS, NORTHROP JL, GREENE EA, FRASCA P (1985) Covalently-crosslinked matrices in sekeletal muscle fibers. Ann New York Acad Sci 455: 699–702

MARUYAMA K, MATSUBARA S, NATORI R, NONOMURA Y, KIMURA S, OHASHI K, EGUCHI G (1977) Connectin an elastic protein of muscle. J Biochem 82: 317–337

PAINE PL (1984) Diffusive and nondiffusive proteins *in vivo*. J Cell Biol 99: 188s–195s

ROBERTS TM (1987) Fine (2–5 nm) filaments: new types of cytoskeletal structures. Cell Motil Cytoskeleton 8: 130–142

SCARTH GW (1927) The structural organization of plant protoplasm in the light of micrurgy. Protoplasma 2: 189–205

SCHLIWA M, VAN BLERKOM J (1981) Structural interactions of cytoskeletal components. J Cell Biol 90: 222–235

WANG K, McCLURE J, TU A (1979) Titin: major myofibrilar components of striated muscle. Proc Natl Acad Sci USA 76: 3698–3702

— RAMIREZ-MICHELL R, PALTER D (1984) Titin is an extraordinarily long, flexible, and slender myofibrillar protein. Proc Natl Acad Sci USA 81: 3685–3689

WOLOSEWICK JJ, PORTER KR (1979) Microtrabecular lattice of the cytoplasmic ground substance. Artifact or reality. J Cell Biol 82: 114–139

Protoplasma (1988) [Suppl. 2]: 145–157



# Role of Microtubules in Hormone Secretory Function of the Rat Anterior Pituitary

O. Shimada[1,*], H. Ishikawa[1], and K. Wakabayashi[2]

[1] Department of Anatomy, School of Medicine, and [2] Hormone Assay Center, Institute of Endocrinology, Gunma University, Maebashi, Gunma

Received May 6, 1988
Accepted June 17, 1988

Dedicated to Professor Dr. Noburo Kamiya on the occasion of his 75th birthday

## Summary

Correlative morphological and physiological analysis was made in order to clarify the role of microtubules in secretory activity in the rat anterior pituitary cells *in vivo*. Immunofluorescence microscopy showed a relatively intense tubulin staining in all secretory cells, although the staining intensity varied among different cell types. By thin-section electron microscopy, microtubules were distributed preferentially in perinuclear regions. They radiated from Golgi areas toward the cell periphery, and were closely associated with secretory granules and mitochondria. Microtubules were more abundant in ACTH-, and some of prolactin (PRL-) and TSH-cells than in other types of cells.

The injections of anti-mitotic drugs, colchicine and vinblastine, reversibly disrupted most of microtubules in secretory cells. These drugs also induced the disturbances in the distribution patterns of secretory granules, mitochondria, Golgi apparatus and rough ER which was dilated considerably. Radioimmunoassay indicated a significant decrease in hormone levels of plasma TSH, PRL and LH after colchicine and vinblastine treatment. When such drug-injected rats were further stimulated with hypothalamic releasing hormones, TRH and LHRH, a marked increase in plasma levels of TSH, PRL and LH was observed. Moreover, exocytotic figures of secretory granules were often observed in the secretory cells of these hormone-stimulated rats.

These results altogether suggest that microtubules may not participate directly in the release of secretory granules at the cell surface. Indeed, they may aid in hormone secretion through their possible roles in intracellular transport of secretory granules toward the cell surface and by their role in maintaining an orderly distribution of various cellular organelles such as Golgi apparatus, rough ER and mitochondria.

*Keywords:* Anterior pituitary cells; Colchicine; Cytoskeleton; Exocytosis; Hormone secretion; Microtubules.

---

* Correspondence and Reprints: Department of Anatomy, Gunma University, School of Medicine, Maebashi, Gunma, 371 Japan.

## 1. Introduction

The hormone secretion is the complex process which comprises several successive steps from synthesis to release of secretory materials (Slot *et al.* 1974, Kelly 1985, Samuel and Flickinger 1987). It has been believed that cytoskeletal components such as microtubules, actin-containing microfilaments and intermediate filaments play some roles in the secretory process of various endocrine as well as exocrine cells (Schneyer *et al.* 1972, Temple *et al.* 1972, Marchand *et al.* 1974, Simson *et al.* 1974, Slot *et al.* 1974, Drenckhahn *et al.* 1977, Aunis *et al.* 1979, Segawa *et al.* 1985, Savino and Dardenne 1986). Since the pioneering work by Lacy *et al.* (1968) on insulin-secreting B cells of Langerhans islets, much attention has been drawn to possible roles of microtubules in the secretion of insulin (Malaisse *et al.* 1971, Slot *et al.* 1974), thyroxine (Williams and Wolff 1970, Neve *et al.* 1972), and various anterior pituitary hormones (Kraicer and Milligan 1971, Labrie *et al.* 1973, Sherline *et al.* 1977). Microtubules are indeed abundant and take a characteristic arrangement in such secretory cells (Shiino and Rennels 1975, Shiino *et al.* 1975, Warchol *et al.* 1975, Ravindra and Grosvenor 1986). Experimental studies using microtubule poisons have established the requirement of intact microtubules for the secretory activities (Labrie *et al.* 1973, Sheterline *et al.* 1975). However, the precise role of microtubules in hormone secretion has not been yet decided.

The anterior pituitary consists of mainly five types of

hormone secreting cells, which produce growth hormone (GH), prolactin (PRL), thyroid-stimulating hormone (TSH), adrenocorticotropin (ACTH), and follicle-stimulating hormone and luteinizing hormone (FSH-LH) respectively. In order to elucidate the involvement of microtubules in the secretion of those hormones from the respective cell types, many studies have been performed analysing isolated tissues *in vitro*. The analysis on such *in vitro* systems seemed to be favoured by many investigators because of the expected simplicity and ease in experiments and interpretation. However, the controversy still exists as to whether or not microtubules are actively involved in the secretory activities (KRAICER and MILLIGAN 1971, MACLEOD *et al.* 1973, KOVACS *et al.* 1974, YOUNG *et al.* 1985). Such inconsistency seems to arise from different experimental designs or from different cell types used. Then, the question may be asked if the degree of microtubule involvement in hormone secretion varies in different cell types, though the basic mechanism of secretion is expected to be common to all. More works are needed carefully correlating physiological phenomena with morphological observations.

The present study has focussed on the role of microtubules in hormone secretion of the rat anterior pituitary using whole animals. Particular emphasis has been placed on morphological and physiological correlates of each hormone secretion. Interestingly, upon stimulation with hypothalamic hormones significant amounts of hormones could be released from the cells in which microtubules had been almost completely disrupted by colchicine and vinblastine. These results together with other data indicate that microtubules are not directly involved in the process of release, namely exocytosis (AUNIS *et al.* 1979, BRECKENRIDGE *et al.* 1987), but many participate in intracellular translocation of secretory granules from the Golgi region to the cell surface besides their roles in the orderly distribution of various membranous organelles.

## 2. Materials and Methods

### 2.1. Animals

Male Wistar rats weighing 150–200 g were caged individually in a room with controlled temperature (24 ± 1 °C) and artificially illuminated (7 am–7 pm). Rats were fed with pellets (Oriental Yeats Co.) and water *ad libitum*. Untreated, drug-injected or saline-injected control rats were sacrificed by decapitation between 2–5 pm. Blood samples were collected in heparinized tubes for radioimmunoassay of TSH, PRL, GH, LH, FSH, thyroxine (T 4), and testosterone. Subsequently, anterior pituitary glands were dissected out for morphological observations and biochemical analysis.

### 2.2. Thin-Section Electron Microscopy

Pituitary glands were quickly removed and cut into small pieces. The tissue was fixed for 1–2 hr at room temperature with 2.5% glutaraldehyde and 2% paraformaldehyde in 0.1 M sodium cacodylate buffer (pH 7.2) without or with 0.2% tannic acid, and then overnight at 4 °C (TSUKITA and ISHIKAWA 1980). Tissues were rinsed extensively with cacodylate buffer, and post-fixed in cold 1% $OsO_4$ in the same buffer for 2 hr followed by *en bloc* staining with cold 1% aqueous uranyl acetate for 1 hr. Tissues were dehydrated in an ethanol series and embedded in Epon 812. Thin sections were cut and stained with uranyl acetate and lead citrate, and then examined with a Hitachi H-800 electron microscope.

### 2.3. Light Microscopic Immunohistochemistry

Tissue samples were fixed in a modified Zamboni's fixative (3% paraformaldehyde and 20% of saturated picric acid in 1/15 M phosphate buffer, pH 7.4) for 24 hr at room temperature and then immersed in 30% sucrose in phosphate buffer at 4 °C for 24 hr. Tissues were then frozen on dry ice and 4 μm serial sections were cut and mounted in 1% gelatin-coated glass slides. Sections were dried at room temperature, incubated with 1% normal goat serum (NGS) in phosphate buffer saline (PBS, pH 7.4) for 20 min, and then incubated with a primary antibody for 1–3 hr at room temperature. After rinsing several times, sections were incubated for 1 hr with an appropriate secondary antibody (anti-rabbit or anti-mouse IgG sheep or goat sera conjugated with either fluoresceine or rhodamine). Sections were rinsed throughly with PBS, and mounted in 50% glycerol in PBS to be viewed with an Olympus BH 2-RFK fluorescence microscopy. The primary antibodies against cytoskeletal proteins used in this study were rabbit anti-tubulin (Cat. No. 17543, Polyscience, U.S.A.), rabbit anti-actin (Cat. No. AB-1150, Advance, Japan), rabbit anti-cytokeratin (Cat. No. 1007, Transformation Res. U.S.A.), and mouse anti-β-tubulin monoclonal antibody (gift from Dr. K. Obata, Gunma University). Antibodies against anterior pituitary hormones were the same as used for radioimmunoassay (see below).

### 2.4. Drug and Hormone Treatment

For anti-mitotic drug treatments, rats received an initial intraperitoneal injection of either 0.7 mg of colchicine or 1 mg of vinblastine sulfate per kg body weight. Six hr later, rats were given an identical second injection. When cells were examined by thin-section electron microscopy 3 days following the drug injection, degenerative cells exhibiting an electron dense cytoplasm were rarely found in the anterior pituitary, and the number and distribution pattern of microtubules were almost the same as those observed in control hormone-secretory cells, indicating the reversible effects of colchicine and vinblastine sulfate.

In order to stimulate hormone release from the anterior pituitary, rats were injected intraperitoneally with a 20 μg/kg body weight solution of the hypothalamic releasing hormone, TRH (Cat. No. 24-0175, Funakoshi Co. Tokyo) or LHRH (Cat. No. 24-0172, Funakoshi Co.). Before releasing hormone injection, and at 5, 15, 30, 60, and 120 min after the injection, the rats were killed by decapitation to collect blood samples for measurement of plasma hormone levels, and pituitary tissues were then quickly dissected out for electron microscopic observations. To see the effects of repeated stimulation with hypothalamic hormones, the second injection of the same dose of TRH or LHRH was made at 120 min after the first injection.

Blood samples and tissues were collected at 15, 30 and 60 min after the second injection.

### 2.5. Radioimmunoassay for Hormones

To assay the hormone content of the anterior pituitary, each tissue sample was placed in a plastic microcentrifuge tube containing 1 ml of ice-cold extracted buffer (1% urea, 1% Triton X-100 in 0.1 M PBS, pH 7.4) and homogenized for 20 sec using an ultra-sonic disintegrating apparatus. After centrifugation at 4 °C the tissue extract was diluted to 1 : 1,000–1 : 10,000 with PBS and processed for radioimmunoassay. To assay the plasma content of each hormone, a blood sample from each rat was collected in a heparinized tube, and then centrifuged at 3,000 rpm for 15 min at 4 °C. The supernatant was collected and used for radioimmunoassay.

According to the method previously described (Ishikawa *et al.* 1987), radioimmunoassay of various hormones was performed using antiserum specific to each hormone, and the specificity of each anti-GH (HAC-RT 25-01 RBP 85), anti-TSH (HAC-RT 29-01 RBP 86), anti-PRL (HAC-RT 26-01 RBP 85), anti-LH, and anti-FSH sera was determined.

### 2.6. Classification of Rat Anterior Pituitary Cells

For the fine structural identification of secretory cells we used two different immunocytochemical methods; either preembedding peroxidase-anti-peroxidase (PAP) or colloidal gold staining as previously described with minor modifications (Polak 1986). The characteristics of each type of secretory cells as follows. (1) Gonadotropin-producing (Gn) cells were identified as cells containing two different types of secretory granules: small electron dense granules (about 200 nm in diameter) and large less dense granules (about 700 nm in diameter). (2) Typical prolactin-producing (PRL) cells were usually situated in close proximity to Gn cells, and contained irregular polymorphic granules. There were, however, some of PRL positive cells which contained uniformly round shaped granules. The latter type of PRL cells were not included in the present analysis, since it was often difficult to identify them without immunocytochemistry. (3) Growth hormone-producing (GH) cells contained large secretory granules (300–500 nm in diameter) which were evenly distributed throughout in the cytoplasm. (4) Both adrenocorticotropin- (ACTH) and thyrotropin-producing (TSH) cells were identified with their small (150–250 nm in diameter) cytoplasmic secretory granules. Granule morphology alone was not sufficient for distinguishing between these two cell types. However, many of the ACTH cells were distinguished by their stellate shape and the presence of long cytoplasmic processes which extended between neighbouring cells. On the other hand, TSH cells tended to be larger than ACTH cells, and they did not exhibit long cytoplasmic processes. These criteria are in good agreement with the previous ones described by other investigators (Nakane 1975, Childs 1986, Kurosumi and Inoue 1986). Nevertheless, precise morphological classification of the rat anterior pituitary cells is often problematic. Thus, in the present study, we examined only those cells which were morphologically identifiable by our criteria, and neglected those cells whose characteristics were obscure in thin sections.

### 2.7. Statistical Analysis

Statistical analysis of the hormone levels is expressed as the mean ± SE, and the Students' *t*-test (p < 0.05) was used to determine significant differences between two groups. To calculate the number

of secretory granules which were closely associated with the plasma membrane in Gn cells, we chose those cells which were sectioned almost through the cell center judging from the shape and size of the nuclei, and drew a 700 nm line apart from the plasma membrane on micrographs magnified at × 6,000, and then counted the number of secretory granules within the peripheral zone bounded by this line including those intersected by this line. A small number of cells, however, contained mainly large granules or small ones, which were so-called FSH-rich cells or LH-rich cells, so in this calculation we examined only the cells in which a ratio of small granules to large ones was from of 5 : 1 to 20 : 1.

## 3. Results

### 3.1. Cytoskeletal Components in Secretory Cells

By immunofluorescence microscopy, most cells of the anterior pituitary showed a strong anti-tubulin immunoreactivity, although there were some cells which stained weakly (Fig. 1 *a*). Interestingly, similarly stained cells showed a tendency to grouping. Anti-actin antibodies stained virtually all cells, in which actin was shown to be localized preferentially at the cell periphery (Fig. 1 *b*). On the other hand, staining with anti-cytokeratin was variable among cells, some cells being completely negative in immunoreactivity (Fig. 1 *c*). Double-staining of pituitary hormones and tubulin indicated that many of ACTH, PRL and TSH cells possessed strong tubulin-like immunoreactivity (Fig. 1 *d–g*). Cells intensely stained with an anti-tubulin monoclonal antibody were 81 cells/102 ACTH cells, 81 cells/120 PRL cells, 42 cells/80 TSH cells, 51 cells/188 GH cells and 7 cells/63 Gn (LH-positive) cells. Similarly, anti-cytokeratin staining tends to be intense in most of ACTH and GH cells as well as in S-100 protein containing non-secretory cells among other types of cells.

By electron microscopy, cytoskeletal fibrous structures such as microtubules, intermediate filaments and microfilaments were observed in every anterior pituitary cell. Microfilaments were prominent at the cell periphery just beneath the plasma membrane of all secretory cells (Fig. 1 *h*). Intermediate filaments were distributed throughout the cytoplasm, though they were prominent in the perinuclear region (Fig. 1 *i*). Microtubules were the most prominent component of the cytoskeleton in secretory cells, although the abundance of microtubules varied among cell types. Microtubules were most abundant in ACTH cells and in many PRL and TSH cells (Fig. 2 *a*), while fewer microtubules were observed in Gn and GH cells (Fig. 2 *b*). These observations were well consistent with the immunofluorescence microscopy. Nevertheless, the distribution pattern of microtubules was similar in all secretory cell types. A number

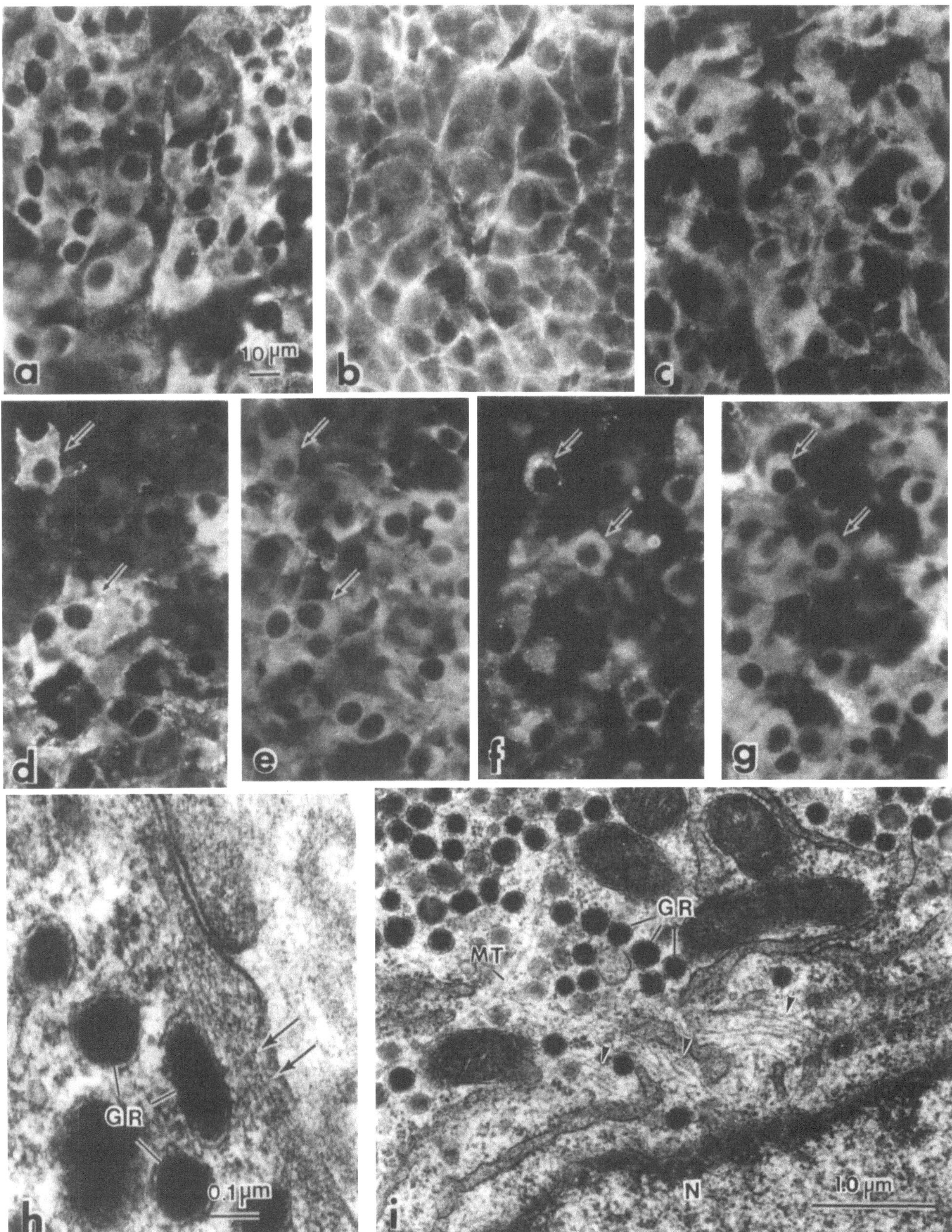

Fig. 1. Immunofluorescence and electron micrographs of the rat anterior pituitary cells. *a–c* Immunofluorescence for cytoskeletal proteins; anti-tubulin (*a*), anti-actin (*b*), and anti-cytokeratin (*c*). *d–g* Double immunostaining for hormones and tubulin; anti-TSH (rhodamine) (*d*) and anti-β-tubulin monoclonal antibody (fluoresceine) (*e*), and anti-PRL (rhodamine) (*f*) and anti-β-tubulin monoclonal antibody (fluoresceine) (*g*). Note that most of the TSH- and PRL-cells show an intense tubulin-immunoreactivity (arrows). *h–i* Electron micrographs of TSH cells. Microfilaments (arrows) are found preferentially at the cell periphery, while intermediate filaments (arrowheads) are distributed throughout the cytoplasm particularly around the nucleus (*N*). *GR* Secretory granules. *MT* Microtubule

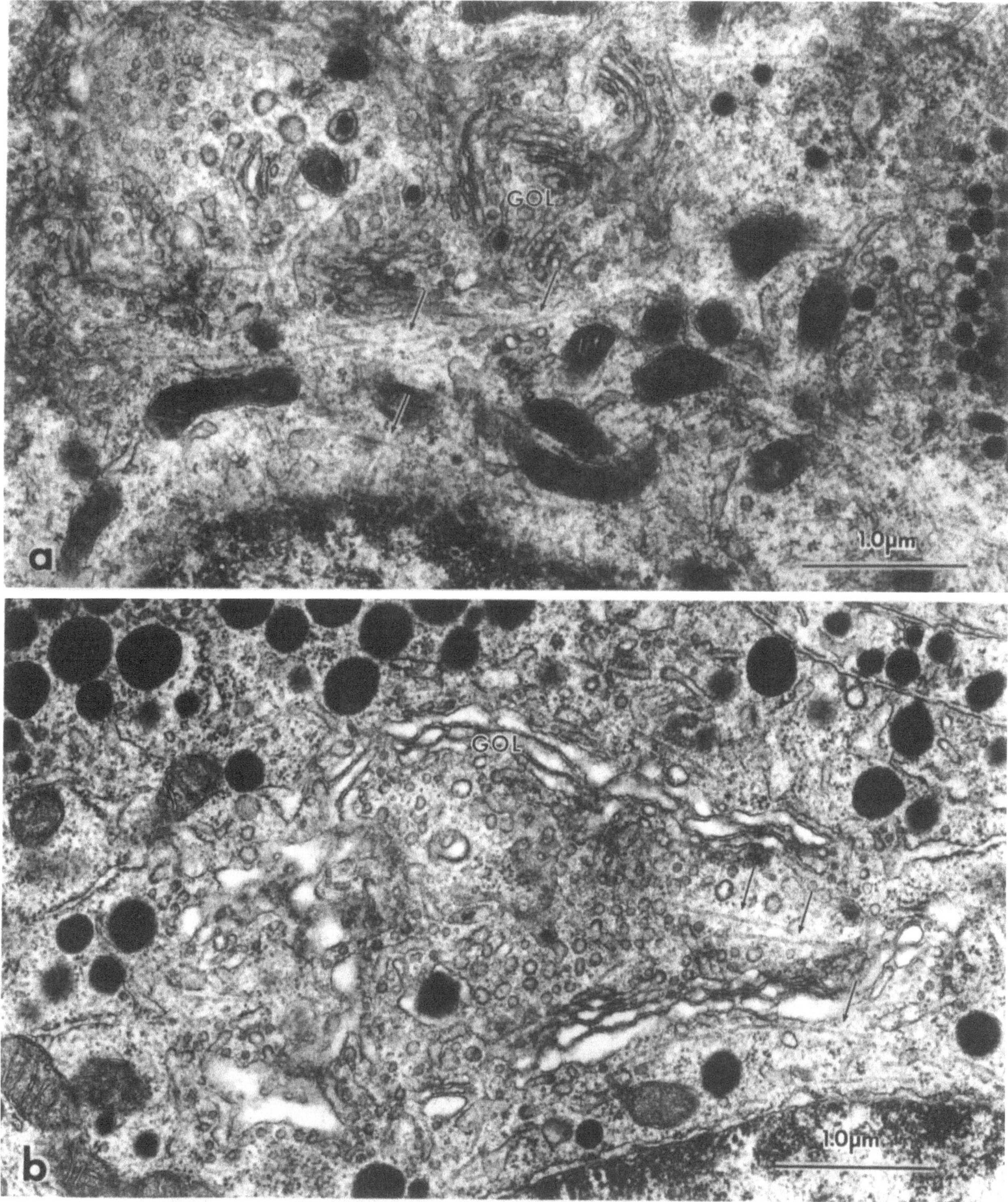

Fig. 2. Electron micrographs of two different types of secretory cells of the rat anterior pituitary. *a* ACTH cell. *b* GH cell. In the similar Golgi regions (*GOL*), microtubules (arrows) are more abundant in an ACTH cell than in a GH cell

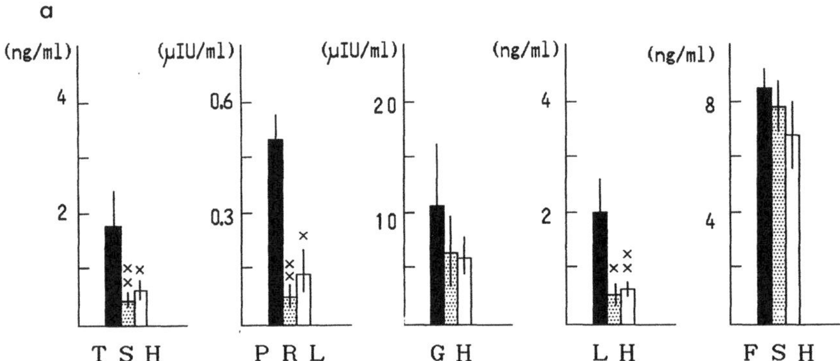

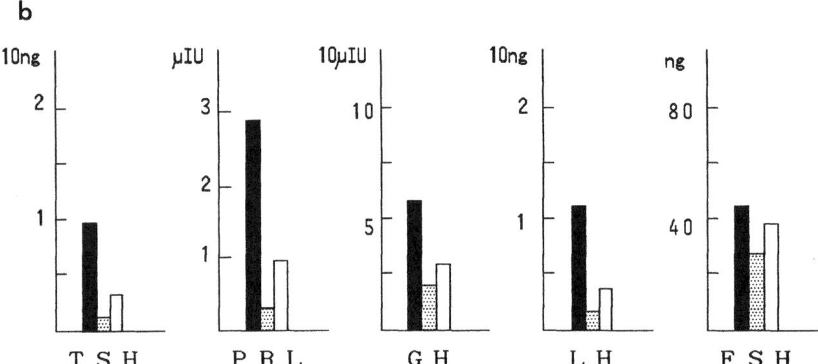

Fig. 3. Effect of anti-mitotic drugs (■ control, ▨ colchicine, and □ vinblastine) on plasma hormones. *a* The basal levels of plasma TSH, PRL, GH, LH, and FSH are shown by the means ± SE of 5–10 animals. × p < 0.05, × × p < 0.01, significantly different compared to the saline-injected control, respectively. *b* The total amount of each hormone in the whole blood. The circulating blood volume is determined as described in Results

of microtubules were seen in the Golgi region, from which they tended to radiate toward the cell periphery. Microtubules were also observed to encircle the nucleus, running parallel to the nuclear envelope. Mitochondria and secretory granules were often associated with microtubules. Microtubules were also prominent in the long cytoplasmic processes of ACTH cells, where they formed parallel bundles.

### 3.2. Basal Hormone Levels Following Anti-Mitotic Drug Injection

Seven hours after the first intraperitoneal injection of either saline or an anti-mitotic drug, the basal levels of plasma TSH, PRL, GH, LH, and FSH were measured by radioimmunoassay. As compared to the saline-injected controls, the drug-injected rats showed a significant decrease of over 70% in the basal levels of plasma TSH, PRL, and LH, while the plasma levels of GH and FSH did not change significantly (Fig. 3 *a*). Although the mean GH level was lowered after anti-mitotic drug injection, the standard deviation was too

large to regard this decrease as significant (Fig. 3 *a*). This result might be explained by a possible decrease in the total volume of circulating blood in the drug-injected rat. Then, the circulating blood volume was determined by the Evans blue (T-1824) method (KANAI and KANAI 1983). The total volume of circulating blood indeed decreased to 72.9% and 89.3% of the control and accordingly hematocrit increased after the injections of colchicine and vinblastine, respectively. When calculated according to the blood condensation ratio (Fig. 3 *b*), the total amounts of all hormones in blood showed a significant decrease in drug-injected rats, though the decreases of TSH, LH, and PRL were more prominent than those of GH and FSH.

Hormone contents of the whole anterior pituitary glands in the rats injected with either saline or an anti-mitotic drug were also analyzed for TSH, PRL, GH, LH, and FSH by radioimmunoassay. The Students' *t*-test did not show a significant difference in hormone levels between control and drug-injected rats. However, by Wilcoxons' Signed Rank test, there was a small, but significant increase in the levels of tissue TSH and PRL after the colchicine injection (p < 0.05).

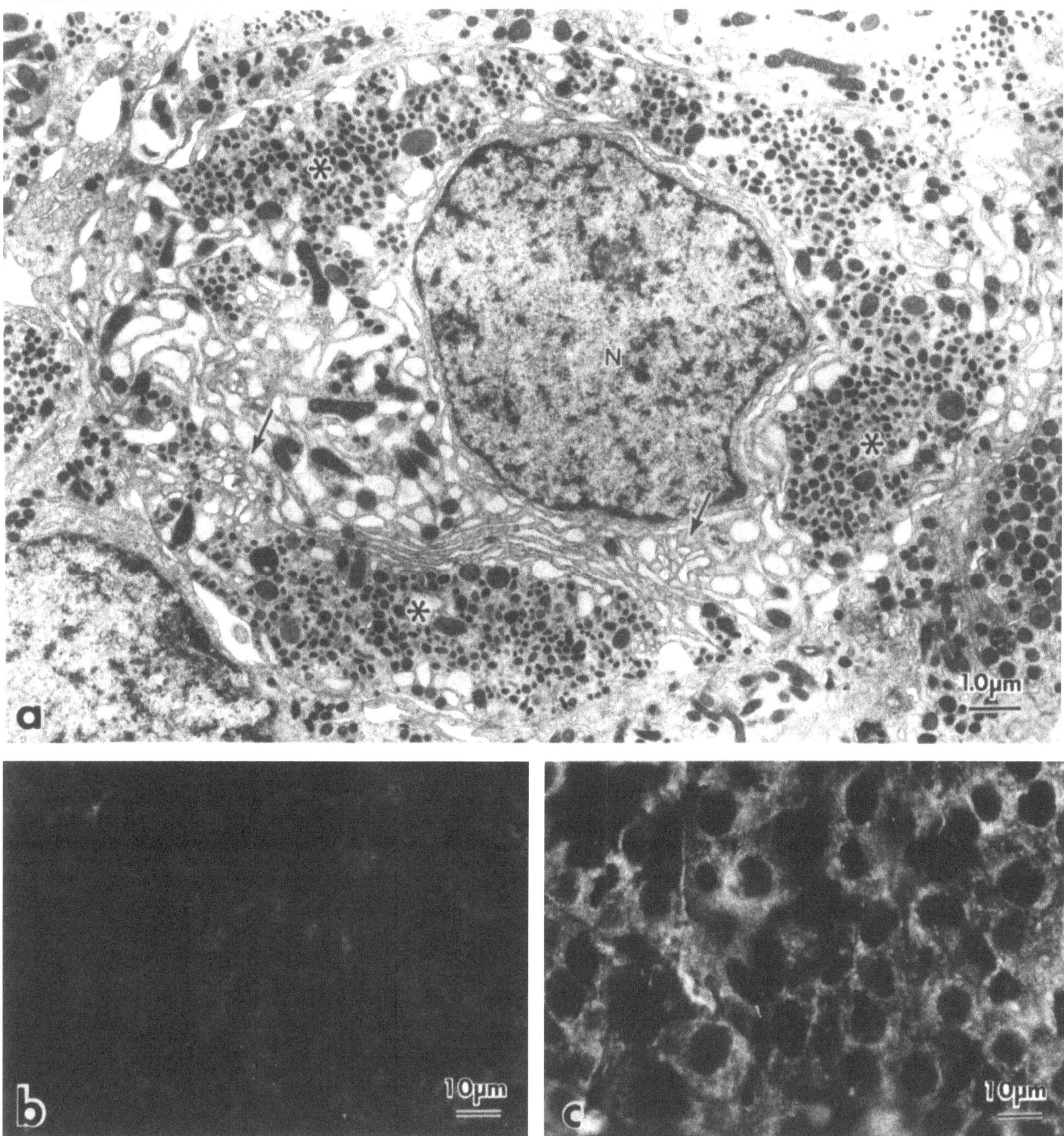

Fig. 4. *a* Electron micrograph of an Gn cell treated with colchicine for 7 hr. Secretory granules tend to aggregate (asterisks), Golgi apparatus is fragmented (arrows) and endoplasmic reticulum slightly dilates. *b* and *c* Immunofluorescence micrographs for tubulin of the colchicine-treated (*b*) and saline-treated control (*c*). *N* Nucleus

### 3.3. *Ultrastructural Features Following Colchicine Treatment*

Seven hours after the first injection of colchicine, almost all microtubules in the anterior pituitary secretory cells had disappeared, as observed by thin-section electron microscopy (Fig. 4 *a*). Centrioles remained almost intact and were situated near the Golgi apparatus, and some of the secretory cells were arrested in metaphase. Immunofluorescence microscopy also showed that immunostaining for tubulin in the colchicine-treated secretory cells was considerably weak (Fig. 4 *b*) as compared to that in the saline-injected control cells (Fig. 4 *c*).

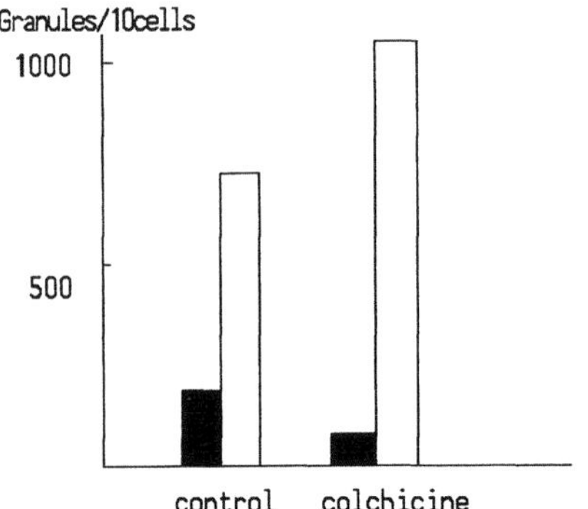

Fig. 5. Distribution of small (□) and large (■) secretory granules beneath the plasma membrane in Gn cells. Counting the numbers of granules and statistical analysis are described in Materials and Methods. The differential distribution of two kinds of secretory granules is evaluated to be significant ($p < 0.05$) by $x^2$-test

Colchicine-treated secretory cells exhibited many characteristic features, which differed from those of saline-injected control cells (Fig. 4 *a*). The distribution of secretory granules and mitochondria were clearly dis-

turbed, and the number of secretory granules was slightly increased. The rough ER dilated and was less orderly arranged. Occasionally, granular materials similar to the contents of secretory granules were found in the cisternae of the dilated rough ER. The Golgi apparatus was a little irregular in shape with slightly dilated lamellae. Interestingly, coated vesicles which were observed in close proximity to the Golgi apparatus in control cells were distributed randomly and aggregated irregularly in drug-treated cells. Similar, but less dramatic ultrastructural changes were observed in the rats which had been injected with vinblastine.

Colchicine treatment resulted in an interesting redistribution of secretory granules within Gn cells. The small dense granules tended to aligned beneath the plasma membrane, while the larger less dense granules were distributed rather diffusely throughout the cytoplasm. When statistical analysis of the distribution patterns of both large and small secretory granules were made (see Materials and Methods), the differential distribution of secretory granules was evaluated to be significant by $x^2$-test ($p < 0.05$) (Fig. 5).

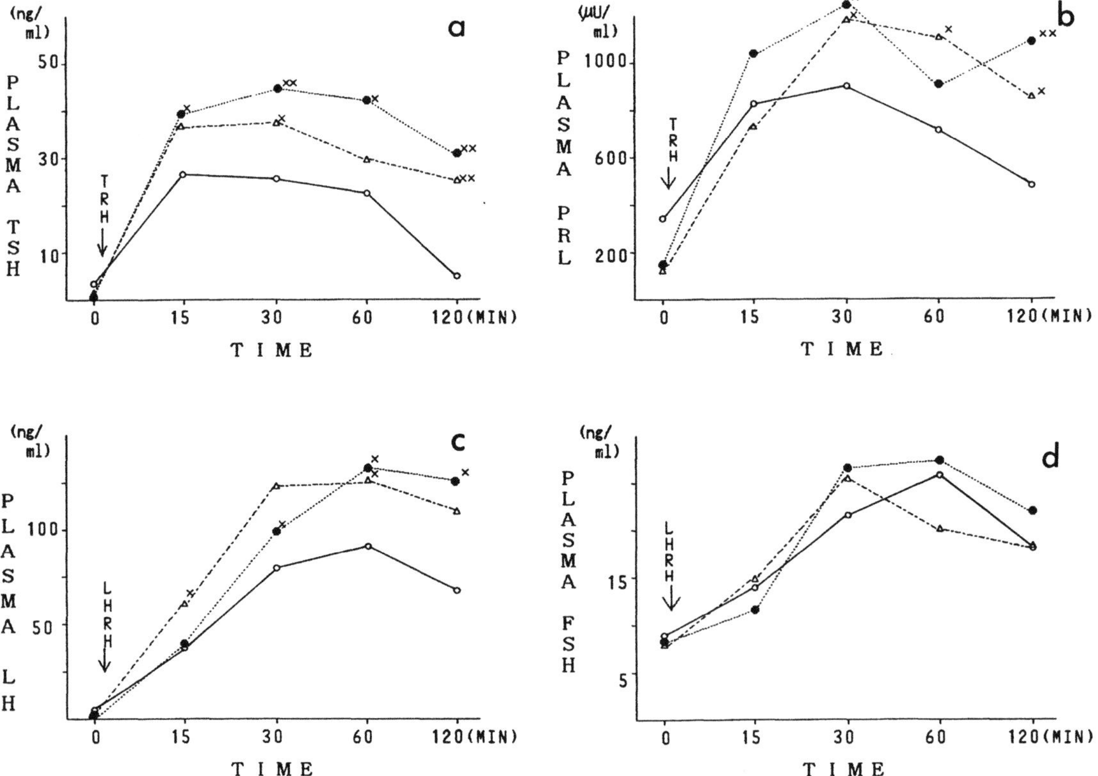

Fig. 6. Time course of effects of the stimulation with hypothalamic hormones on plasma levels of saline (O—O), colchicine (●···●), and vinblastine (△—·—△) pre-treated animals. All results are the means ± SE of 5–7 animals. × $p < 0.05$, × × $p < 0.01$, significantly different compared to the saline pre-treated control, respectively. *a* and *b* Changes of plasma TSH and PRL levels after TRH injection. *c* and *d* Changes of plasma LH and FSH levels after LHRH injection. Note a significant increase in the plasma hormone levels after stimulation

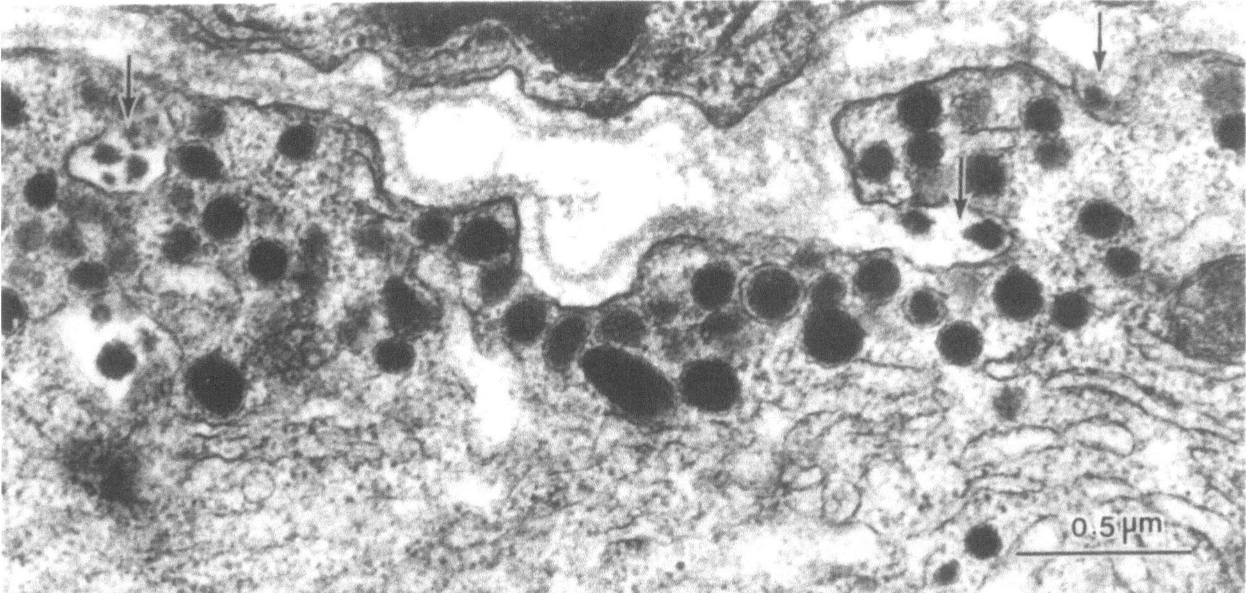

Fig. 7. Part of a colchicine-treated TSH cell 15 min after TRH stimulation. Note the increased exocytotic figures of secretory granules (arrows)

### 3.4. Hormone Level Following Stimulation

Changes in plasma hormone levels were induced by intraperitoneal injection of hypothalamic-releasing hormones, TRH or LHRH. Following injection of TRH, the plasma levels of TSH and PRL were markedly increased in saline- as well as drug-injected rats (Fig. 6 *a, b*). Interestingly, the induced release of both TSH and PRL was significantly higher in colchicine- and vinblastine-treated rats, as compared with the saline-injected control rats.

Injection of LHRH also induced a significant increase in the plasma levels of LH and FSH in both control and drug-injected rats (Fig. 6 *c, d*). The increase of plasma LH in colchicine-treated rats was significantly higher than that in saline-treated controls. However, the increase in plasma FSH was similar in both drug-treated and control rats.

Thin-section electron microscopy of secretory cells fixed at from 5 to 15 min after the injection of TRH of LHRH revealed an increased number of exocytotic figures of secretory granules in the respective cell types in colchicine- and vinblastine-treated rats as well as in saline-treated controls (Fig. 7). Interestingly, immature secretory granules were found in the lamellae of the Golgi apparatus and the deposition of such granules was much more prominent in colchicine-treated rats than in saline-treated control ones. To see the effects of repeated stimulation with hypothalamic hormones, the second injection of the same dose of releasing hormones, TRH or LHRH, was made two hr after the first injection of TRH or LHRH. The plasma levels of hormones were measured as shown in Fig. 8. The increments of plasma TSH, PRL and LH were significantly less in colchicine-treated animals than in saline-treated ones.

### 4. Discussion

The present study has demonstrated that anti-mitotic drugs disrupt virtually all microtubules in secretory cells of the anterior pituitary *in vivo* and concurrently the basal plasma levels of anterior pituitary hormones decrease to varying extents. Apparently, the disruption of microtubules causes the decrease of hormone secretion from the anterior pituitary cells. Nevertheless, certain hypothalamic releasing hormones can induce a marked increase in the plasma levels of the corresponding hormones in such drug-treated animals. The results indicate that these hormones can be released from the microtubule-disrupted secretory cells. Intact microtubules are not neccessarily required for hormone release. Then, what is the function of microtubules in secretory cells?

Although there are numerous reports regarding the effects of anti-mitotic drugs on secretory cells, the involvement of microtubules in secretion remains controversial (KRAICER and MILLIGAN 1971, MACLEOD *et al.* 1973, KOVACS *et al.* 1974, YOUNG *et al.* 1985). Most of the previous works have dealt with the overall effects of microtubule disruption on hormone secretion (LABRIE *et al.* 1973, SHETERLINE *et al.* 1975). The *in vivo*

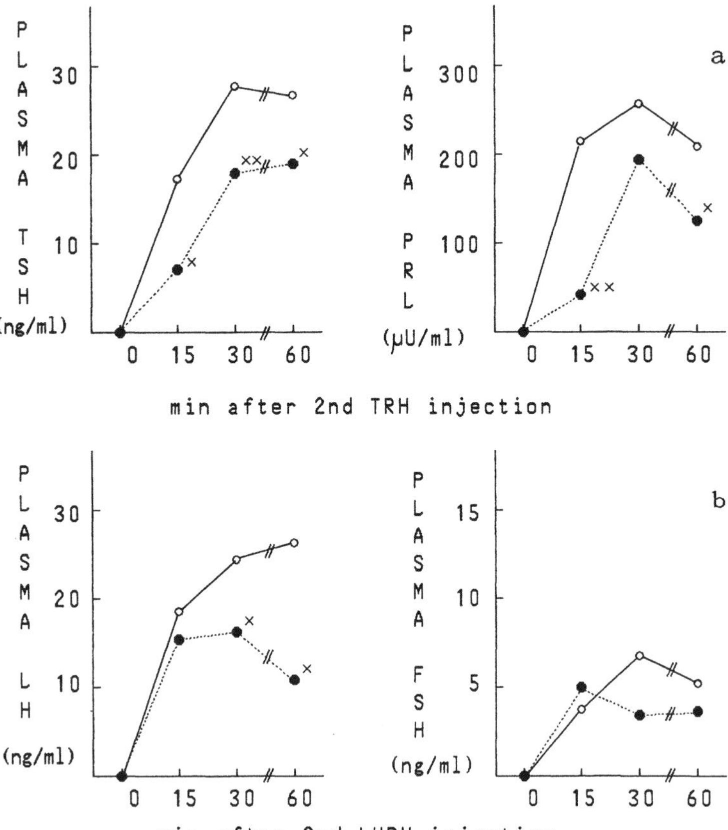



Fig. 8. Time course of the increment of plasma hormone levels after the second injections of TRH (*a*) or LHRH (*b*) in saline (O—O) or colchicine (●···●) pre-treated rats. The plasma level of each hormone 2 hr after the first injection is shown to be zero in value at 0 min. All results are the means ± SE of 5 or 6 animals in each group. × p < 0.05, × × p < 0.01, significantly different compared to the control, respectively

system has been less favored mainly because of the complexity of regulation for hormone secretion such as the presence of feedback action and because of the possible uncertainty of effects of various drugs and stimuli on the secretory cells. However, such difficulties were not very much serious as far as the present analyses are concerned. The feedback action is negligible, for the plasma content of $T_4$ and testosterone did not change significantly within a limited duration after the drug treatment (data not shown). Furthermore, the TRH level in the median eminence and its vicinity remained constant (data not shown). Anti-mitotic drugs were effective in microtubule disruption without any signs of cell degeneration and death. On the contrary, the cell degeneration and damage are often inevitable *in vitro* systems especially after the anti-mitotic drug treatment.

Since the secretion is a complex process comprising several distinct steps from protein synthesis to release of synthesized proteins (SLOT *et al.* 1974, KELLY 1985, SAMUEL and FLICKINGER 1987). One should ask in which step microtubules may be actually involved. In thin sections, microtubules are the most prominent

component of the cytoskeleton on the secretory cells of the rat anterior pituitary, though their abundance varied in different cell types. Microtubules are most abundant in ACTH cells, which are characterized by stellate cell shape with long cytoplasmic processes. The increase in microtubules was reported after thyroidectomy, which induced TSH cells to be in a hypersecretory state (SHIINO and RENNELS 1975). On the other hand, anti-mitotic drugs cause the disturbances of the distribution pattern of secretory granules, Golgi apparatus and rough ER. In many other types of cells microtubules have been demonstrated to be involved in intracellular distribution and translocation of membranous organelles (PORTER 1966, 1973, TSUKITA and ISHIKAWA 1980). The similar functions can be conceivable in hormone secretory cells (NEVE *et al.* 1970, 1972, WILLIAMS and WOLFF 1970, MALAISSE *et al.* 1971, SLOT *et al.* 1974). Indeed, many microtubules run radially from Golgi region toward the cell surface and a close association between microtubules and secretory granules is also often observed in secretory cells. The present observations strongly suggest that microtubules may aid in hormone secretion through their major roles in the orderly distribution of various membranous organelles and in the intracellular translocation of secretory granules toward the cell surface. Recently, two kinds of microtubule-associated proteins, kinesin (Vale *et al.* 1985 a) and MAP 1 C (Paschal *et al.* 1987), have been identified as the candidates for translocators which may play a role in the transport of vesicular organelles in neuronal cells. The direction of force production mediated by kinesin corresponds to antero-grade axonal transport (Vale *et al.* 1985 b) and the opposite direction may be ascribed to MAP 1 C (Paschal and Vallee 1987). Similar translocator proteins seem to occur widely in non-neural cells (Cohen *et al.* 1987). It is highly possible that such translocators may also be involved in microtubule-based translocation of various organelles including secretory granules in a variety of cell types. Studies are urgently needed to identify and localize such translocators in the anterior pituitary secretory cells.

The anterior pituitary is advantageous system in a sense to study the secretory mechanism. This organ consists of five major types of cells which secrete different hormones. As also shown in the present analysis, different cell types respond differently to various stimuli and drugs. Detailed analyses of such differences may lead to a better understanding of the mechanism and regulation of hormone secretion. The present study has also cleary shown that when treated with anti-mitotic drugs the extent of inhibition of hormone secretion varied in different cell types. Several possibilities can be considered for such differences. Firstly, the different responses to drugs might be attributed to different amounts of microtubules and thus to different dependency of secretion on microtubules. However, this is not likely, for the abundance of microtubules does not correlate with the effect of colchicine treatment and microtubules are almost completely disrupted in all types of cells after drug treatment. Secondly, the kind and size of secretory granules may be responsible for such different responses. Interestingly, gonadotropic cells contain two kinds of secretory granules, large less dense FSH granules and small dense LH ones in the same cells. After the drug treatment, the plasma LH level decrease to more extent than the FSH level. Moreover, when stimulated by a hypothalamic hormone (LHRH), LH was released more than FSH. A mechanism must exist for more release of one kind of secretory granules than the other. Apparently, small LH granules are sensitive to the drugs and the releasing hormone. Morphologically, small granules are distributed preferentially at the cell periphery closer to the

plasma membrane than large granules in the drug-treated animals. The different releasing responses of two kinds of secretory granules may be related to such differential distribution of secretory granules. The mechanism of redistribution of secretory granules after the drug treatment is another interesting problem to be clarified. Thirdly, different regulatory systems may exist in different cell types. Indeed, some hormones appear to be secreted continuously and steadily, while others are released regulatively and/or rhythmically in response to various stimuli. Such secretory manners may resemble the concept of constitutive and regulated secretion proposed for the secretory activity in general (Kelly 1985).

Actin filaments are also rich in secretory cells especially at the cell periphery, as revealed by immunofluorescence and electron microscopy. Our observations are consistent with the previous works (Ordronneau and Petrusz 1986). It is very likely that actin filaments may be involved in hormone release through exocytosis (Malaisse 1971, Drenckhahn 1977, Selden and Pollard 1983). Further studies are needed to clarify the exact roles of microtubules, actin filaments, and intermediate filaments in hormone secretion, emphasizing the division of labor among those cytoskeletal components in successive steps of secretion.


## Acknowledgements

We thank Prof. Kunihiko Obata for kindly providing a monoclonal antibody against β-tubulin. This work was supported by research grants from Ministry of Education, Science and Culture and from the Ministry of Health and Welfare, Japan.


## References

Auni D, Hesketh JE, Devilliers G (1979) Freeze-fracture study of the chromaffin cell during exocytosis: Evidence for connections between the plasma membrane and secretory granules and the movements of membrane-associated particles. Cell Tissue Res 197: 433–441

Breckenridge LJ, Almers W (1987) Final steps in exocytosis observed in a cell with giant secretory granules. Proc Natl Acad Sci USA 84: 1945–1949

Childs GV (1986) Functional ultrastructure of gonadotrophs: a review. In: Tixier-Vidal AT, Fahrquhar MG (eds) Morphology of hypothalamus and its connection. Springer, Berlin Heidelberg New York Tokyo, pp 49–98

Cohn SA, Ingold AL, Sholey JM (1987) Correlation between the ATPase and microtubule translocating activities of sea urchin egg kinesin. Nature 328: 160–163

Drenckhahn D, Groschel-Stewart U, Unsicker K (1977) Immunofluorescence-microscopic demonstration of myosin and actin in salivary glands and exocrine pancreas of the rat. Cell Tissue Res 183: 273–279

Kanai I, Kanai M (1983) Rinsho-Kensaho-Teiyo (in Japanese). Kanehara, Tokyo Osaka Kyoto

Ishikawa J, Fuse Y, Wakabayashi K (1987) Choice of extraction procedure for estimation of anterior pituitary hormone content. Endocrinol Japon 34: 755–767

Kelly RB (1985) Pathways of protein secretion in eukaryotes. Science 230: 25–32

Kraicer J, Milligan JV (1971) Effect of colchicine on *in vitro* ACTH release induced by high K and by hypothalamus-stalk-median eminence extract. Endocrinology 89: 408–412

Kovacs K, Horvath E, Szabo S, Dzau VJ, Feldman D, Chang Y, Reynolds ES (1974) Effect of vinblastine on neurohypophysial and adenohypophysial microtubules. Steroids Lipids Res 5: 167–172

Kurosumi K, Inoue K (1986) Ultrastructure of anterior pituitary cells. Curr Top Neuroendocrinol 7: 99–134

Labrie F, Gauthier M, Pelletier G, Borgeat P, Lemay A, Gouge JJ (1973) Role of microtubules in basal and stimulated release of growth hormone and prolactin in rat adenohypophysis *in vitro*. Endocrinology 93: 903–914

Lacy PE, Hawell SL, Young DA, Fink CJ (1968) New hypothesis of insulin secretion. Nature 291: 1177–1180

Macleod RM, Lehmeyer JE, Bruni C (1973) Effect of anti-mitotic drugs on the *in vitro* secretory activity of mammotrophs and somatotrophs and on their microtubules. Proc Soc Exp Biol Med 144: 259–267

Malaisse WJ, Malaisse-Lagae F, Walker MO, Lacy PE (1971) The stimulus-secretion coupling of glucose-induced insulin release V. The participation of a microtubular-microfilamentous system. Diabetes 20: 257–265

Marchand YL, Patzelt C, Assimacopoulos-Jeannet F, Loten EG, Jeanrenaud B (1974) Evidence for a role of microtubular system in the secretion of newly synthesized albumin and other proteins by the liver. J Clin Invest 53: 1512–1517

— Singh A, Assimacopoulos-Jeannet F, Orci L, Rouiller C, Jeanrenaud B (1973) A role for the microtubular system in the release of very low density lipoproteins by perfuse mouse livers. J Biol Chem 248: 6862–6870

Nakane PK (1975) Identification of anterior pituitary cells by immunoelectron microscopy. In: Ganten D, Pfaff D (eds) The anterior pituitary. Academic Press, London, pp 45–61

Neve P, Ketelbant-Balasse P, Willems C, Dumont JE (1972) Effect of inhibitors of microtubules-microfilaments on dog thyroid slices *in vitro*. Exp Cell Res 74: 227–244

— Williams C, Dumont JE (1970) Involvement of the microtubule-microfilament system in thyroid secretion. Exp Cell Res 63: 457–460

Ordronneau P, Petrusz P (1986) Non-hormonal markers in the pituitary. In: Polak JM, Noorden SV (eds) Immunocytochemistry: modern methods and applications, 2nd edn. Wright, Bristol, pp 425–437

Paschal BM, Shpetner H, Vallee RB (1987) MAP 1C is a microtubule-activated ATPase which translocates microtubules *in vitro* and has dynein-like properties. J Cell Biol 105: 1273–1282

— Vallee RB (1987) Retrograde transport by the microtubule-associated protein MAP 1C. Nature 330: 181–183

Porter KR (1966) Cytoplasmic microtubules and their functions. In: Wolstenholme GEW, O'Connor M (eds) Principles of biomolecular organization. Churchill, London, pp 308–345

— (1973) Microtubules in intracellular locomotion. In: Locomotion of tissue cells. Associated Scientific Publishers, New York, pp 149–169 (Chiba Foundation Symposium, vol 5)

Ravindra R, Grosvenor CE (1986) Microtubules are mobilized in the lactating rat anterior pituitary gland during suckling. Endocrinology 118: 1194–1198

— — (1988) Soluble and polymerized tubulin levels in the anterior pituitary lobe of the lactating rat during suckling. Endocrinology 122: 114–119

Samuel LH, Flickinger CJ (1987) The relationship between the morphology of cell organelles and kinetics of the secretory process in male sex accessory glands of mice. Cell Tissue Res 247: 203–213

Savino W, Dardenne M (1986) Thymic hormone-containing cells. VIII. Effects of colchicine, cytochalasin B, and monensin on secretion of thymulin by cultured human thymic epithelial cells. J Histochem Cytochem 34: 1719–1723

Schneyer LH, Young JA, Schneyer CA (1972) Salivary secretion of electrolytes. Physiol Rev 52: 720–777

Segawa A, Sahara N, Suzuki Y, Yamashina S (1985) Acinar structure and membrane regionalization as a prerequisite for exocrine secretion in the rat submandibular gland. J Cell Sci 78: 67–85

Selden SC, Pollard TD (1983) Phosphorylation of microtubule-associated proteins regulates their interaction with actin filaments. J Biol Chem 258: 7064–7071

Sherline P, Lee YC, Jacobs LS (1977) Binding of microtubules to pituitary secretory granules and secretory granule membranes. J Cell Biol 72: 380–389

Sheterline P, Schofield JG, Mira F (1975) Colchicine binding to bovine anterior pituitary slices and inhibition of growth hormone release. J Biol Chem 148: 435–459

Shiino M, Warchol JB, Rennels EG (1974) Microtubules in prolactin cells of the rat anterior pituitary gland. Proc Soc Exp Biol Med 147: 361–365

— Rennels EG (1975) Microtubules in thyroidectomy cells of the rat anterior pituitary gland. Proc Soc Exp Biol Med 149: 380–383

Simson JAV, Spicer SS, Hall BJ (1974) Morphology and their alteration by isoproterenol. J Ultrastruct Res 48: 465–482

Slot JW, Geuze JJ, Poort C (1974) Synthesis and intracellular transport of proteins in the exocrine pancreas of the frog (*Rana esculenta*). Cell Tissue Res 155: 135–154

Temple R, Williams JA, Wilber JF, Wolff J (1972) Colchicine and hormone secretion. Biochem Biophys Res Comm 46: 1454–1461

Tsukita S, Ishikawa H (1980) The movement of membranous organelles in axons: electron microscopic identification of antero-gradely and retrogradely transported organelles. J Cell Biol 84: 513–530

Vale RD, Reese TS, Sheetz MP (1985 a) Identification of a novel force-generating protein, kinesin, involved in microtubule-based motility. Cell 42: 39–50

— Schnapp BJ, Mitchison T, Steuer E, Reese TS, Sheetz MP (1985 b) Different axoplasmic proteins generated movement in opposite. Cell 43: 623–632

Varndell IM, Polak JM (1986) Electron microscopical immunocytochemistry. In: Polak JM, Noorden SV (eds) Immunocytochemistry: modern methods and applications, 2nd edn. Wright, Bristol, pp 146–166

WARCHOL JB, DAMON CH, WILLIAMS MG, RENNELS EG (1975) Distribution of microtubules in prolactin cells of lactating rats. Cell Tissue Res 159: 205–212

WILLIAMS JA, WOLFF J (1970) Possible role of microtubules in thyroid secretion. Proc Natl Acad Sci USA 67: 1901–1908

YOUNG LS, NAIK SI, CLAYTON RN (1985) Pituitary gonadotropin-releasing hormone receptor up-regulation *in vitro*: dependence on calcium and microtubule function. J Endocrinol 107: 49–56

# Index of Keywords


Actin  27, 65, 104
Actin-binding protein  9, 22
Actin-binding sites  37
Actin cytoskeleton  116
Actin filaments  95
Actomyosin  3
*Amoeba proteus*  76
Anterior pituitary cells  145
ATPase active site  37

Calmodulin  9
*Chara corallina*  116
Chemical stimulation  76
Colchicine  145
Crayfish abdominal nerve  137
Cytochalasin  65, 116
Cytoplasmic membrane  27
Cytoskeleton  65, 104, 145
Cytoskeleton mutant  22

36 kDalton actin binding
  protein  48
Dark field microscopy  27

Development  22
*Dictyostelium discoideum*  22
Diffusion rate in cytoplasm  88
Dimethyllysine  37
Domain structure  37
Dynamic cooperativity  3

Egg cells  88
Endocytotic activity  76
Energy conversion  3
Exocytosis  48, 145

Fertilization  104
Fixactive  131
Fluorescence microscopy  88
Freeze substitution  65

Hormone secretion  145

Immunoelectron microscopy  48
Immunocytochemistry  131
Insoluble cytomatrix  137
Intermediate filament  131

*In vitro* elongation  27
Isolated cortex  104

Light chain-binding sites  37
Limited proteolysis  37

Membrane-microfilament inter-
  action  76
Microfilaments  9, 65
Microinjection  88
Microtubules  95, 131, 145
Morphodynamic changes  76
Motility  22
Myosin crossbridge  3
Myosin heavy chain  37

$N^\varepsilon$-($\gamma$-glutamic)lysine
  crosslinks  137
Nerve cytoskeleton  137
*Nitella*  27
*Nitellopsis obtusa*  116
Network fibers  27

Nonmuscle caldesmon  9

Permeability of nuclear
  envelope  88
Photoactivation  131
Phragmoplast (isolated)  95
*Physarum* myosin  37
*Physarum* plasmodium  48
Polarity  116
Pollen tubes  65

Rapid freeze fixation  65
Reconstituted system  3

Sea urchin egg  104
Sulfhydril residues  37

Tissue cultured cells  9
Tobacco BY-2-cell  95

Vesicle fusion  48

Water transport  116

PROTOPLASMA
*Supplementum 1*

*M. Tazawa (ed.)*

# Cell Dynamics 1

*Cytoplasmic Streaming*
*Cell Movement—Contraction and Migration*
*Cell and Organelle Division*
*Phototaxis of Cell and Cell Organelle*

1988. 147 figures. VIII, 193 pages.
Cloth DM 200,—, öS 1400,—
Reduced price for subscribers to "Protoplasma"
Cloth DM 180,—, öS 1260,—
ISBN 3-211-82094-9

The 37 original papers in "Cell Dynamics" are dedicated to Prof. Dr. N. Kamiya on the occasion of his 75th birthday. Prof. Kamiya is still active as the president of the International Federation for Cell Biology. He started his studies on cell dynamics in 1940 with the publication of his famous double-chamber method for measuring the motive force of shuttle streaming of *Physarum* plasmodium. With these measurements he opened the way to study the physico-chemical basis of cytoplasmic streaming. Since then, for nearly five decades, he spearheaded biophysical researches on cytoplasmic streaming by making full use of his own ingenious methods. Reflecting Prof. Kamiya's interests in cell dynamics, the papers in these two volumes include various aspects of primitive and basic motile systems. They are classified in six topics. The topics in volume 1 treat mostly physiological and biophysical aspects of cell, cytoplasm and cell organelle motility and include "Cytoplasmic Streaming", "Cell Movement: Contraction and Migration", "Cell and Organelle Division", "Phototaxis of Cell and Cell Organelle".
The contributors include Prof. Kamiya's pupils and colleagues and researchers who have been influenced by his science and personality. They stand foremost in the field of cell motility of both plant and animal kingdoms. These books thus offer us a synopsis of the current status of research on cell motility, while also representing a tribute to one of the greatest of all cell biologists.

## Springer-Verlag Wien New York

Moelkerbastei 5, A-1010 Wien ● Heidelberger Platz 3, D-1000 Berlin 33 ● 175 Fifth Avenue, New York, NY 10010, USA ●
37-3, Hongo 3-chome, Bunkyo-ku, Tokyo 113, Japan

# The Cytoskeleton

By Professor **Manfred Schliwa,**
Department of Zoology, University of California,
Berkeley, Calif., USA

(Cell Biology Monographs, Vol. 13)

1986. 88 figures. XI, 326 pages.
Cloth DM 172,—, öS 1200,—
Reduced price for subscribers to the series "Cell
Biology Monographs" and subscribers to the journal
"Protoplasma":
Cloth DM 137,60, öS 960,—
ISBN 3-211-81884-7
*Prices are subject to change without notice*

The last 15 years have seen the emergence and explosive development of a new branch of cell biological research that deals with the fibrous components of the cytoplasm: Microtubules, microfilaments, intermediate filaments, and their associated proteins. These cellular components are known collectively as "the cytoskeleton". This book constitutes a concise, introductory survey of the structure, biochemistry, and function of the major components of the cytoskeleton. Particular attention is paid to their interactions with each other and with the cell membrane. The survey's emphasis on recent findings and key references allows the reader to gain rapid access to a vast body of literature without being overwhelmed. The book is richly illustrated with some of the finest micrographs taken by experts around the world. To researchers in the field, this book will be useful as a summary of existing information. To students and researchers outside the field, it provides an up-to-date introduction into the contemporary knowledge of a complex and rapidly developing area of modern cell biological research.

# The Nucleolus and Ribosome Biogenesis

By Professor **Asen A. Hadjiolov,** M. D., Ph. D., Department of Molecular Genetics, Institute of Molecular Biology, Bulgarian Academy of Sciences, Sofia, Bulgaria

(Cell Biology Monographs, Vol. 12)

1985. 46 figures. XI, 267 pages.
Cloth DM 138,—, öS 970,—
Reduced price for subscribers to the series "Cell
Biology Monographs" and subscribers to the journal
"Protoplasma":
Cloth DM 110,40, öS 776,—
ISBN 3-211-81790-5
*Prices are subject to change without notice*

The monograph draws together the vast body of knowledge on the structure and function of the cell nucleolus, with primary emphasis on the major nucleolar role: the production of ribosomes destined for their central role in protein synthesis carried on in the cytoplasm. The book is more than an assembly of information, however, as the author provides a critical synthesis of cytological, biochemical, and genetic data into a coherent picture of the common molecular denominators of the nucleolus that lead to the complete production of mature ribosomes.

The monograph deals in depth with such subjects as: the great multiplicity of ribosomal RNA and ribosomal protein genes; their chromosomal and extra-chromosomal location and organization; the transcription of ribosomal RNA and ribosomal protein genes; the maturation (processing) of ribosomal RNA precursors and of "pre-ribosomes"; the microscopic and molecular architecture of the nucleolus; the regulation of ribosome production, including transcriptional and post-transcriptional controls; and, finally, the patterns of ribosome biogenesis in the life cycles of normal, senescent, and cancer cells. Throughout the book, evidence from a broad range of organisms — including, but not limited to, yeast, protozoa, plants, and mammals — is considered.

# *Springer-Verlag Wien New York*

Moelkerbastei 5, A-1010 Wien · Heidelberger Platz 3, D-1000 Berlin 33 ·
175 Fifth Avenue, New York, NY 10010, USA · 37-3, Hongo 3-chome, Bunkyo-ku, Tokyo 113, Japan